HYDROSTATIC EXTRUSION

Theory and Applications

HYDROSTATIC EXTRUSION

Theory and Applications

Edited by

N. INOUE

Science University of Tokyo, Tokyo, Japan

and

M. NISHIHARA

Kobe Steel Ltd, Kobe, Japan

ELSEVIER APPLIED SCIENCE PUBLISHERS
LONDON and NEW YORK

ELSEVIER APPLIED SCIENCE PUBLISHERS LTD
Crown House, Linton Road, Barking, Essex IG11 8JU, England

Sole Distribution in the USA and Canada
ELSEVIER SCIENCE PUBLISHING CO., INC.
52 Vanderbilt Avenue, New York, NY 10017, USA

WITH 20 TABLES AND 266 ILLUSTRATIONS

British Library Cataloguing in Publication Data

Hydrostatic extrusion: theory and applications.
1. Hydrostatic extrusion
I. Inoue, N. II. Nishihara, M.
621.8′9   TS255

Library of Congress Cataloging in Publication Data

Hydrostatic extrusion

1. Hydrostatic extrusion. I. Inoue, Nobuo,
1919–   . II. Nishihara, M. (Masao), 1920–
TS255.H94 1985   671.3′4   85-15884

ISBN-13: 978-94-010-8696-7   e-ISBN-13: 978-94-009-4954-6
DOI: 10.1007/978-94-009-4954-6

Preface

This book is intended to be a reference text on hydrostatic extrusion, a multidisciplinary technology involving the forming process of materials, tribology, high pressure engineering and so forth.

Until now only one book bearing the title of hydrostatic extrusion, by Prof. Alexander and Dr. Lengyel, has been published since 1971. Although there are chapters on hydrostatic extrusion in such books as THE MECHANICAL BEHAVIOUR OF MATERIALS UNDER PRESSURE edited by Dr. Pugh, METAL FORMING by Prof. Avitzur and HIGH PRESSURE TECHNOLOGY by Drs. Spain and Paauwe, it is regrettable that no up-to-date reference books on hydrostatic extrusion are available.

As is well known, hydrostatic extrusion is a nearly-ideal lubricated extrusion. Its advantages have been demonstrated by laboratory research in the past two decades, yet many manufacturers, however, still hesitate to adopt the technology in their plants. Their hesitation is certainly due to the lack of exact information on the process and its equipment and also to their unfamiliarity with the actual method of operation.

In order to provide a useful introduction to the subject for engineers who work in industries which plan to employ this technique and also to give exact and reliable information on the durability and performance of production facilities, as well as the capabilities of the process and the properties of extruded products, we decided to publish this book.

Starting with theories and computational methods, the processes of cold, warm and hot hydrostatic extrusion are described by experts in their respective fields. Then follows a chapter devoted to industrial hydrostatic extrusion plant equipment and the book concludes with an up-to-date account of hydrostatic extrusion of new materials, such as composites, special alloys, fine wires, or polymers.

Nobuo Inoue
Masao Nishihara

CONTENTS

LIST OF CONTRIBUTORS

R. J. Fiorentino, Program Manager, Metalworking Section, Engineering and Manufacturing Technology Department, Battelle Columbus Laboratories, 505 King Avenue, Columbus, Ohio, U. S. A.   43201

N. Inoue, Professor, Department of Mechanical Engineering, Faculty of Engineering, Science University of Tokyo, 1 - 3 Kagurazaka, Shinjuku-ku, Tokyo, JAPAN   162

A. Kobayashi, Assistant Manager, Toyoura Works, Hitachi Cable, Ltd., 1500 Kawajiri-cho, Hitachi, JAPAN   319-14

T. Matsushita, Senior Researcher, Central Research Laboratory, Kobe Steel, Ltd., 1 - 3 - 18 Wakinohama-cho, Chuo-ku, Kobe, JAPAN   651

P. B. Mellor, Professor, Department of Mechanical Engineering, University of Bradford, Bradford, West Yorkshire, BD7 1DP, U. K.

S. Mitsugi, Senior Researcher, Metals Research Laboratory, Hitachi Cable, Ltd., 3 - 1 - 1 Sukegawa-cho, Hitachi, JAPAN   317

M. Nishihara, Adviser, Kobe Steel, Ltd., 1 - 3 - 18 Wakinohama-cho, Chuo-ku, JAPAN   651

K. Osakada, Professor, School of Mechanical Engineering, Faculty of Engineering, Hiroshima University, Shitami, Saijo, Higashi-hiroshima, JAPAN   724

M. Seido, Senior Researcher, Metals Research Laboratory, Hitachi Cable, Ltd., 3550 Kidamari-cho, Tsuchiura, JAPAN   300

W. R. D. Wilson, Professor, Department of Mechanical and Nuclear Engineering, Northwestern University, The Technological Institute, Evanston, Illinois, U. S. A.   60201

M. Yamaguchi, General Manager, Mechanical Engineering Research Laboratory, Kobe Steel, Ltd., 1 - 3 - 18 Wakinohama-cho, Chuo-ku, Kobe, JAPAN 651

Y. Yamaguchi, Chief Researcher, Central Research Laboratory, Kobe Steel, Ltd., 1 - 3 - 18 Wakinohama-cho, Chuo-ku, Kobe, JAPAN   651

CHAPTER 1

INTRODUCTION

N. INOUE

SCIENCE UNIVERSITY OF TOKYO

The idea of employing hydrostatic pressure for metalworking was conceived as early as in the nineteenth century and a British patent was
granted to Robertson in 1893 [1]. Experimental proofs of the pressure-induced ductility in metals and other materials, however, were not obtained
until Bridgman made an extensive study of subject along with metalworking
under pressure and published the result in the form of a monograph [2] in
1949. He invented a high-pressure seal, now called the 'Bridgman Seal',
which was so effective that he found himself in a uniquely advantageous
position to do high-pressure work. He spent the rest of his life studying
the fascinating world of high pressure and left a mountain of research
papers on the subject, which were later compiled in the form of a seven-
volume monograph entitled COLLECTED EXPERIMENTAL PAPERS OF P. W. BRIDGMAN
and published by Harvard University Press in 1964 [3]. An extremely versa-
tile high-pressure unit bearing his name has long been manufactured by the
Harwood Engineering Company of Newhall, with the assistance of Abbot,
former Research Assistant of Bridgman, and has benefited countless researchers in the high-pressure field. A similar type of press was also produced by the Pressure Technology Corporation of America, headed by
Bobrowsky.

In Great Britain, on the other hand, systematic work on the hydrostatic extrusion of metals and alloys was started by Pugh at the National
Engineering Laboratory in Glasgow. He and his collaborators developed the
technique of hydrostatic extrusion to such an extent that it is now established as a useful industrial process. In the Soviet Union high-
pressure study was started by Vereshchagin at the High Pressure Laboratory
of the Academy of Sciences and reports on hydrostatic extrusion began
to appear in 1957 [4]. For a historical review of high-pressure work the
reader is referred to an excellent paper by Vodar and Kieffer of the High
Pressure Laboratory, C.N.R.S., Bellevue, France, included in the book, THE

MECHANICAL BEHAVIOUR OF MATERIALS UNDER PRESSURE, edited by Pugh [5]. Great progress in high-pressure engineering was made in the 1960's, as can be seen by the fact that the first conference exclusively on this subject was held at Imperial College in London and the first seminar on hydrostatic metalworking processes at Battelle Columbus Laboratories, both in 1967. During that decade the United Kingdom Atomic Energy Authority made an important contribution to the development of the technique of hydrostatic extrusion.

In Japan the early recognition of the importance of the study on the effects of hydrostatic pressure on the mechanical properties of materials by Nishihara, then at Doshisha University in Kyoto, led to a series of research at room temperature [6,7] and also at elevated temperatures [8,9, 10]. The strength and ductility of carbon steel, magnesium, titanium, and zinc were determined at pressures up to 5 kilobars and temperatures up to 600°C. Later on he moved to Kobe Steel and was instrumental in organizing a powerful group of engineers for the production application of hydrostatic extrusion [11]. In industry, the Western Electric Company in the United States, ASEA in Sweden, Kobe Steel in Japan, and some other companies started to use the new technology on a commercial basis. The first international meeting devoted solely to this subject was held at University of Sterling in Scotland in June 1973 under the joint sponsorship of NEL and AIRAPT.

In the early period of development the main effort in the application of hydrostatic extrusion seems to have been expended on work with the difficult-to-form materials. It goes without saying that the deformation of materials is caused by tangential stresses while fracture is caused by normal ones. By applying hydrostatic pressure fracture will be suppressed without imparting a substantial change in the yield and flow characteristics of the materials. It is not without reason therefore that the brittle materials were the prime target for the application of the techniques of hydrostatic extrusion. In actuality, however, ductile metals were the first to be hydrostatically extruded commercially. Thus various sizes of copper tubing were produced using the ASEA manufactured Quintus hydrostatic extrusion plant by Lips BV, Holland. The extremely high reduction ratios obtainable in a single pass were an advantage but the process still had the limitation of being a batch one. The decade of the 1970's saw an increase in the commercial use of the batch-type production unit on one hand and of research and development of a continuous process on the other.

Quite a variety of ideas on the continuous process were conceived and some of them were subjected to experimentation. Beginning with 'Continuous Extrusion with Viscous Drag' by Fuchs [12], novel methods of 'Continuous Extrusion Forming' often called 'Conforming' by Green [13], 'Linear Continuous Extrusion' or 'Linex' by Black and Voorhes [14], 'Helical Extrusion' by Green [15], 'Extrolling' by Avitzur [16], and 'Hydrostatic Extrusion with Continuous Feed' by Kobe Steel [17] were proposed. Among them 'Conforming' seems to have been most developed at this time [18].

The theory of hydrostatic extrusion has also made great progress. A free-body approach, often called a slab method, was most frequently employed in the study of metalworking processes since the days of von Karman, and hydrostatic extrusion was not an exception. In hydrostatic extrusion, however, if certain conditions are fulfilled, a situation of hydrodynamic lubrication is realized between the billet and die, resulting in extremely low friction. By making use of the Reynolds equations the way to obtain this desirable state of lubrication has been clarified. The application of slip line field theory, extremum principles, especially the upper bound approach, and other energy methods, have been quite successful. Useful solutions were obtained and made it possible to predict the pressure required to effect the extrusion with a prescribed amount of reduction in area, to determine the optimum shape of the die, or to select process parameters, such as temperature or the rate of extrusion. Visioplasticity has been used to analyze displacement, velocity, strain, strain rate, stress, and temperature in the billet. More recently, the methods of finite elements, finite differences, or boundary elements have been applied to the mathematical analysis of stress and strain distributions in the billet. These mathematical tools, however, require the accurate constitutive equations of the billet material to be known as a basis of the calculation. It is a formidable job to spell out the constitutive relations for large strains with temperature, pressure, and strain rate as the process parameters. Furthermore, if the mechanical behavior of the billet material is dependent on time, as is the case with polymers or the metals at elevated temperatures, the work involved is almost insurmountable. Much work still remains to be done.

After two decades of research and development the techniques of hydrostatic extrusion have made progress to such a level that they are now established as an industrial process. The process has been successfully applied to copper tubings, copper-clad aluminum wire, fine wires of noble

metals, tubings of aluminum alloys, Nb-Ti superconductors, honeycomb-shaped
gamma-ray collimeters of scintillation cameras for medical use, aluminum-
copper transition pieces of refrigerators and coolers, and so forth. With
the application of high pressure alone metallic materials do not necessarily
gain in ductility [19]. Under the combined action of pressure and temperature, however, any metals and alloys can be expected to behave in a ductile
manner [20]. Hydrostatic extrusion at elevated temperatures has begun as
early as in the mid 70's for aluminum and copper alloys [21] and steels
[22] at Kobe Steel, where a new plant for hot hydrostatic extrusion is now
in operation. Its application to a wider class of materials is yet to come.

When the recent development of the techniques of hydrostatic extrusion
is reviewed, their application to polymeric materials cannot be overlooked.
During the last decade a number of primarily crystalline polymers, especially high-density polyethylene, have been extruded with large reductions in area to the effect that their moduli as well as strengths may
be raised to the level of steels. It is still at a research and development
stage, but the outlook for obtaining novel materials of unprecedented modu-
lus-to-density or strength-to-density ratios is bright. A new field for
the commercial application of hydrostatic extrusion is being opened. Much
work should be done, however, before these highly oriented polymers are
produced commercially.

New studies on hydrostatic extrusion are going on at numerous research
institutions throughout the world. Papers on the subject are published in
the proceedings of the AIRAPT International High Pressure Conference, which
is held approximately every other year, and many other technical journals.
Although there are chapters on hydrostatic extrusion in books by Pugh [23],
Avitzur [24], or Spain and Paauwe [25], monographs relevant to this subject
have not been published in the last thirteen years since Alexander and
Lengyel wrote the first and last book bearing the title of hydrostatic
extrusion [26]. In the present volume, which is the second book ever
published bearing the title of hydrostatic extrusion, theories and
practices in this important field of plastic forming are presented in six
chapters. The present chapter having provided a brief review of the development of hydrostatic extrusion, Chapter 2 contains mathematical theories presented by active researchers who have been and are still publishing
interesting and important papers in this field. Available mathematical
tools to analyze the process are reviewed and their application to lubrication and dynamics is described by college professors and factory manag-

ers. Chapters 3 and 4 are devoted to an explanation of the process by highly experienced researchers, managers, and technical consultants from Kobe Steel and Hitachi Cable who have joined forces to give a detailed description of cold, warm and hot hydrostatic extrusion processes. These chapters will be especially useful to factory managers who are contemplating introducing the techniques of hydrostatic extrusion into their production plants. Chapter 5 describes the plant equipment of hydrostatic extrusion and is based on the experience obtained by manufacturing the press and tooling at Kobe Steel and also by running the plant at Hitachi Cable. Chapter 6, the final chapter, describes the properties of the products along with the process characteristics which are affected by the properties of the billet materials. The considerable experience obtained by Hitachi Cable, in extruding composites on a production base, and by Battelle Columbus Laboratories, in working with a variety of metallic materials, is included. The selected studies conducted at Battelle include the thin-film hydrostatic extrusion process, products made by this process, hydrostatic extrusion of brittle materials with double-reduction dies, and hydrostatic extrusion of extra long billets with a stepped-bore container. Last but not least, the application of techniques of hydrostatic extrusion to fine wires of metallic materials is described along with the recent work on polymers.

The book is intended to be practical and of use to those employing high pressure to form materials at departments of planning, research, development, and in factories producing non-ferrous material, steel, and polymers. Engineers and scientists working in the aircraft, nuclear reactor, electrical, electronic, automobile, armament, and other heavy industries will find it useful to refresh their knowledge. Most of the materials covered are presented here in a book form for the first time. At the end of each chapter or section updated references are appended. These will be especially useful to graduate students and researchers at universities and research institutions.

REFERENCES

1 J. Robertson, British Patent No. 19,356, October 14, 1893; U.S. Patent No. 524,504, August 24, 1894.

2 P.W. Bridgman, Physics of High Pressure, International Textbooks of Exact Sciences, G. Bells & Sons, Ltd., London, 1949.

3 P.W. Bridgman, Collected Experimental Papers, Harvard University Press, 1964.

4 B.I. Beresnev, L.R. Vereshchagin, and Yu.N. Ryabinin, Izv. Akad. Nauk SSSR, Otdl. Tech. Nauk, 5(1957)48-55.

5 B. Vodar and J. Kieffer in H.Ll.D. Pugh (Ed.), The Mechanical Behaviour of Materials Under Pressure, Applied Science Publishers, 1971, pp.1-53.

6 M. Nishihara, K. Tanaka and T. Muramatsu, Proc. 7th Japan Cong. Test. Mat., 7(1964)154.

7 M. Nishihara et al., Proc. 8th Japan Cong. Test. Mat., 8(1965)73.

8 M. Nishihara, K. Tanaka, S. Yamamoto and Y. Yamaguchi, J. Soc. Materials Science, Japan, 16(1967)169-173.

9 M. Nishihara, K. Tanaka, S. Yamamoto and Y. Yamaguchi, Proc. 10th Japan Cong. Test. Mat., 10(1967)50-53.

10 M. Nishihara and K. Izawa, Science and Engineering Review of Doshisha University, 9(1968)41-64.

11 M. Nishihara, Preprints, NEL/AIRAPT International Conference on Hydrostatic Extrusion, Univ. Stirling, Scotland, (1973)33-81.

12 F.J. Fuchs, The Wire Journal, Oct.(1960)105-113.

13 D. Green, Journal of the Institute of Metals, 100(1972)295-300.

14 J.T. Black and W.G. Voorhes, Journal of Engineering for Industry, Trans. ASME, 100(1978)37-42.

15 D. Green, TRG Report 2332(s), Reactor Fuel Element Laboratories, UKAEA, Springfield, December (1972).

16 B. Avitzur, The Wire Journal, July(1975)73-80.

17 M. Nishihara, T. Matsushita and Y. Yamaguchi, Preprints, 14th Annual Meeting of European High Pressure Research Group, Dublin, Ireland, Sept. 6-8(1976).

18 Y. Sugiyama and T. Hiramatsu, Journal of the Japan Society for Technology of Plasticity, 18(1977)1006-1011.

19 A. Oguchi, ibid., 22(1981)210-214.

20 A. Oguchi, M. Nobuki, Y. Kaieda, M. Ohtaguchi and T. Tagashira, Reports, National Research Institute for Metals, Japan, 4(1983)280-289.

21 M. Nishihara, M. Noguchi, T. Matsushita and Y. Yamaguchi, Proc. 18th MTDR Conference, London, (1977)91-96.

22 M. Nishihara, T. Matsushita, Y. Yamaguchi, T. Yamazaki and M. Noguchi, Proc. 20th MTDR Conference, Birmingham, (1979)87-92.

23 H.Ll.D. Pugh (Ed.), The Mechanical Behaviour of Materials Under Pressure, Applied Science Publishers, 1971.

24 B. Avitzur, Handbook of Metal-Forming Processes, Wiley Interscience, 1983.

25 I.L. Spain and J. Paauwe (Ed.), High Pressure Technology, Marcel Dekker, 1977.

26 J.M. Alexander and B. Lengyel, Hydrostatic Extrusion, Mills & Boon, 1971.

CHAPTER 2

THEORY OF HYDROSTATIC EXTRUSION

Section 1

FUNDAMENTALS OF HYDROSTATIC EXTRUSION

K.OSAKADA

HIROSHIMA UNIVERSITY

P.B.MELLOR

BRADFORD UNIVERSITY

1. Introduction

The most significant feature of the hydrostatic extrusion process is
that there is no friction between the billet and the container wall and
thus the length of the billet is not limited as it is in conventional
extrusion. Also, the frictional forces between the die and the billet tend
to be lower in hydrostatic extrusion because the lubricant is forced, by
the high fluid pressure, to flow into the interface. A low frictional
stress on the die surface permits a smaller die angle to be used which, in
turn, results in more uniform deformation in the extruded product. On the
other hand, the indirect driving of the billet through the compressible
fluid sometimes makes it difficult to control the motion of the billet.

Since the working limit of hydrostatic extrusion is set mainly by the
fatigue strength of the container it is essential to be able to predict the
extrusion pressure for each application. The extrusion pressure is affected
by geometrical factors such as extrusion ratio and die angle, as well as by
the flow stress of the material and the friction forces between die and
billet. The most effective way of reducing the extrusion pressure is by
increasing the temperature of the billet.

At an early stage in the development of hydrostatic extrusion it was
expected that the high fluid pressure would prevent fracture in materials
having poor ductility. This is not the case since, under the same conditions of die angle, reduction and material properties, almost the same
hydrostatic stress is generated in conventional extrusion as in hydrostatic

extrusion. However, lower frictional forces in hydrostatic extrusion permit the use of lower die angles and higher extrusion ratios, both of which lead to higher hydrostatic stress and therefore to conditions which supress fracture. The uniform deformation near to the surface, caused by low friction, reduces the danger of surface cracking. Fracture can be suppressed by extruding into a high fluid pressure but then the hydrostatic extrusion pressure has to be raised accordingly [1].

It is found that the hardness of a product of hydrostatic extrusion is uniform over its cross-section. However, at large deformations hardness measurements are insensitive to variations in the amount of strain and residual stresses will still occur in the product.

In the next section the above fundamental aspects are explained further.

2. Flow Stress and Extrusion Pressure
(1) Equivalent Strain and Extrusion Pressure

To evaluate the strain in large plastic deformation the equivalent plastic strain $\bar{\varepsilon}$ is used. In simple compression or tension, the equivalent strain is equal to the absolute value of the logarithmic plastic strain,

$$\bar{\varepsilon} = \ln (l_1 / l_0) . \tag{1}$$

The equivalent strain rate $\dot{\bar{\varepsilon}}$ is the equivalent strain caused in unit time,

$$\dot{\bar{\varepsilon}} = d\bar{\varepsilon}/dt . \tag{2}$$

(2) Flow Stress

The resistance of material to further plastic flow is measured by the flow stress, $\bar{\sigma}$. The flow stress in uniform compression or tension is the absolute value of the true stress,

$$\bar{\sigma} = |P/A| , \tag{3}$$

where P is the applied force and A is the current cross-sectional area. The flow stress is equal to the equivalent stress defined by the von Mises yield criterion in complex stress systems.

(3) Flow Curve

The curve of equivalent plastic strain vs. flow stress illustrated in Fig.1 is called a flow curve. To simplify the analysis the flow stress is often assumed to remain constant with increasing plastic strain. The mate-

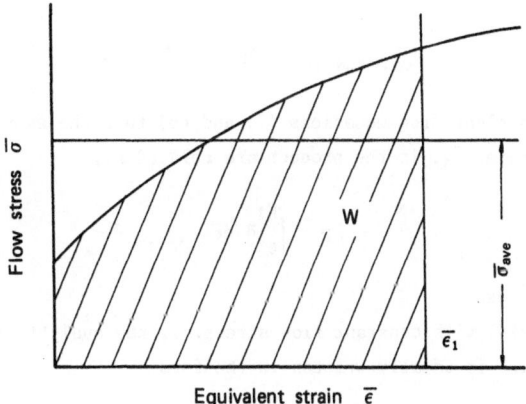

Fig.1 Flow curve

rial is then said to be "perfectly plastic".

(4) Plastic Work

The plastic work per unit volume W up to an equivalent strain $\bar{\varepsilon}_1$ is,

$$W = \int_0^{\bar{\varepsilon}_1} \bar{\sigma} \, d\bar{\varepsilon} \quad , \tag{4}$$

which is given by the cross-hatched area under the flow curve in Fig.1.

(5) Extrusion Ratio

The extrusion ratio, R, is the ratio of the cross-sectional area A_b of the billet to that of the product, A_p.

$$R = A_b/A_p \tag{5}$$

If the deformation in extrusion is ideally uniform, the equivalent strain in the product is given by,

$$\bar{\varepsilon}_{ideal} = \ln R \quad . \tag{6}$$

(6) Extrusion Pressure and Plastic Work

The work done by the fluid pressure, p, in extruding a billet of length l_b and cross-sectional area A_b is,

$$W_{pressure} = p \, l_b \, A_b \quad . \tag{7}$$

If it is assumed that all of this work is dissipated in plastic deformation and causes an equivalent strain $\bar{\varepsilon}_1$ in the product,

$$W_{deformation} = \ell_b A_b \int_0^{\bar{\varepsilon}_1} \bar{\sigma} \, d\bar{\varepsilon} \quad . \tag{8}$$

It is then clear from equations (7) and (8) that the extrusion pressure and the strain, $\bar{\varepsilon}_1$, in the product are related by,

$$p = \int_0^{\bar{\varepsilon}_1} \bar{\sigma} \, d\bar{\varepsilon} \quad . \tag{9}$$

(7) Mean Flow Stress

For a material with constant flow stress, Y, the analytical result for extrusion pressure is usually written in the form,

$$p = Y \, f(\, \alpha, \mu, \, \cdots) \tag{10}$$

where α is the semi-angle of the conical die and μ is the coefficient of friction. For a constant flow stress, Y, equation (9) becomes $p = Y\bar{\varepsilon}_1$ and identifying this with equation (10) it is seen that,

$$\bar{\varepsilon}_1 = f(\, \alpha, \mu, \cdots) \quad . \tag{11}$$

In applying equation (10) to actual metals which work-harden, Y is replaced by a mean flow stress $\bar{\sigma}_{ave}$ which is defined by,

$$\bar{\sigma}_{ave} = \frac{1}{\bar{\varepsilon}_1} \int_0^{\bar{\varepsilon}_1} \bar{\sigma} \, d\bar{\varepsilon} \tag{12}$$

as shown in Fig.1

(8) Factors Affecting Flow Stress

The flow stress usually decreases as the temperature is increased. Fig.2 illustrates the flow curves of 18-8 stainless steel [SUS 304] at various temperatures up to $1100°C$, [2].

Although the flow stress increases as the strain-rate increases, the degree of change is not as significant as for temperature changes. The flow stress is increased only by 20-50% when the strain-rate is increased by 10^3 at room temperature.

The effect of pressure on the flow stress is negligibly small for ordinary metals under extrusion conditions.

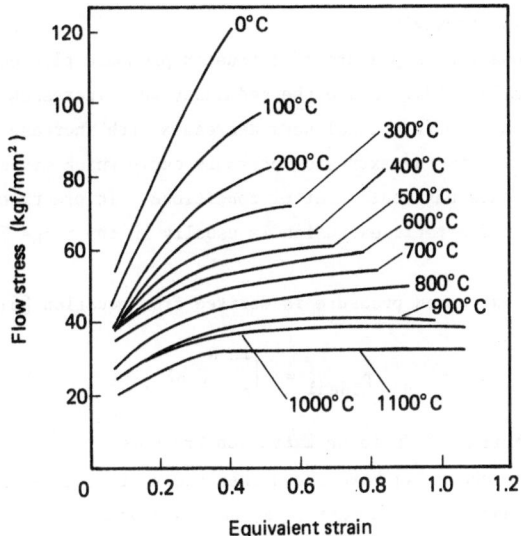

Fig.2 Flow curve of 18-8 stainless steel at various temperatures
(strain rate =450/sec)

(9) Adiabatic Temperature Rise

Almost all the plastic work is converted into heat. The temperature
rise ΔT due to complete conversion of plastic work is calculated from,

$$\Delta T = W \,/\, J \rho \, c \tag{13}$$

where W is the plastic work, J the mechanical equivalent of heat, ρ and c
are the density and specific heat of the material respectively.

3. Geometrical Factors Affecting Extrusion Pressure

(1) Components of Extrusion Pressure

The extrusion pressure is considered to consist of three components:
the pressure for uniform [ideal] deformation, the pressure for redundant
deformation and the pressure to overcome the frictional force between
billet and die. The uniform deformation is the minimum deformation
necessary to change the outside dimensions of the billet. The redundant
deformation is the additional internal shearing that occurs and which
increases with die angle. The frictional force depends on the contact area

between billet and die surface and therefore, for a constant reduction
ratio and constant coefficient of friction, the frictional force increases
with decrease in die angle.

Fig.3 shows the components of extrusion pressure plotted against the
included die angle, [3]. Since the redundant work increases with increase
in die angle and the frictional work decreases with increase in die angle
there is a minimum in the extrusion pressure curve which gives the optimum
die angle for this particular set of conditions. In practice the optimum
die angle in hydrostatic extrusion is usually in the range 30 to 60 degrees.

The ideal extrusion pressure is derived from equation [6] as,

$$P_{ideal} = \int_0^{\ln R} \bar{\sigma} \, d\bar{\varepsilon} \, . \tag{14}$$

(2) Effect of Extrusion Ratio on Extrusion Pressure

Past researchers [4] have shown that the variation between extrusion
pressure, p, and extrusion ratio, R, can be represented by the empirical
expression,

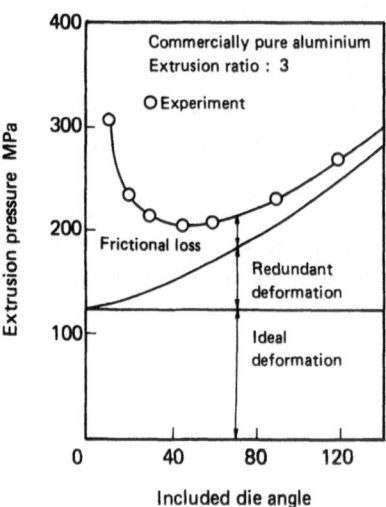

Fig.3 Components of extrusion pressure
(pure aluminium, extrusion ratio =3, MoS_2 paste lubricant)

13

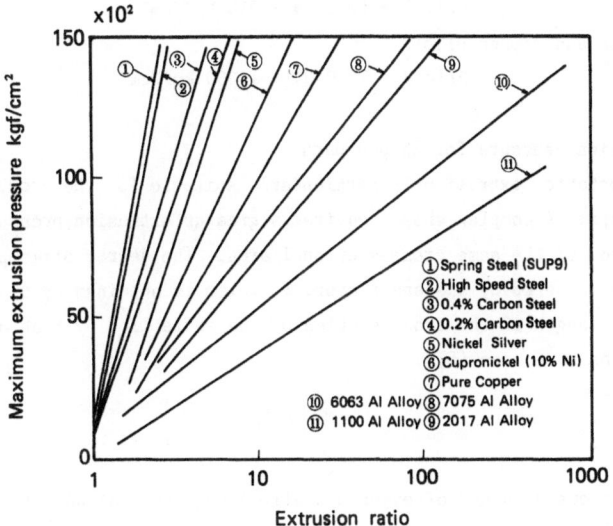

Fig.4 Extrusion pressure plotted against log R

$$p = (a + b \ln R) Y \qquad (15)$$

where a and b are constants. In hydrostatic extrusion, where the die angles are small, the value of a is small compared with $b\ln R$ at large reductions. In this case equation [15] may be approximated by,

$$p \simeq c \ln R \qquad (16)$$

where c is a constant for a particular material.

Fig.4 shows a diagram of extrusion ration vs. extrusion pressure for several materials and alloys plotted on a semi-logarithmic scale [5]. It can be observed that these experimental results can be represented by equation [16].

Since Vickers and Brinell hardness numbers denote the contact pressure in kg/mm^2 of the identors with the material they are related to the flow stress. Taking this fact into account, Nishihara (5) obtained the following empirical equations for specified groups of materials.

aluminium and aluminium alloys
$$p/\ln R = 39.4 \; Hv + 920 \; kgf/cm^2 \qquad (17)$$

steels

$$p/\ln R = 56.5 \ Hv + 210 \ kgf/cm^2 \qquad (18)$$

copper and copper alloys

$$p/\ln R = 75.0 \ Hv + 1,000 \ kgf/cm^2 \qquad (19)$$

(3) Extrusion Pressure for Shaped Bars

Hydrostatic extrusion is particularly suitable for the production of shaped bars. A complex shape requires a greater extrusion pressure than a circular rod of the same cross-sectional area. The degree of complexity of a shape is measured by a shape factor, β, which is obtained by dividing the peripheral length of the cross-section of the shape (S_s) by that of a round bar with the same area (S_r).

$$\beta = S_s \ / \ S_r \qquad (20)$$

Table 1 shows a number of experimentally determined ratios of extrusion pressure of a shaped bar (p_s) to that of a round bar (p_r), [6].

Table 1 Measured ratio of extrusion ratio of shaped bar (p_s)
to that of round bar (p_r)
(1100 Al, included die angle=45 deg., beeswax)

Shape	Extrusion ratio	S_s/S_r	p_s/p_r
	8.9	1.17	1.03
	12.0	1.52	1.12
	7.6	1.06	1/10
	3.8	1.76	1.33
	3.0	4.22	1.61

(4) Effect of Applied Force at Exit from Die

The extrusion pressure can be changed by the application of a tensile or a compressive force on the extruded product, as shown in Fig.5. If the frictional stress is not varied by the normal stress, the change in the extrusion pressure is equal to the applied stress. Thus, if p is the extrusion pressure for normal extrusion, Fig.5(a), the extrusion pressure. p' when a tensile stress σ_t is applied to the product from the exit, Fig.5(b), is given by,

$$p' = p - \sigma_t \ . \tag{21}$$

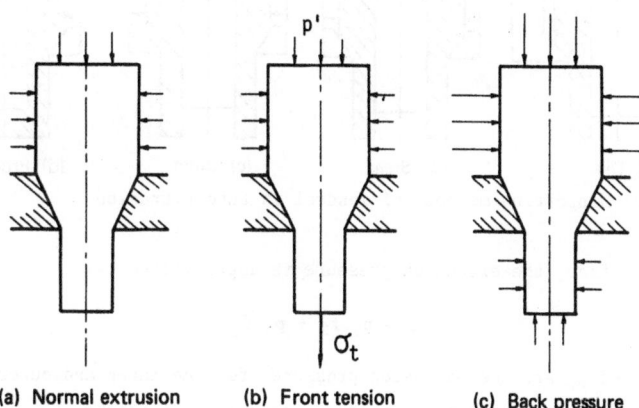

(a) Normal extrusion   (b) Front tension   (c) Back pressure

Fig.5 Extrusion with additional force from die exit

(5) Extrusion of tubes

There are several forms of tube extrusion which are distinguished by the method adopted for supporting the mandrel, see Fig.6. When the mandrel is fixed to the container or die the friction between the mandrel and the billet increases the required extrusion pressure. If the mandrel moves with the stem the frictional loss is reduced. A floating mandrel resting on the rear of the billet intensifies the axial force, thus decreasing the extrusion pressure, but the additional axial force can cause unfavourable barrelling or buckling of the billet. Finally, for the case of a floating mandrel moving with the product, the extrusion pressure is further reduced because the tensile stress occurring at the exit and the frictional stress between the mandrel and billet both aid the extrusion of the billet. However, there is a possibility of fracture of the product due to the large tensile stress.

(6) Clad materials

In the hydrostatic extrusion of a clad billet the mode of deformation is altered by the combination of the outer and inner materials [7]. When a sound product is obtained, without failure of the harder material or slip

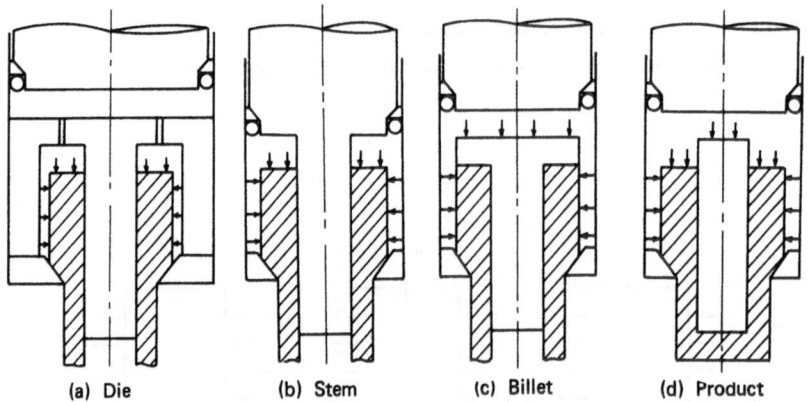

| (a) Die | (b) Stem | (c) Billet | (d) Product |

Fig.6 Supporting methods of mandrel in tube extrusion

at the interface, the extrusion pressure is approximated by,

$$p = p_i \ f_i + p_o \ f_o \qquad (22)$$

where p_i and p_o are the extrusion pressures for the inner and outer mate-rials, and f_i and f_o are the corresponding volume fractions.

4. Ductile Fracture

(1) Fracture Mode in Extrusion

In the extrusion of metals having insufficient ductility, some of the types of cracking caused by ductile fracture are as shown in Fig.7. Central cracking is the most common defect in high carbon steels. Surface cracking is sometimes observed when, for example, bismuth or magnesium alloys are extruded [1].

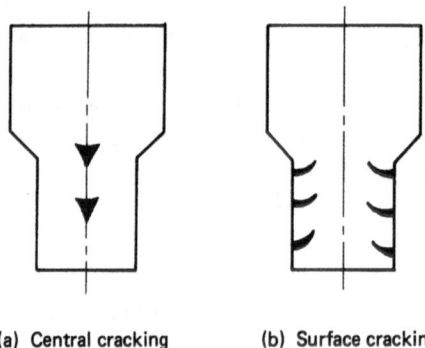

(a) Central cracking   (b) Surface cracking

Fig.7 Cracking defects in extrusion

(2) Stress State

In extrusion the compressive hydrostatic stress is low in the central region, especially when the extrusion ratio is small [8]; this is the cause of central cracking. A large tensile stress is also seen near the surface just after the die exit and the surface cracking is induced by the stress state if the extruded product has lost ductility.

(3) Criterion of Ductile Fracture

Since a compressive hydrostatic stress is known to be effective in increasing the strain to fracture, the fracture strain $\bar{\epsilon}_f$ may be expressed as a function of the hydrostatic stress component σ_m, [9],

$$\bar{\epsilon}_f = a + b \, \sigma_m \tag{23}$$

where a and b are constants of the material.

When the stress state is changed significantly during deformation, the effect of stress history must be taken into consideration. Some criteria for ductile fracture of this type are given below.

Cockcroft et al.[10]

$$\int_0^{\bar{\epsilon}_f} < \frac{\sigma_{max}}{\sigma} > d\bar{\epsilon} = c_1 \tag{24}$$

Sekiguchi et al.[11], Oyane [12]

$$\int_0^{\bar{\epsilon}_f} (a_2 + \sigma_m / \bar{\sigma}) \, d\bar{\epsilon} = c_2 \tag{25}$$

Osakada et al.[13]

$$\int_0^{\bar{\epsilon}_f} (a_3 + b_3\bar{\epsilon} + \sigma_m/\bar{\sigma}) \, d\bar{\epsilon} = c_3 \tag{26}$$

where σ_{max} denotes the maximum principal stress and
$$<x> = x \text{ when } x \geq 0 \text{ and } <x> = 0 \text{ when } x < 0.$$

5. Property of product

(1) Hardness Distribution

Because of redundant work the equivalent strain in the product at the outer surface is greater than at the centre where it is equal to the ideal strain defined by equation (14). Fig.8 shows the hardness distribution for a hydrostatically extruded copper rod [3]. The hardness becomes more uni-

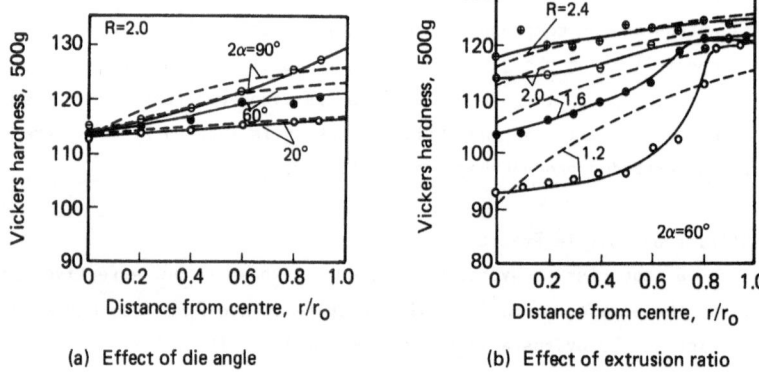

(a) Effect of die angle (b) Effect of extrusion ratio

Fig.8 Hardness distribution in extruded copper rod

form as the extrusion ratio increases, since at large strains the hardness
is not sensitive to changes in strain. The hardness also becomes more
uniform as the die angle decreases, Fig. 8(a).

(2) Mechanical Properies

It is well known that the strength of a deformed product increases
with equivalent strain, whereas the elongation and ductility decrease with
strain. If recovery or recrystallization takes place the material is softened and elongation and ductility increase. The adiabatic temperature rise
in aluminium alloy is about 410°C at an extrusion pressure of 10,000
kg/cm^2, and in copper it is 280 C°at the same pressure, and thus recovery
is expected.

Fig.9 shows the mechanical properties of hydrostatically extruded
copper rod [14]. Up to an extrusion ratio of about 10 the strength increases as the extrusion ratio increases. At larger reductions the material
is softened by the adiabatic temperature rise.

In the case of aluminium the effect of temperature rise is much more
significant. The strength of the extruded wire recovers to almost the same
level as the initial state when the extrusion ratio exceeds 100.

(3) Residual Stress

Because of non-uniform elastic recovery some residual stress is left
in the extruded material. Fig.10 shows an example of a residual stress
distribution [15]. In general, tensile residual stress is left near the
surface in the tangential and longitudinal directions [16,17]. The residual
stress can be cancelled by raising the temperature.

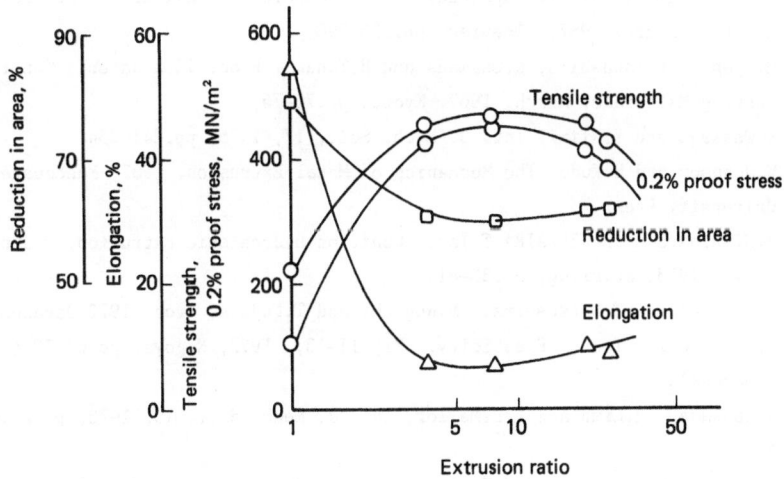

Fig.9 Mechanical properties of copper after hydrostatic extrusion

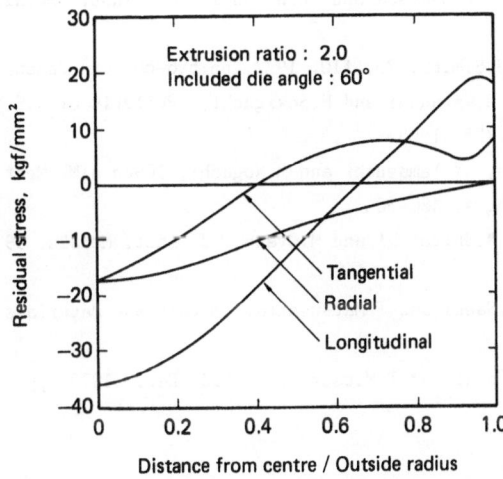

Fig.10 Residual stress in copper rod after hydrostatic extrusion
(extrusion ratio = 2, included die angle = 60 deg.)

REFERENCES

1 H.Ll.D.Pugh, in H.Ll.D.Pugh (Ed.), The Mechanical Behaviour of Materials under Pressure, 1987, Elsevier, pp.525-590.

2 M.Oyane, F.Takashima, K.Osakada and H.Tanaka, Proc. 10th Japan Congr. Testing Materials, March, 1967, Kyoto, pp.72-76.

3 K.Osakada and Y.Niimi, Int. J. Mech. Sci., 17, 1975, pp.241-254.

4 W.Johnson and H.Kudo, The Mechanics of Metal Extrusion, 1962, Manchester University Press.

5 M.Nishihara, in NEL/AIRAPT Int. Conf. on Hydrostatic Extrusion, June 13-15, 1973, Stirling, pp.33-81.

6 Y.Yamaguchi, T.Matsushita, M.Noguchi and T.Fujita, Proc. 1972 Japanese Spring Conf. Tech. Plasticity, May 11-13, 1972, Nagoya, pp.69-72 (in Japanese).

7 K.Osakada, M.Limb and P.B.Mellor, Int. J. Mech. Sci., 15, 1973, pp.291-307.

8 K.Iwata, K.Osakada and S.Fujino, Trans. ASME, Ser.B, 94-2,1972,pp.697-703.

9 P.W.Bridgman, Studies in Large Plastic Flow and Fracture, McGraw-Hill, 1952.

10 M.G.Cockcroft and D.J.Latham, J. Inst. Metals, 96, 1968, pp.33-39.

11 H.Sekiguchi, K.Osakada and H.Hayashi, J. Inst. Metals, 101, 1973, pp.167-171.

12 M.Oyane, J.J.S.M.E., 75, 110, 1972, pp.596-601 (in Japanese).

13 K.Osakada, A.Watadani and H.Sekiguchi, Bulletin of J.S.M.E., 20-150, 1977, pp.1557-1565.

14 T.Matsushita, Y.Yamaguchi and M.Noguchi, 109th AIME Meeting, Feb.24-28, 1980, Las Vegas, A80-30.

15 K.Osakada, N.Shiraishi and M.Oyane, J. Inst. Metals, 99, 1971, pp.341-344.

16 S.Miura, Y.Saeki and T.Matsushita, Metals and Materials, Oct., 1973, pp.441-447.

17 S.Miura, Y.Saeki and T.Matsushita, ibid, Dec., 1973, pp.547-551.

SECTION 2
MECHANICS OF HYDROSTATIC EXTRUSION

K. OSAKADA

HIROSHIMA UNIVERSITY

P.B. MELLOR

BRADFORD UNIVERSITY

1. Analytical Methods for Extrusion

The slab method was developed to calculate the stress and load by integrating the differential equation of equilibrium under a simplified stress state. Because of the simplicity in handling, this method has been applied to many forming processes. When extruding an ideally plastic material with a yield stress Y through a conical die of an included angle α at a extrusion ratio R is given as follows when the friction coefficient between the die and billet is μ,

$$p = Y\{(1 + \frac{1}{\mu \cot \alpha})(R^{\mu \cot \alpha} - 1) + \frac{4\alpha}{3\sqrt{3}}\}. \tag{1}$$

For $\mu \cot \alpha \ll 1$, the above equation is simplified to;

$$p = Y\{(1 + \mu \cot \alpha)\ln R + \frac{4\alpha}{3\sqrt{3}}\}. \tag{2}$$

The last term in brackets takes into account the redundant deformation.

The slip line field method presents an accurate solution for plane strain problems with rigid-perfectly plastic materials. In this method, exact stress and velocity distributions are derived by the Hencky and Geiringer equations respectively. The book by Johnson and Kudo [1] introduces various slip line fields for extrusion. Although the series method by Ewing [2] and the matrix method by Collins [3] were proposed later, these methods do not seem to be used in practical situations because they are limited to plane strain problems. Thomason [4] tried to extend the slip line field solution to axisymmetric extrusion by assuming linear variation from a uniaxial stress state at the axis to a plane strain condition at the surface contacting the die. This method is also of limited value in predicting the state of deformation in extrusion.

The upper and lower bound methods were developed on the basis of the variational principle and some constraining conditions were removed. In the upper bound analysis, an overestimate for the exact load is obtained from a kinematically admissible velocity field which satisfies the condition of incompressibility and the boundary conditions of velocity. The lower bound analysis provides an underestimate for exact load from a statically admissible stress field. The upper bound technique has been extensively applied to hydrostatic extrusion, because kinematically admissible velocity fields for extrusion can be obtained easily. Avitzur [5] assumed a special velocity field with the material within the die flowing towards the apex of the die. For a constant friction stress $mY/\sqrt{3}$ (m: friction shear factor $0\leq m\leq 1$) the extrusion pressure is given by,

$$p = Y\{(g(\alpha) + \frac{m}{\sqrt{3}}\cot\alpha)\ln R + \frac{2}{\sqrt{3}}(\frac{\alpha}{\sin^2\alpha} - \cot\alpha)\} ,$$

where

$$g(\alpha) = \frac{1}{\sin^2\alpha}[1 - \cos\alpha\sqrt{1 - \frac{11}{12}\sin^2\alpha}$$
$$+ \frac{1}{\sqrt{132}}\ln\{(1 + \sqrt{\frac{11}{12}})/(\sqrt{\frac{11}{12}}\cos\alpha + \sqrt{1 - \frac{11}{12}\sin^2\alpha})\}].$$

(3)

The finite element method is one of the numerical methods for solving differential equations governing engineering problems by dividing the material into many hypothetical elements interconnected at nodal points. For analysis of metal forming, there are two schemes of finite element method, i.e., an elastic-plastic one and a rigid-plastic one. In the later sections, the upper bound method and the finite element method are introduced in more detail.

2. Stress and Strain

(1) Stress

To express the equations in the Cartesian coordinates in simple forms, it is convenient to use the stress tensor;

$$\sigma_{ij} = \begin{bmatrix} \sigma_{xx} & \sigma_{xy} & \sigma_{xz} \\ \sigma_{yx} & \sigma_{yy} & \sigma_{yz} \\ \sigma_{zx} & \sigma_{zy} & \sigma_{zz} \end{bmatrix} = \begin{bmatrix} \sigma_x & \tau_{xy} & \tau_{xz} \\ \tau_{yx} & \sigma_y & \tau_{yz} \\ \tau_{zx} & \tau_{zy} & \sigma_z \end{bmatrix} .$$

(4)

The first subscript of a stress component refers to the direction of the normal to the area over which the stress is transmitted, and the second subscript refers to the direction of the force component.

The stress components should satisfy the equations of equilibrium in the body:

$$\frac{\partial \sigma_x}{\partial x} + \frac{\partial \tau_{xy}}{\partial y} + \frac{\partial \tau_{xz}}{\partial z} + F_x = 0 \ ,$$

$$\frac{\partial \tau_{yx}}{\partial x} + \frac{\partial \sigma_y}{\partial y} + \frac{\partial \tau_{yz}}{\partial z} + F_y = 0 \ , \qquad (5)$$

$$\frac{\partial \tau_{zx}}{\partial x} + \frac{\partial \tau_{zy}}{\partial y} + \frac{\partial \sigma_z}{\partial z} + F_z = 0 \ .$$

where F_x, F_y and F_z are components of body forces per unit volume. These equations are expressed by tensor simply by;

$$\frac{\partial \sigma_{ij}}{\partial x_j} + F_i = 0 \ . \qquad (6)$$

From the equilibrium of moment, it can be shown that the first and the second subscripts are exchangeable:

$$\sigma_{ij} = \sigma_{ji} \ . \qquad (7)$$

In the finite element method, the stress is often presented by a matrix;

$$\{\sigma\} = \begin{Bmatrix} \sigma_x \\ \sigma_y \\ \sigma_z \\ \tau_{yz} \\ \tau_{zx} \\ \tau_{xy} \end{Bmatrix} = \{\sigma_x \quad \sigma_y \quad \sigma_z \quad \tau_{yz} \quad \tau_{zx} \quad \tau_{xy}\}^T \ . \qquad (8)$$

The hydrostatic stress component is the average value of normal stress components:

$$\sigma_m = (\sigma_x + \sigma_y + \sigma_z)/3 = \sigma_{ii}/3 \ . \qquad (9)$$

In the theory of plasticity, the deviatoric stress tensor defined in the following is often used.

$$\sigma'_{ij} = \sigma_{ij} - \delta_{ij}\,\sigma_m \ , \qquad (10)$$

where δ_{ij} is the Kronecker delta which is 1 when i=j, and 0 when i≠j.

(2) Strain and Strain Rate

The small strain at a point of the body is defined by tensor as,

$$\varepsilon_{ij} = \frac{1}{2}\left(\frac{\partial w_i}{\partial x_i} + \frac{\partial w_i}{\partial x_j}\right), \qquad (11)$$

where w_i and w_j are the components of displacement.

The strain rate is the increasing rate of strain in terms of time;

$$\dot{\varepsilon}_{ij} = \frac{d\varepsilon_{ij}}{dt} = \frac{1}{2}\left(\frac{\partial v_j}{\partial x_i} + \frac{\partial v_i}{\partial x_j}\right) \tag{12}$$

where v_i and v_j are the components of velocity. It should be noted that the engineering shear strain rate, e.g., $\dot{\gamma}_{yz}$ is double the tensor component of shear strain, e.g., $\dot{\varepsilon}_{yz}$:

$$\dot{\gamma}_{yz} = \dot{\varepsilon}_{yz} + \dot{\varepsilon}_{zy} = 2\dot{\varepsilon}_{yz} \ , \quad \dot{\gamma}_{zx} = 2\dot{\varepsilon}_{zx} \ , \quad \dot{\gamma}_{xy} = 2\dot{\varepsilon}_{xy} \ . \tag{13}$$

The matrix presentation of strain rate is given by;

$$\{\dot{\varepsilon}\} = \{\dot{\varepsilon}_x \ \ \dot{\varepsilon}_y \ \ \dot{\varepsilon}_z \ \ \dot{\gamma}_{yz} \ \dot{\gamma}_{zx} \ \dot{\gamma}_{xy}\}^T \ . \tag{14}$$

The volumetric strain rate is defined by;

$$\dot{\varepsilon}_v = \dot{\varepsilon}_x + \dot{\varepsilon}_y + \dot{\varepsilon}_z = \dot{\varepsilon}_{ii} \ . \tag{15}$$

The rotating velocity in the counter clockwise direction is defined by;

$$\omega_{ij} = \frac{1}{2}\left(\frac{\partial v_j}{\partial x_i} - \frac{\partial v_i}{\partial x_j}\right) \ . \tag{16}$$

(3) Mises Yield Criterion

In many of the analyses of metal forming problems, the material is assumed to obey von Mises yield criterion:

$$(\sigma_y - \sigma_z)^2 + (\sigma_z - \sigma_x)^2 + (\sigma_x - \sigma_y)^2$$

$$+ 6(\tau_{yz}^2 + \tau_{zx}^2 + \tau_{xy}^2) - 2\bar{\sigma}^2 = 0 \ , \tag{17}$$

where $\bar{\sigma}$ is the yield or flow stress. This yield criterion can be expressed by the deviatoric stress tensor as;

$$3\sigma'_{ij} \ \sigma'_{ij} - 2\bar{\sigma}^2 = 0 \ . \tag{18}$$

(4) Flow Rule for Rigid-Plastic Deformation

If the plastic deformation is large, the elastic deformation may be neglected in the analyses. In this case, the strain or strain rate components are only plastic ones, and this material property is called "rigid-plastic". By assuming the left-hand side of equation (17) to be a plastic potential, the relation between deviatoric strain rate and stress components is given by the Lévy-Mises equation:

$$\dot{\varepsilon}_{ij} = \frac{3}{2} \frac{\dot{\bar{\varepsilon}}}{\bar{\sigma}} \sigma'_{ij} \ . \tag{19}$$

where $\dot{\bar{\varepsilon}}$ is the equivalent strain rate:

$$\dot{\bar{\varepsilon}} = \sqrt{\frac{2}{9}\{(\dot{\varepsilon}_y - \dot{\varepsilon}_z) + (\dot{\varepsilon}_z - \dot{\varepsilon}_x) + (\dot{\varepsilon}_x - \dot{\varepsilon}_y) + \frac{3}{2} (\dot{\gamma}_{yz} + \dot{\gamma}_{zx} + \dot{\gamma}_{xy})\}}$$

$$= \sqrt{(3/2)\dot{\varepsilon}_{ij}\dot{\varepsilon}_{ij}} \ . \tag{20}$$

This flow rule implicitly includes the condition of incompressibility;

$$\dot{\varepsilon}_v = \dot{\varepsilon}_x + \dot{\varepsilon}_y + \dot{\varepsilon}_z$$

$$= \frac{3}{2}\frac{\dot{\bar{\varepsilon}}}{\bar{\sigma}}(\sigma'_x + \sigma'_y + \sigma'_z) = 0 \ . \tag{21}$$

The equivalent strain is obtained by integrating the equivalent strain rate in terms of time:

$$\bar{\varepsilon} = \int \dot{\bar{\varepsilon}} \ dt \ , \tag{22}$$

and it is used as a measure of the degree of work-hardening.

$$\bar{\sigma} = H(\bar{\varepsilon}) \ . \tag{23}$$

(5) Flow Rule for Elastic-Plastic Deformation

In elastic-plastic deformation, there is a relation between strain rate and stress rate called the Prandtl-Reuss relation. A strain rate component is expressed in terms of elastic strain rate, which is related with stress rate by the Hooke equation, and plastic strain rate, which is not related with the stress rate components but with the stress components as the Lévy-Mises equation.

$$\dot{\varepsilon}_{ij} = \frac{\dot{\sigma}'_{ij}}{2G} + \delta_{ij}\frac{\dot{\sigma}_m}{3K} + \frac{3\dot{\bar{\varepsilon}}}{2\bar{\sigma}} \sigma'_{ij}$$

$$= \frac{\dot{\sigma}'_{ij}}{2G} + \delta_{ij}\frac{\dot{\sigma}_m}{3K} + \frac{3\dot{\bar{\sigma}}}{2\bar{\sigma}H'} \sigma'_{ij} \tag{24}$$

where

$$H' = \frac{d\bar{\sigma}}{d\bar{\varepsilon}} \ . \tag{25}$$

The inversion of the relation was carried out by Hill [6] as,

$$\dot{\sigma}_{ij} = 2G(\dot{\varepsilon}_{ij} + \frac{\nu}{1 - 2\nu}\delta_{ij}\dot{\varepsilon}_{ii} - \sigma'_{ij}\frac{\sigma'_{kl}\dot{\varepsilon}_{kl}}{S}) \tag{26}$$

where

$$S = \frac{2}{3} \bar{\sigma}^2 (1 + \frac{1}{3G} H') \quad . \tag{27}$$

In the elastic-plastic finite element method, the above result is used in the form of matrix as demonstrated by Yamada et al. [7]:

$$\{\sigma\} = [D^P]\{\dot{\varepsilon}\} \quad , \tag{28}$$

where

$$[D^P] = \begin{bmatrix} \frac{1-\nu}{1-2\nu} - \frac{\sigma_x'^2}{S} & & & & & \\ \frac{\nu}{1-2\nu} - \frac{\sigma_x'\sigma_y'}{S} & \frac{1-\nu}{1-2\nu} - \frac{\sigma_y'^2}{S} & & \text{SYMMETRY} & & \\ \frac{\nu}{1-2\nu} - \frac{\sigma_x'\sigma_z'}{S} & \frac{\nu}{1-2\nu} - \frac{\sigma_y'\sigma_z'}{S} & \frac{1-\nu}{1-2\nu} - \frac{\sigma_z'^2}{S} & & & \\ -\frac{\sigma_x'\tau_{yz}}{S} & -\frac{\sigma_y'\tau_{yz}}{S} & -\frac{\sigma_z'\tau_{yz}}{S} & \frac{1}{2} - \frac{\tau_{yz}^2}{S} & & \\ -\frac{\sigma_x'\tau_{zx}}{S} & -\frac{\sigma_y'\tau_{zx}}{S} & -\frac{\sigma_z'\tau_{zx}}{S} & -\frac{\tau_{yz}\tau_{zx}}{S} & \frac{1}{2} - \frac{\tau_{zx}^2}{S} & \\ -\frac{\sigma_x'\tau_{xy}}{S} & -\frac{\sigma_y'\tau_{xy}}{S} & -\frac{\sigma_z'\tau_{xy}}{S} & -\frac{\tau_{yz}\tau_{xy}}{S} & -\frac{\tau_{zx}\tau_{xy}}{S} & \frac{1}{2} - \frac{\tau_{xy}^2}{S} \end{bmatrix}$$

$$\tag{29}$$

3. Upper Bound Method

(1) Upper Bound Theorem

Let us consider that the whole body of a rigid-plastic material is deforming. The traction T on the surface of the body is equilibrated with the internal stress;

$$T_i = \sigma_{ij} n_j \quad , \tag{30}$$

where n_j is the outward unit normal of the surface. The surface of the body may be divided into two parts S_t and S_v from the viewpoint of the boundary conditions. The traction is prescribed on the surface S_t as follows;

$$\bar{T}_i - T_i = 0 \quad . \tag{31}$$

The velocity v_i is prescribed on the surface S_v,

$$\bar{v}_i - v_i = 0 \quad . \tag{32}$$

Let σ_{ij}, $\dot{\varepsilon}_{ij}$ and v_i denote the exact stress, strain rate and velocity. The rate of energy dissipation in a unit volume is given by,

$$\bar{\sigma} \, \dot{\bar{\epsilon}} = \sigma_{ij} \dot{\epsilon}_{ij} = \sigma'_{ij} \dot{\epsilon}_{ij} \; . \tag{33}$$

Also, let v_i^* be a kinematically admissible velocity field which satisfies the incompressibility condition and the velocity boundary conditions on S_v and $\dot{\epsilon}_{ij}^*$ be the strain rate derived from v_i^*.

By using the principle of virtual work:

$$\int_V \sigma_{ij} \dot{\epsilon}_{ij}^* \cdot dV = \int_{S_t} \bar{T}_i v_i^* dS + \int_{S_v} T_i \bar{v}_i dS \quad , \tag{34}$$

and Schwarz's inequality:

$$\sigma'_{ij} \dot{\epsilon}_{ij} \leq \sqrt{\sigma'_{ij}\sigma'_{ij}} \; \sqrt{\dot{\epsilon}_{kl}\dot{\epsilon}_{kl}} \quad , \tag{35}$$

Markov's principle [8] is derived: among admissible velocity fields, the actual field renders the following functional an absolute minimum.

$$\Phi_1 = \int_V \bar{\sigma} \, \dot{\bar{\epsilon}} *dV - \int_{S_t} \bar{T}_i v_i^* dS \tag{36}$$

This functional cannot be extended to a strain rate sensitive material, because the flow stress of the material is not determined without knowing the strain rate beforehand.

For an ideally plastic material, a kinematically admissible velocity field can include velocity discontinuity surface S_s. Further, the interfacial friction with constant frictional stress τ_f can also be considered on tool surface S_f. Then, the following inequality is derived;

$$\int_{S_v} T_i \bar{v}_i dS \leq \int_V \bar{\sigma} \, \dot{\bar{\epsilon}} *dV - \int_{S_t} \bar{T}_i v_i^* dS + \int_{S_s} \frac{\bar{\sigma}}{\sqrt{3}} \Delta v* dS + \int_{S_f} \tau_f \Delta v_f^* dS \tag{37}$$

where Δv^* and Δv_f are the absolute values of velocity discontinuity along S_s and of sliding velocity on S_f, respectively. The right-hand side of equation (37) gives an upper bound for the rate of work of the unknown tractions acting on S_v.

(2) Velocity Fields for Axisymmetric Extrusion

In 1960s and early 1970s, extensive attempts were made to apply the upper bound theorem to axisymmetric extrusion, which was an important subject of cold forging. Ring-shaped elements of parallel flow having a rectangular and a triangular cross section were first introduced by Kudo [9]. Halling and Mitchell [10] proposed a hodograph method which made

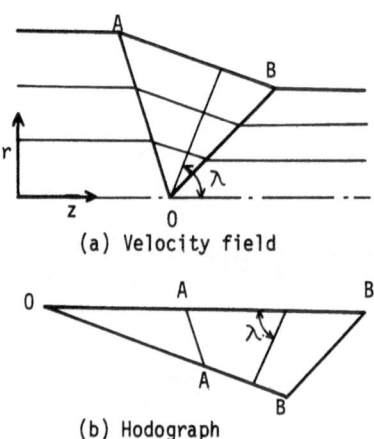

(a) Velocity field

(b) Hodograph

Fig.1 Parallel flow field for axisymmetric extrusion

possible the simple determination of the velocity distribution for the triangular ring elements. They applied the technique to extrusion as shown in Fig.1.

A flow field which has spherical boundaries was used by Avitzur [5]. Later, generalization of the boundary shape was carried out by Osakada and Niimi [11] (see Fig.2). They showed that optimizing the boundary shape resulted in extrusion pressures which were at most 10% lower than for the spherical boundary, especially when the die angle was large. Fig.3 shows the effect of frictional shear factor, m, on the upper bound value of extrusion pressure calculated from flow fields with optimized boundaries.

The stream function method was introduced to derive admissible velocity fields for axisymmetric steady-state forming processes by Lambert and Kobayashi [12] in their study of extrusion and drawing of bar and tube through a conical die. A stream function ϕ of axisymmetric flow is defined in the cylindrical system by,

$$v_z = \frac{1}{r}\frac{\partial \phi}{\partial r} \quad \text{and} \quad v_r = -\frac{1}{r}\frac{\partial \phi}{\partial r} \ . \tag{38}$$

Bar extrusion through a cosine, elliptic and hyperbolic dies, and tube drawing or extrusion through a conical die are treated by the this method. Nagpal [13] discussed the stream functions to be applied to steady-state flow processes through an arbitrarily contoured die. The use of the stream function enabled the selection of an admissible velocity field from a wider

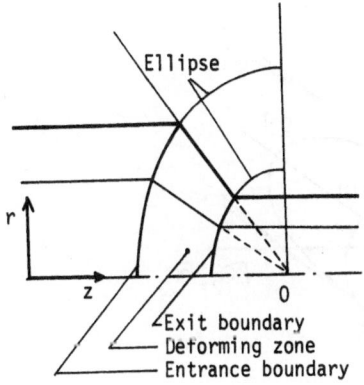

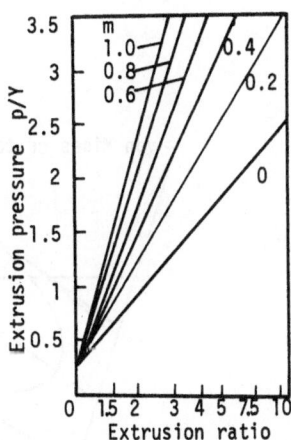

Fig.2 Converging flow field with
      elliptic boundaries

Fig.3 Upper bound value of extrusion
      pressure calculated with flow
      field shown in Fig.2 ($\alpha=15°$)

class of velocity fields.

4. Rigid-Plastic Finite Element Method

(1) Yield Criterion and Flow Rule

The stress cannot be determined uniquely from the strain rate components by the upper bound method for incompressible rigid-plastic material.
This property is brought about by the incompressibility condition of the
material. This suggests that direct stress calculation from the strain rate
components may be carried out if compressibility is allowed.

We consider a slightly compressible rigid-plastic material which obeys
the yield criterion [14]:

$$3\sigma'_{ij}\sigma'_{ij} + 2g\sigma_m - 2\bar{\sigma}^2 = 0 \tag{39}$$

where g is a small positive constant which represents the degree of compressibility, σ_m is the hydrostatic stress and $\bar{\sigma}$ is the equivalent stress
for this material. A cross section of the yield surface of this criterion
in the principal stress space is shown in Fig.4. The plane of cross section
is in parallel with the σ_z-axis and is inclined 45 degrees to the σ_x- and
σ_y-axes. The yield surface is an ellipsoid which contacts internally the
circular cylinder representing the von Mises yield criterion. The ellipsoid
approaches the Mises cylinder as g decreases.

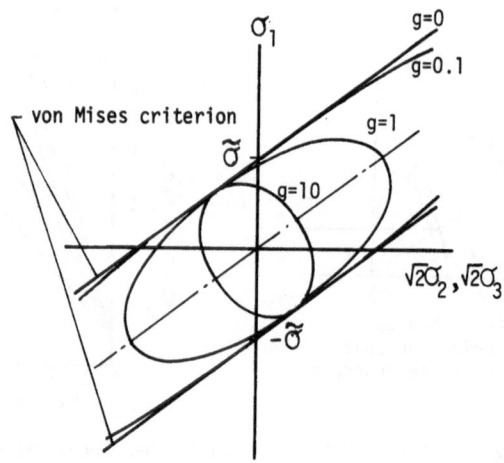

Fig.4 Yield surface for compressible material

By assuming the left-hand side of equation (39) to be a plastic potential, the stress-strain rate relations are derived as,

$$\sigma_{ij} = \frac{\tilde{\sigma}}{\dot{\varepsilon}} \{ \frac{2}{3} \dot{\varepsilon}_{ij} + \delta_{ij}(\frac{1}{g} - \frac{2}{9}) \dot{\varepsilon}_v \} , \tag{40}$$

where $\dot{\varepsilon}_v$ is the volumetric strain rate, The equivalent strain rate $\dot{\varepsilon}$ for this material is given by,

$$\dot{\varepsilon} = \sqrt{\frac{2}{3} \dot{\varepsilon}'_{ij}\dot{\varepsilon}'_{ij} + \frac{1}{g} \dot{\varepsilon}_v^2} , \tag{41}$$

where $\dot{\varepsilon}'_{ij}$ is the deviatoric strain rate defined by,

$$\dot{\varepsilon}'_{ij} = \dot{\varepsilon}_{ij} - \delta_{ij}\dot{\varepsilon}_{ii}/3 . \tag{42}$$

The material undergoes a volumetric change during plastic deformation because of the pressure sensitivity of the yield criterion. The property enables the direct calculation of stress from the strain rate components. The energy dissipation in a unit volume of this material is expresed by equation (33) in which $\bar{\sigma}$ and $\bar{\dot{\varepsilon}}$ are exchanged by σ and $\dot{\varepsilon}$ respectively.

(2) Variational Principle

Let σ_{ij}, $\dot{\varepsilon}_{ij}$ and v_i denote the exact solution for the slightly compressible material under the prescribed tractions $\bar{T}_i$ over the surface S_v and the prescribed velocities $\bar{v}_i$ on the surface S_v. Also, let v_i^* be a

kinematically admissible velocity field satisfying only the velocity bound-
ary conditions on S_v , and $\dot{\varepsilon}_{ij}*$ the strain rate obtained from the velocity
field v_i*.

For the yield criterion introduced above, a variational principle can
be derived. The exact solution renders the following functional a minimum
under the prescribed boundary conditions.

$$\Phi_2 = \int_V \bar{\sigma}\, \dot{\bar{\varepsilon}}* \, dV - \int_V \bar{T}_i v_i^* \, dS \tag{43}$$

Since incompressibility is not a prerequisite condition for this material
property, the functional is comparatively easy to minimize.

The volume change for the compressible material varies with the value
of g. Table 1 shows the degree of deviation of this yield criterion from
the von Mises criterion and the volume change at 10% reduction in height
for some values of g in simple compression. σ_{zC} and σ_{zM} are the compressive
stresses in the materials yielding under equations (39) and (17) respectively. It is seen that when the value of g is smaller than 0.01 equation
(39) provides a good approximation for the incompressibility material.

Table 1 Deviation of yield criterion equation (39) from
von Mises yield criterion in simple compression

g	$\dfrac{(\sigma_{zM} - \sigma_{zC})}{\sigma_{zM}}$ %	Volume change at 10% reduction in height %
0.1	0.6	−0.3
0.01	0.06	−0.03
0.001	0.006	−0.003
0.0001	0.0006	−0.0003

(3) Discretization into Elements

In the case of axisymmetric deformation, the meridian plane of a
billet is divided into number of elements as shown in Fig.5. The velocity
components of each nodal point u_r and u_z in r and z directions, respectively, are the unknown variables. The velocity components in an element is
interporated using the velocities of the nodal points belonging to the
element. Then the functional equation (43) can be expressed in terms of the
nodal velocities. If a combination of nodal velocities which minimizes the
functional is found, it is the best approximation of the actual velocity

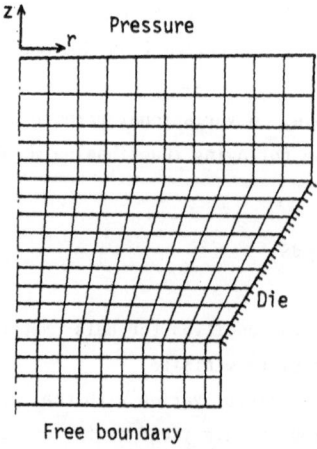

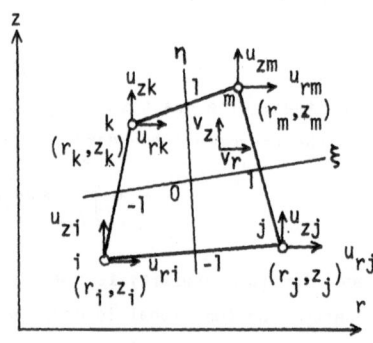

Fig.5 Finite element discretization
of extrusion billet

Fig.6 Isoparametric quadrilateral
element

field for the mesh under consideration. In general, as the mesh becomes
finer the accuracy of computation is improved but the computing time and
the necessary size of memory in CPU increase sharply. Thus, there should be
a compromize between accuracy and economy of computation.

By taking the isoparametric quadrilateral element ijmk, Fig.6, as an
example, the method of interpolation is explained. The velocity components
v_r and v_z within the element are expressed by,

$$\left.\begin{aligned}
v_r &= \tfrac{1}{4}\{(1-\xi)(1-\eta)u_{ri}+(1+\xi)(1-\eta)u_{rj}+(1+\xi)(1+\eta)u_{rm}+(1-\xi)(1+\eta)u_{rk}\} \;, \\
v_z &= \tfrac{1}{4}\{(1-\xi)(1-\eta)u_{zi}+(1+\xi)(1-\eta)u_{zj}+(1+\xi)(1+\eta)u_{zm}+(1-\xi)(1+\eta)u_{zk}\} \;,
\end{aligned}\right\} \quad (44)$$

where ξ and η are the normalized coordinates having values from 1 to -1.
u_{ri} and u_{zi} denote the velocity components of the nodal point i. The
coordinates within the element are related with ξ and η by,

$$\left.\begin{aligned}
r &= \tfrac{1}{4}\{(1-\xi)(1-\eta)r_i+(1+\xi)(1-\eta)r_j+(1+\xi)(1+\eta)r_m+(1-\xi)(1+\eta)r_k\} \;, \\
z &= \tfrac{1}{4}\{(1-\xi)(1-\eta)z_i+(1+\xi)(1-\eta)z_j+(1+\xi)(1+\eta)z_m+(1-\xi)(1+\eta)z_k\} \;,
\end{aligned}\right\} \quad (45)$$

where r_i and z_i are the coordinates at node i. In general, the strain rates
components are given by the nodal velocities of the element $\{u_e\}$,

$$\{\dot{\varepsilon}\} = [B]\{u_e\} \; . \tag{46}$$

In the case of the axisymmetric quadrilateral element,

$$
\begin{Bmatrix} \dot{\varepsilon}_r \\ \dot{\varepsilon}_\theta \\ \dot{\varepsilon}_z \\ \dot{\gamma}_{rz} \end{Bmatrix}
=
\begin{bmatrix}
B_i & 0 & B_j & 0 & B_m & 0 & B_k & 0 \\
D_i & 0 & D_j & 0 & D_m & 0 & D_k & 0 \\
0 & C_i & 0 & C_j & 0 & C_m & 0 & C_k \\
C_i & B_i & C_j & B_j & C_m & B_m & C_k & B_k
\end{bmatrix}
\begin{Bmatrix} u_{ri} \\ u_{zi} \\ u_{rj} \\ u_{zj} \\ u_{rm} \\ u_{zm} \\ u_{rk} \\ u_{zk} \end{Bmatrix} , \tag{47}
$$

where

$$
\begin{aligned}
J = &\{(-r_i+r_j+r_m-r_k) + (r_i-r_j+r_m-r_k)\eta\}\{(-z_i-z_j+z_m+z_k) \\
&+(z_i-z_j+z_m-z_k)\xi\} -\{(-r_i-r_j+r_m+r_k) + (r_i-r_j+r_m-r_k)\xi\} \\
&\times\{(-z_i+z_j+z_m-z_k) + (z_i-z_j+z_m-z_k)\eta\} \; ,
\end{aligned}
$$

$$
\begin{Bmatrix} B_i \\ B_j \\ B_m \\ B_k \end{Bmatrix}
=
\begin{Bmatrix} z_{jk}-z_{mk}\xi-z_{jm}\eta \\ -z_{im}+z_{mk}\xi+z_{ik}\eta \\ -z_{jk}+z_{ij}\xi-z_{ik}\eta \\ z_{im}-z_{ij}\xi+z_{jm}\eta \end{Bmatrix} ,
\quad
\begin{Bmatrix} C_i \\ C_j \\ C_m \\ C_k \end{Bmatrix}
=
\begin{Bmatrix} -r_{jk}+r_{mk}\xi+r_{jm}\eta \\ r_{im}-r_{mk}\xi-r_{ik}\eta \\ r_{jk}-r_{ij}\xi+r_{ik}\eta \\ -r_{im}+r_{ij}\xi-r_{jm}\eta \end{Bmatrix} ,
$$

$$
\begin{Bmatrix} D_i \\ D_j \\ D_m \\ D_k \end{Bmatrix}
=
\frac{J}{8r}
\begin{Bmatrix} (1-\xi)(1-\xi) \\ (1+\xi)(1-\eta) \\ (1+\xi)(1+\eta) \\ (1-\xi)(1+\eta) \end{Bmatrix} ,
\quad
\begin{aligned} r_{ij} &= r_i - r_j , \\ z_{ij} &= z_i - z_j . \end{aligned}
\tag{48}
$$

The equivalent strain rate $\dot{\bar{\varepsilon}}$ is a function of the strain rate components as equation (41), and thus the first term of the right hand side of equation (43) can be expressed in terms of the nodal velocities. By representing the traction on the boundary by the nodal force {F}, the second term of equation (43) becomes a fuction of the nodal velocities {u},

$$\Phi_2 = \sum_{\text{element}}(\bar{\sigma}\,\dot{\bar{\varepsilon}}\,V_e) - \{\dot{F}\}\{u\} \tag{49}$$

where V_e is the volume of element.

(4) Minimization of Functional

When the functional is minimized, the stationary condition is fulfilled; equations obtained by differentiating equation (49) with respect to the nodal velocities are equal to zero.

$$\frac{\partial \Phi 2}{\partial u_{r1}} = q_{r1}(u_{r1}, u_{z1}, u_{r2}, u_{z2}, \dots\dots) = 0 \; ,$$

$$\frac{\partial \Phi 2}{\partial u_{z1}} = q_{z1}(u_{r1}, u_{z1}, u_{r2}, u_{z2}, \dots\dots) = 0 \; ,$$

$$\left. \frac{\partial \Phi 2}{\partial u_{r2}} = q_{r2}(u_{r1}, u_{z1}, u_{r2}, u_{z2}, \dots\dots) = 0 \; , \right\} \quad (50)$$

$$\dots\dots\dots\dots\dots\dots\dots\dots\dots\dots$$

$$\dots\dots\dots\dots\dots\dots\dots\dots\dots\dots \quad .$$

Since these equations are not linear, special techniques are required to solve them.

The most commonly used method for solving non-linear equations is the Newton-Raphson method: $q_{r1}, \; q_{z1}, \; \dots, q_{ri}, \; q_{zi}, \dots$ are linearized by using trial nodal velocities $(u_{r1})_0, \; (u_{z1})_0, \dots$ which approximate the exact velocity field, and the corrective nodal velocities $du_{r1}, \; du_{z1}, \dots$ are the values to be determined. An example of the linearized equation is,

$$q_{ri} = (q_{ri})_0 + (\frac{\partial q_{ri}}{\partial u_{r1}})_0 du_{r1} + (\frac{\partial q_{ri}}{\partial u_{z1}})_0 du_{z1} + \dots\dots \quad (51)$$

where $(q_{ri})_0$ and $(\partial q_{ri}/\partial u_{r1})_0$, $(\partial q_{ri}/\partial u_{z1})_0$, $\dots$ are the values calculated from the trial velocities.

Since the corrective velocities are not always properly obtained when the trial velocity field is not very close to the actual field, the calculated corrective velocities are multiplied by a modification factor which is determined to minimize the functional, and the products are added to the trial velocities. The modification factor is usually much smaller than unity at early stages of iteration and becomes nearly unity at final stages. By employing the renewed velocity field as the trial velocity field, the iteration is continued until a convergence is attained.

The strain rate components in each element are calculated using equation (47), and then the stress components are determined by equation (40).

(5) Steady State Problem

In the analysis of non-steady state processes, the increments of the nodal points are determined by multiplying the calculated nodal velocities with the incremental time, and the nodal points are located at new positions after each incremental deformation.

Since the flow stress distribution is not determined beforehand in the case of steady state deformation, the functional expressed in equation (43) cannot be directly adopted for a work-hardening material. The procedure for an approximate analysis of the steady state problem is given below.

1) The functional is minimized by assuming the material to be rigid-perfectly plastic, and the distributions of velocity and equivalent strain rate in the whole region are calculated.

2) Stream lines are located by using the velocity field and the distribution of equivalent strain is calculated by integrating the equivalent strain rate along the stream lines from the entry boundary.

3) By using the flow curve and the equivalent strain distribution so determined, the equivalent stress distribution is given, and the functional is minimized again.

4) The above procedures are repeated until the calculated result converges.

(6) Results

Continuous extrusion of an aluminium billet is analysed. The process geometries and working conditions are as follows;

Extrusion ratio $\qquad$ R = 2

Half Die Angle $\qquad$ α = 30°

Coefficient of Friction $\qquad$ μ = 0.05

Flow Curve $\qquad$ $\bar{\sigma} = 127\bar{\varepsilon}^{0.27}$ MPa

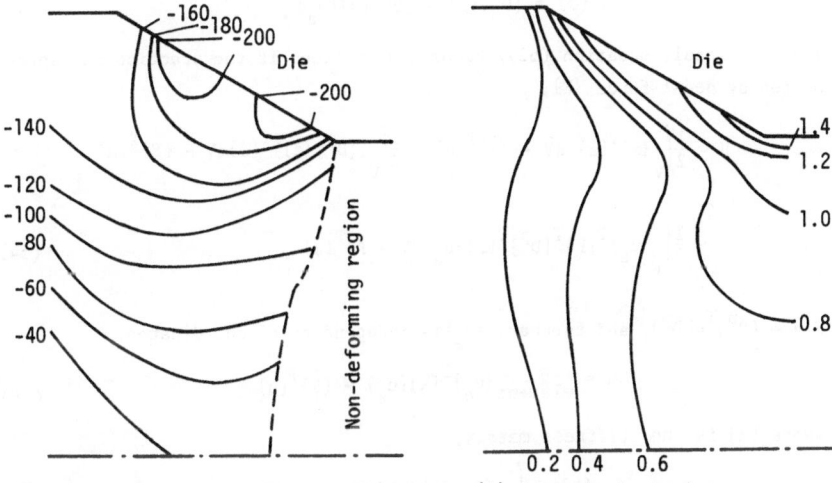

(a) Hydrostatic stress MPa   (b) Equivalent strain

Fig.7 Stress and strain distributions in extrusion billet calculated by rigid-plastic finite element method

Fig.7 (a) is the hydrostatic stress distribution in the deforming region. High contact pressure regions are observed near the entrance and exit corners. The hydrostatic compressive pressure is lowest at the centre. Fig.7 (b) shows the distribution of equivalent strain in the billet. The strain at the centre is equal to ln R after extrusion and the equivalent strain increases towards the surface.

5. Elastic-Plastic Finite Element Method

(1) Variational Principle

For elastic-plastic deformation, the following variational principle was proposed by Hill[6]. Among kinematically admissible velocity fields v*, which allow elastic compressibility, the exact solution renders functional

$$\Pi = \frac{1}{2}\int_V \dot{\sigma}^*_{ij} \; \dot{\varepsilon}^*_{ij} \; dV \; - \; \int_{S_t} \dot{\overline{T}}_i v^*_i \; dS \; , \tag{52}$$

a minimum. It should be noted that the stress rate $\dot{\sigma}_{ij}^*$, not stress σ_{ij}^* associated with the strain rate $\dot{\varepsilon}_{ij}^*$ is used in this expression.

(2) Finite Element Formulation

Let us divide the body into a number of elements, and consider that the nodal velocities {u} are unknown variables. The strain rate is expressed by the nodal velocities $\{u_e\}$ of the element using equation (46),

$$\{\dot{\sigma}\} = [D^P]\{\dot{\varepsilon}\} = [D^P][B]\{u_e\} . \tag{53}$$

The functional, equation (52), becomes as follows if the traction is represented by nodal force {F},

$$\Pi = \frac{1}{2}\int_V \{\dot{\sigma}\}^T\{\dot{\varepsilon}\} \; dV - \{\dot{F}\}^T\{u\} = \frac{1}{2}\int_V ([D^P]\{\dot{\varepsilon}\})^T\{\dot{\varepsilon}\}dV - \{\dot{F}\}^T\{u\}$$

$$= \frac{1}{2}\int_V \{u_e\}^T[B]^T[D^P]^T[B]\{u_e\}dV - \{\dot{F}\}^T\{u\} . \tag{54}$$

Since $[D^P]^T = [D^P]$, and further, $\{u_e\}$ is independent of coordinates,

$$\Pi = \sum_{\text{element}} \{u_e\}^T[k]\{u_e\} - \{\dot{F}\}^T\{u\} , \tag{55}$$

where [k] is the stiffness matrix,

$$[k] = \int_{V_e} [B]^T[D^P][B]dV . \tag{56}$$

To minimize the functional, equation (55) is differentiated by each of the nodal velocities u_m;

$$\frac{\partial \Pi}{\partial u_m} = \sum_{element} \begin{Bmatrix} 0 \\ 0 \\ \vdots \\ 0 \\ 1 \\ 0 \\ \vdots \\ 0 \\ 0 \end{Bmatrix} [k]\{u_e\} - \dot{F}_m = 0 . \tag{57}$$

The full equations are given by,

$$[K]\{u\} - \{\dot{F}\} = 0 . \tag{58}$$

By solving the linear equations, the nodal velocities are obtained. Then the strain rate and stress rate in each element are calculated using equations (46) and (53). More detailed explanation of the calculation method is given in reference [15]. The nodal velocities and the equivalent plastic strain rate are multiplied by the increment of time and added to the nodal coordinates and the equivalent strain, respectively, to obtain the new nodal positions and the equivalent strain. Further, the stress rate components are also multiplied by the increment of time and is added to to the stress components. Using these results, the next step of calculation is carried out.

The above explained method is based on the infinitesimal deformation theory, and allows accurate calculation of the initial state of extrusion in which the plastic strain is in the same order as the elastic strain. In large plastic strains, however, error of stress calculation is accumulated due to 1) the neglect of rotation and shape change in the stiffness matrix, and 2) the finite step of time interval. These problems can be overcome by applying the method based on the finite deformation theory [16], and by using very small time interval. By employing these methods, however, actual metal forming problems are still difficult to solve by the elastic-plastic finite element method. This is because there are corners where the billet begins to contact the tool surface and the stress state is changed sharply; serious error which is inherited to subsequent stages tends to be caused when an element is passing the corner. For the same reason, steady state analysis of extrusion is difficult. Thus, it may be practical to prepare both the rigid-plastic and elastic-plastic finite element methods and use an appropriate one or combine them depending on the purpose of analysis.

38

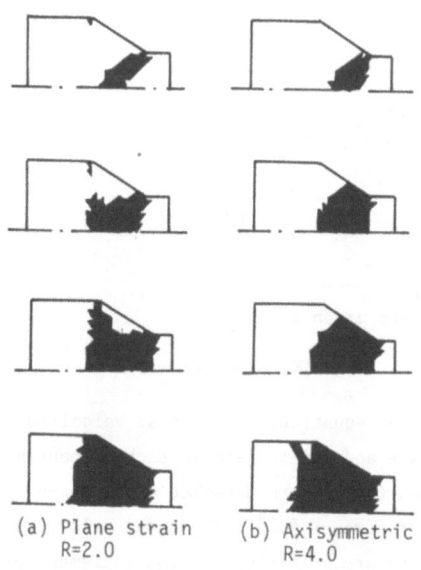

(a) Plane strain   (b) Axisymmetric
    R=2.0   R=4.0

Fig.8 Spread of plastic zone in plane-strain
and axisymmetric hydrostatic extrusion

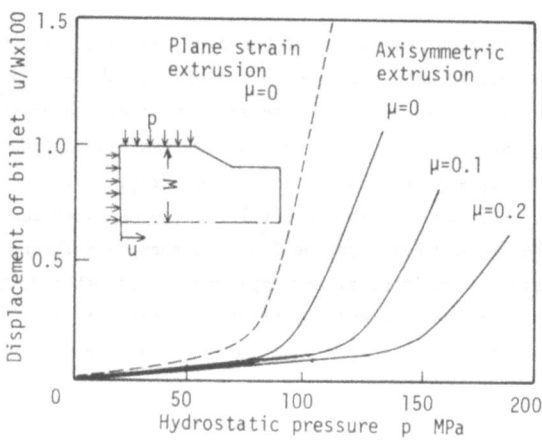

Fig.9 Effect of friction coefficient on pressure-displacement curve
(R=2, α=30°)

(3) Results [17]

The material considered is annealed pure copper and the simplified flow curve is employed;

Young's modulus	E	= 1.2	GPa
Poisson's ratio	ν	= 0.34	
Initial yield stress	Y	= 80	MPa
Work-hardening coefficient	H'	= 550	MPa

Fig.8 shows the spread of the plastic zone with increasing displacement in plane-strain and axisymmetric extrusions. The extrusion ratio is 4.0 for the axisymmetric case, whereas it is 2.0 for the plane-strain case.

The effect of friction coefficient on pressure-displacement curve is given in Fig.9. A curve for plane-strain extrusion is also shown for comparison.

REFERENCES

1 W.Johnson and H.Kudo, The Mechanics of Metal Extrusion, Manchester Univ. Press, 1962.

2 D.J.F.Ewing, J. Mech. Phys. Solids, 15(1967), pp.105-114.

3 I.F.Collins, Proc. Roy. Soc. London, A, 303(1968), pp.317-338.

4 P.F.Thomason, Proc. Instn. Mech. Engrs., 184 Pt.1 (1969-70), pp.896-911.

5 B.Avitzur, Trans. ASME, Ser.B, 86(1964), pp.305-316.

6 R.Hill, Mathematical Theory of Plasticity, Oxford Univ. Press, 1950

7 Y.Yamada, N.Yoshimura and T.Sakurai, Int. J. Mech. Sci., 10(1968), pp.343-354.

8 K.Washizu, Variational Methods in Elasticity and Plasticity, Pergamon Press, 1968

9 H.Kudo, Int. J. Mech. Sci., 2(1961), pp.102-127, 3(1961), pp.91-117.

10 J.Halling and L.A.Mitchell, ibid, 7(1965), pp.277-295.

11 K.Osakada and Y.Niimi, ibid, 17(1975), pp.241-254.

12 E.R.Lambert and S.Kobayashi, Proc. JSME Semi Int. Symp., Experimental Mechanics Vol.2, Tokyo, (1967), pp.53-61.

13 V.Nagpal, Trans. ASME, Ser.B, 96(1974), pp.1197- 1201.

14 K.Osakada, J.Nakano and K.Mori, Int. J. Mech. Sci., 24(1982), pp.459-468.

15 O.C.Zienkiewicz, The Finite Element Method in Engineering Science, McGrawhill, 1971.

16 E.H.Lee,et al., Computer Methods Appl. Mech. Engg., 10(1977), pp.339–353.

17 K.Iwata, K.Osakada and S.Fujino, Trans. ASME, Ser.B, 94(1972), pp.697–703.

SECTION 3

LUBRICATION IN HYDROSTATIC EXTRUSION

WILLIAM R. D. WILSON

NORTHWESTERN UNIVERSITY

1. Introduction

As in most metalforming processes, considerations of friction and lubrication are of vital importance in hydrostatic extrusion. An efficient lubrication system will reduce the pressure required for a given reduction ratio by lowering friction at the billet-die interface. This will reduce the container strength or number of steps required for a particular product with attendant reductions in cost.

Friction control is particularly important during the early stages of hydrostatic extrusion because of the tendency to have a pressure peak due to a temporary lubrication breakdown shortly after the billet starts moving. The pressure peak fixes the maximum container loading during the extrusion cycle. The sudden drop in the pressure required for extrusion as the lubrication system recovers may result in a loss of control of billet speed with the result that the billet is violently ejected or enters into stick-slip motion.

Good lubrication is also vital in controlling die wear. While this is relatively unimportant in the laboratory it is of crucial importance in production. This is particularly true where large volumes of a small section product such as wire are being produced.

Product quality is often impacted by the lubrication system. The incidence of surface defects such as scoring which result from metal-to-metal contact is reduced by the presence of a thick lubricant film at the billet-die interface. On the other hand, if the film is too thick a roughened product surface is produced by the inhomogeneous deformation of individual grains. Residual stresses and associated defects such as surface cracking are reduced by effective lubrication which tends to promote more homogeneous deformation and temperature fields in the product.

It is important to understand that lubrication in hydrostatic extrusion must be treated as a system in which a number of factors including the geometry of the conjunction, surface motions, temperatures and micro-topographies and the physical and chemical properties of the billet, die and lubricant all can play a role. An important concept is that the system can operate in any of a variety of regimes where different mechanisms provide the lubrication action and where different combinations of the factors are important. The regimes are the thick film, thin film, mixed and boundary regimes. These are illustrated in Figure 1.

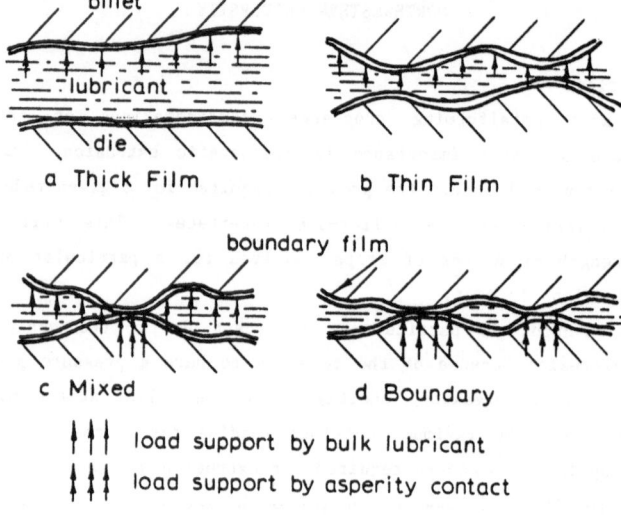

Fig. 1. Lubrication regime in hydrostatic extrusion.

In the thick film regime shown in Figure 1a, the surfaces are separated by a continuous lubricant film which is much thicker than either the roughnesses of the surfaces involved or the molecular size of the lubricant. In most hydrostatic extrusion systems with relatively smooth dies this will require a film thickness of about ten times the RMS roughness of the billet. Under these circumstances the roughness has little influence on the lubrication process and the lubricant can be treated as a continuum between smooth surfaces. Lubrication is essentially a mechanical process and friction is controlled by the local lubricant film thickness and sliding velocity and the mechanical properties of the lubricant under the conditions at the interface. In this regime friction tends to be very small. While the idea of a coefficient of friction is of little value in characterizing this regime, the measured coefficient would generally be less than 0.05 and particularly with liquid lubricants it could be much smaller. Since the surfaces are completely separated by a relatively thick lubricant film, wear is unlikely. However wear can occur due to corrosion, erosion by abrasive particles or surface fatigue. The lubricant film is generally relatively compliant so that the die surface has little constraining influence on the micro-deformation of individual grains in the billet surface. Inhomogeneous deformation on the grain scale thus leads to workpiece surface roughening often to a level similar to that found in free surface deformation to the same strain [1,2,3]. This roughening is used as an indication that the system is operating in the thick film regime. It may result in a product with a surface that is unacceptable for certain applications.

In the thin film regime shown in Figure 1b the lubricant film thickness is typically from three to ten times the RMS billet roughness. In this regime surface roughness influences lubricant flow and generates local ripples in the interface pressure. However the local lubricant film thickness is generally much larger than the molecular size of the lubricant. A negligible fraction of the load is carried by "contacts" between individual roughness elements (asperities). Lubrication and friction is still dominated by mechanics in the thin film regime but surface roughness now plays an important role. Friction levels are generally similar to those in the thick film regime but may be influenced by roughness. Wear is usually slight. Workpiece surface roughening can occur but it is more common for the roughness to remain essentially constant or to decrease during forming. Roughness changes are specially important because they can modify the lubrication process.

In the mixed lubrication regime shown in Figure 1c, the mean film thickness is generally less than three times the RMS roughness of the billet. The interface loading is shared between the pressurized bulk lubricant film in the roughness valleys and "contacts" at the asperity peaks. In fact, if the lubricant is properly formulated, actual metal-to-metal contact will not occur at asperity peaks because of the presence of a tightly adhering "boundary" film. This film is formed as a result of the chemical reaction or physical adsorption of compounds in the lubricant. Even though the boundary films typically have thicknesses of the order of the lubricant molecular size, they can prevent metal-to-metal contact, welding and pick-up during asperity collisions.

The mixed lubrication regime is extremely complicated and difficult to model. Different lubrication mechanisms are active at the asperity valleys and peaks. In the valleys the films are relatively thick and a modified form of mechanics-based thin-film lubrication theory can be used. At the peaks the films are so thin that chemistry and surface physics effects become important in addition to the mechanics of local deformation of the asperities.

In the mixed lubrication regime friction coefficients can vary from values of less than 0.05 associated with thick film lubrication to values of greater than 0.4 associated with poor boundary lubrication, depending on the proportion of load carried by asperity contacts. Since this proportion is sensitive to process conditions and geometry, small changes in these can produce large changes in friction. Workpiece roughness tends to be decreased during forming in the mixed regime. However, if the boundary films are disrupted, workpiece scoring can result in increased roughness. Asperity interactions can also result in significant die wear.

In the boundary lubrication regime shown in Figure 1d, all the interface load is carried by the thin films on the asperity peaks. The mechanics of local deformation of the asperities and the surface physics and chemistry of the surfaces and the active constituents of the lubricant control the lubrication process. Comprehensive quantitative models for the boundary lubrication regime in metalforming processes are yet to be developed.

Under ideal conditions coefficients of friction of about 0.1 can be achieved in the boundary regime. Under the conditions of high load and new surface generation found in hydrostatic extrusion, values are usually larger than 0.1 and values of 0.4 or more can result if conditions are sufficiently severe to partially disrupt the boundary films. Relatively high levels of wear are to be expected. Under conditions of good boundary lubrication workpiece surface smoothing will occur but film breakdown will lead to extensive surface roughening due to scoring.

From the point of view of friction and wear reduction, the thick and thin lubrication regimes (often collectively called the full film regime) are highly desirable. Many hydrostatic extrusion processes can be successfully operated in these regimes if the lubriction system is carefully designed. However in some cases the necessity of controlling surface roughness requires that the process must operate in the mixed or boundary lubrication regimes. Even when surface roughening is not a problem, mixed or boundary lubrication regimes may be necessary at some stages in the extrusion cycle, particularly during starting. Under such circumstances mixed lubrication is more desirable than purely boundary lubrication because the latter will generally result in higher friction and wear and an increased likelihood of surface damage due to metal-to-metal contact. It is probably unwise to rely on boundary lubrication in a process such as hydrostatic extrusion which places very stringent requirements on the lubrication system. Even the best boundary additives will perform better in a mixed lubrication environment where the bulk lubricant in the roughness valleys not only supports part of the load but also provides a fresh supply of active material to replace boundary films as they are worn away.

Lubrication systems in hydrostatic extrusion have used either the pressurizing liquid or a separate billet solid coating as lubricant. In either case the system can operate in all of the various regimes described above. In order to predict what regimes will occur in a particular case, it is necessary to understand the mechanics by which lubricant films are carried into and distributed within the billet-die interface. Small variations in speed, surface roughness, die geometry, etc. can change the regime in which the lubrication system operates and alter its performance dramatically.

An example of the influence of the lubrication regime on system performance is given in Figure 2. This shows the variation of measured coefficient of friction and surface roughness with die semi-angle in strip drawing aluminum with wedge-shaped dies and a wax coating as a lubricant [4]. This process has very similar lubrication mechanics to hydrostatic extrusion. As the die angle is increased the entrained film thickness is decreased and the system passes progressively through the thick film, thin film, mixed and boundary lubrication regimes. Figure 2a shows that as the film thickness is decreased the coefficient of friction increases from the very small values associated with full film lubrication to values characteristic of good boundary lubrication. Figure 2b shows the associated change from roughening due to unconstrained deformation in the thick film regime, via smoothing in the intermediate regimes to roughening due to scoring in the boundary regime.

In studying the lubrication of hydrostatic extrusion as with other metal-forming processes, it is convenient to divide the billet-die conjunction into a number of zones in which different factors govern the flow of the lubricant film. The three main zones, the inlet zone, the work zone and the outlet zone are shown in Figure 3 for the case of full film lubrication. The film thickness and length of the inlet zone have been exaggerated for clarity.

In the inlet zone the lubricant is drawn into the convergent space between the essentially rigid die and billet surfaces by the billet motion. Very large pressure gradients, controlled by lubrication mechanics, occur in the inlet zone where the pressure at the billet-die interface builds up to a high enough level to initiate plastic flow of the billet as it enters the work zone. In the work zone the billet undergoes plastic deformation. The pressures at the billet-die interface are largely controlled by the billet plasticity and have relatively small gradients. The lubricant is carried along by the billet surface motion and its average film thickness decreases as the surface stretches. In the outlet zone the billet becomes rigid again and the lubricant film pressure falls rapidly to the downstream ambient value.

Two main approaches have been used to estimate the lubricant film thickness carried into the billet-die interface. One approach uses the classical methods of lubrication theory to analyze flow in the inlet zone [5,6,7,8]. The other applies a condition of minimum energy to the dissipation in the billet and lubricant in the work zone [9,10]. It is now generally held by researchers in lubrication mechanics that the former approach is correct. As in many other lubricated compliant systems [11,12], lubricant film entrainment in hydrostatic extrusion is governed by conditions in the inlet zone. Film transport in the work zone decides the current distribution of the entrained lubricant and in most processes almost all of the frictional drag on the billet occurs in the work zone.

It is also important to distinguish between steady and unsteady lubrication conditions in hydrostatic extrusion. Steady conditions occur when the billet velocity has been maintained constant over a long enough period so that the local conditions at the billet-die interface no longer vary with time. In general the inlet zone is sufficiently short so that it can be treated as quasi-steady and only the current conditions affect its performance. However the influence of variations in speed or other factors in the inlet zone propagate through the wok zone and this is important in the development of the local lubrication regime.

2.   Inlet Zone-Liquid Lubricants

In many hydrostatic extrusion processes the pressurizing liquid also acts as the lubricant at the billet-die interface. The conditions are as shown in Figure 4 for thick-film lubrication. The motion of the billet drags the liquid against the steep pressure gradient necessary to cause plastic flow where the

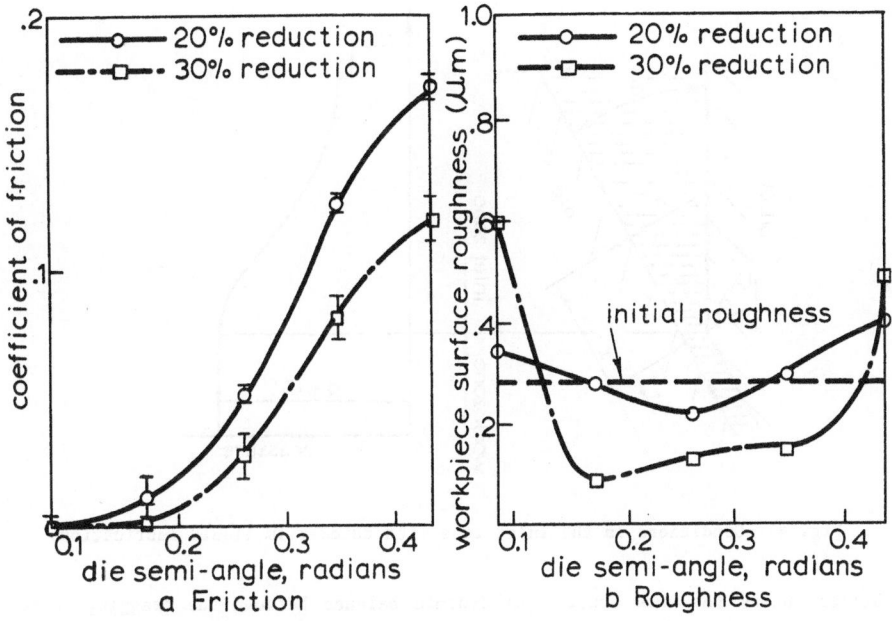

Fig. 2. Friction and roughness in drawing aluminum alloy strip
with wax lubricant (after Wilson and Cazeault [4]).

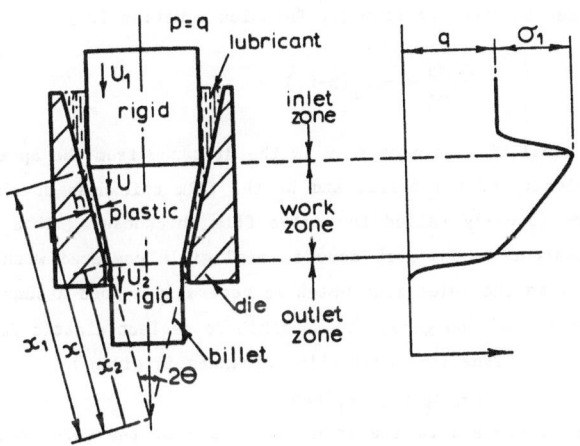

Fig. 3. Zones of lubrication in hydrostatic extrusion.

48

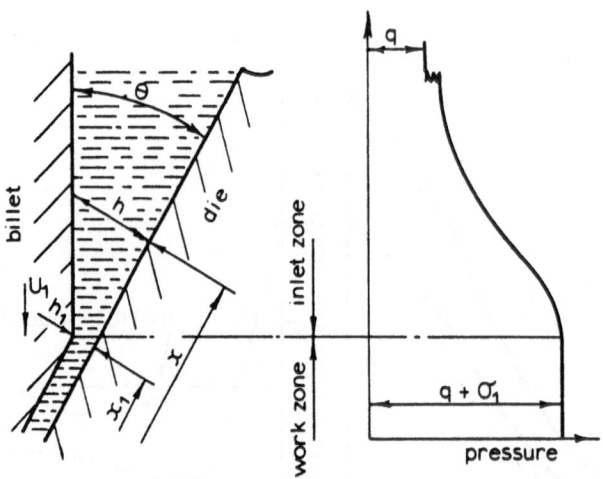

Fig. 4. Conditions in the inlet zone with thick-film liquid lubrication.

billet enters the work zone. The dynamic balance between the dragging action assisting lubricant entrainment and the pressure gradient resisting lubricant entrainment decides how much lubricant passes into the work zone.

If the lubricant can be assumed to be incompressible and Newtonian with a constant viscosity μ, then under steady conditions the local pressure p in the lubricant film can be obtained from the Reynolds equation [13]

$$\frac{h^3}{\mu} \frac{dp}{dx} = -6U_1(h-h_1)$$ (1)

where h is the local film thickness, x is the distance from the apex of the die, U_1 the inlet velocity of the billet and h_1 the film thickness at the inlet edge of the work zone (usually called the inlet film thickness). This assumes that the pressure gradient in the work zone is negligible compared with the typical pressure gradient in the inlet zone which is generally a good assumption.

In practice it is also generally possible to neglect elastic deflections in the inlet zone [14]. Thus the local film thickness is given by

$$h = h_1 + (x-x_1)\tan\theta$$ (2)

where x_1 is the distance from the inlet zone edge of the work zone to the die apex and θ is the die semi-angle.

Eqns. (1) and (2) yield the general solution

$$p = -\frac{6\mu U_1}{\tan\theta}\left(\frac{h_1}{2h^2}\right) + C_1 \tag{3}$$

where C_1 is an arbitrary constant. Upsteam in the extrusion container where h = ∞, p = q the extrusion pressure. Thus C_1 = q and Eq. 3 becomes

$$p - q = -\frac{6\mu U_1}{\tan\theta}\left(\frac{h_1}{2h^2} - \frac{1}{h}\right) \tag{4}$$

From the yield criterion at the inlet edge of the work zone where h = h_1, p = σ_1 + q where σ_1 is the billet flow stress or yield stress at inlet. therefore from Eq. (4)

$$h_1 = \frac{3\mu U_1}{\sigma_1 \tan\theta} \tag{5}$$

it is evident from Eq. 5 that the film thickness increases with increasing lubricant viscosity and billet speed and decreases with increasing workpiece yield stress and die angle. Since Reynolds' equation is only truly valid for small angles, the analysis should be used with caution for large die angles.

The extrusion pressure q does not appear in Eq. 5. This reflects the fact that, contrary to popular opinion, the high upstream pressure does not directly force lubricant into the interface. The film thickness at the interface is generally much too small for this mode of lubrication (leakage hydrodynamics) to play a significant role. It is the motion of the billet dragging the lubricant into the converging space between the billet and die in the inlet zone (wedge hydrodynamics) which tends to carry lubricant into the pressurized interface.

In practice the viscosity of most practical pressurizing liquids (with the notable exception of water-based fluids) increases substantially under pressure [15]. The viscosity of a typical oil may increases by several orders of magnitude under the high pressures characteristic of hydrostatic extrusion and it is important to allow for this effect in lubrication analyses.

Pressure induced viscosity changes are usually included by using the Barus equation

$$\mu = \mu_o e^{\gamma p} \tag{6}$$

where μ is the viscosity at pressure p, μ_o is the viscosity at atmosphere (essentially zero) pressure and γ is the pressure coefficient of viscosity. At room temperature the pressure coefficient γ has values of about 0.5-1.0 × 10^{-8} m^2/N for animal and vegetable oils and about 1.0-2.0 × 10^{-8} m^2/N for mineral oils. Values for a variety of liquids are given by the ASME [15].

If the inlet analysis is repeated using Eq. 6, which assumes isothermal conditions, then Eq. 5 becomes

$$h_1 = \frac{3\mu_o e^{\gamma q} \gamma U_1}{\tan\theta(1-e^{-\gamma\sigma_1})} \qquad (7)$$

In general the increase of viscosity with pressure increases the inlet film thickness h_1. A comparison of Eq. 7 with Eq. 5 indicates that the viscosity μ has been replaced with $\mu_o e^{\gamma q}$. This is the viscosity of the lubricant upstream in the extrusion container under pressure q. The high pressure in the container increases the lubricant film thickness indirectly by increasing the upstream viscosity and not directly by forcing lubricant into the conjunction. In both expressions the base viscosity (μ or μ_o) and the billet speed U_1 appear in the numerator. However in Eq. 7 the pressure coefficient γ also appears in the numerator indicating that this factor is at least as important. The term $\tan\theta$ appears in the denominator of both expressions but the yield stress σ_1 in Eq. 5 has been replaced by the term $(1-e^{-\gamma\sigma_1})$ in Eq. 7. Normally this term is close to unity and the film thickness is relatively insensitive to the yield strength σ_1. The increase in pressure gradient in the inlet zone associated with a higher yield strength tends to be offset by the increased viscosity due to the elevated pressure in the inlet zone.

In addition to increasing under pressure, the viscosity of most liquid lubricants also decreases with temperature. Thus increases in the lubricant temperature can result in a decrease in the inlet film thickness.

Lubricant temperature rises can originate from a number of sources including viscous shear heating within the film, compressive heating and heating from contact with hot billet or die surfaces. Generally the most important of these is viscous heating, particularly under high speed conditions or with high viscosity lubricants.

Viscous shear within the lubricant film in the inlet zone generates heat which is conducted through the film to the adjacent billet and die surface. Even though the films are thin, their relatively low conductivity and the high rate of heat generation can result in a significant temperature rise relative to the surface temperature. This temperature rise reduces the lubricant viscosity and hence the inlet film thickness.

In order to estimate the influence of viscous shear heating on inlet film thickness it is convenient to assume that the lubricant viscosity μ at the pressure p and temperature T (measured relative to the billet temperature) is given by

$$\mu = \mu_o e^{\gamma p - \alpha T} \qquad (8)$$

where μ_o is the viscosity at zero pressure and billet temperature, and γ is the pressure coefficient and α the temperature coefficient of viscosity [15] respectively. With this model it has been shown [8] that the inlet film thickness given in Eq. 7 is reduced by a thermal correction factor C which is given approximately by the semi-empirical equation

$$C = (1 + 0.565 \ G^{0.375} L'^{0.8})^{-1} \qquad (9)$$

where

$$G = \gamma \sigma_1 \qquad (10)$$

$$L' = \frac{\mu_o e^{\gamma q} \alpha U_1^2}{4k} \qquad (11)$$

and k is the lubricant thermal conductivity. G can be thought of as a non-dimensional pressure coefficient of viscosity while L' is the factor which mainly governs the severity of the reduction of inlet film thickness due to viscous heating. With small values of L' (low speed) C tends to unity and the lubrication system behaves isothermally, while at large values of L' (high speed) C is substantially less than unity and the inlet film thickness is reduced by viscous heating.

A clearer picture of the variation of inlet film thickness with processs conditions may be obtained from the example shown in Figure 5. This shows the theoretically predicted variation of inlet film thickness with speed and extrusion pressure for rod-to-rod extrusion of 2011-T3 aluminum alloy through a 45° semi-angle die with castor oil as lubricant. At low speeds the film thickness increases linearly with speed according to isothermal theory. At higher speeds the film thickness goes through a maximum and then decreases with further increase in speed. Increasing the extrusion pressure increases the film thickness because the increase in upstream lubricant viscosity enhances hydrodynamic wedge action but also moves the film thickness peak to lower speeds

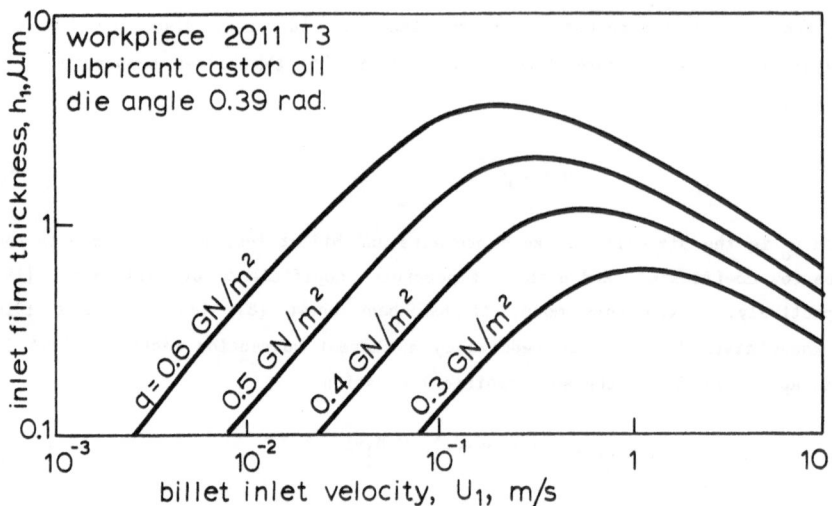

Fig. 5.   Inlet film thickness in a typical hydrostatic extrusion
process (after Wilson and Mahdavian [8]).

because the viscosity increase makes viscous heating more severe. It is evident
that it will be difficult to generate inlet film thickness sufficient for thick-
film lubrication at either very high or very low speeds and under these
conditions the system will likely operate in less favorable regimes.

The theory developed above has been tested by experimental 2.25 area ratio
rod-to-rod extrusions of 2011-T3 aluminum alloy through conical dies with castor
oil as pressurizing fluid and lubricant [8].   Different billet sizes, liquid
volumes and nosing conditions were used to achieve extrusion speeds from about
10 mm/s to 35 m/s. At each end of this range surface damage characteristic of
the mixed or boundary lubrication regimes was noted.   The inlet film thickness
in the experiments was estimated from careful measurements of the product
diameter after extrusion.   The experimental results are compared with theory in
Figure 6, in which the thermal correction factor C (ratio of inlet film
thickness to isothermal value) is plotted on a base of the thermal viscous
variable L'.   It is evident that considering the potential inaccuracies in the
experimental measurements there is quite good agreement between theory and
experiment over the range investigated.   This provides good support for the
validity of the theoretical model.

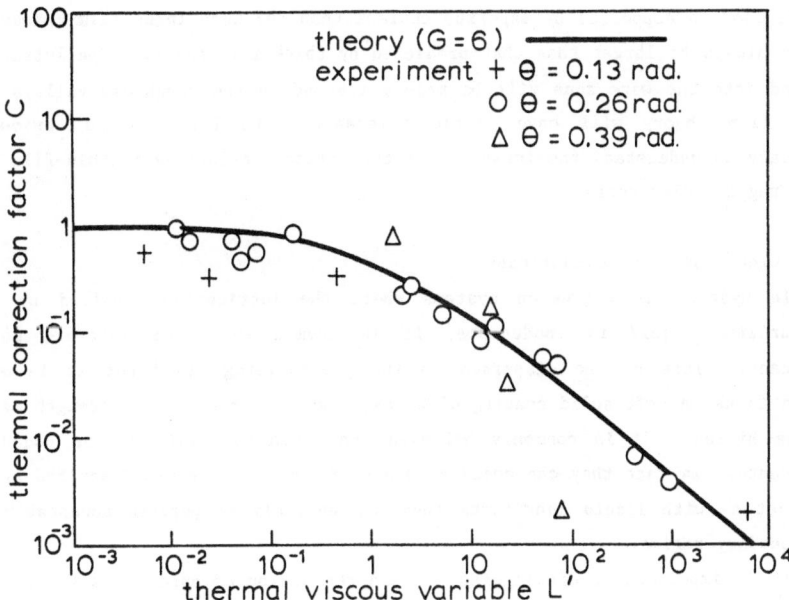

Fig. 6. Comparison of measured film thickness with theory for
hydrostatic extrusion of aluminum alloy with castor
oil as lubricant (after Wilson and Mahdavian [8]).

The theory presented above is for the thick-film regime. The theoretical
modelling of inlet zones in the thin-film and mixed regimes is still in its
infancy. Most of the work has dealt with machine elements rather than
metalforming and with the thin-film rather than the mixed regime. However,
based on work on rolling [16], it appears that thick film theory can be used as
a guide to the mean inlet film thickness formed in the thin film regime with a
correction for the influence of surface roughness. Here the lay of the
roughness is as important as its magnitude. A lay direction parallel to the
direction of billet motio will tend to decrease the inlet film thickness while a
lay normal to the billet motion will tend to increase inlet film thickness in
the thin film regime. A similar influence has been predicted for machine
lubrication but no experimental verification of the effect has emerged.

In the mixed regime part of the interface load will be supported by
asperity contact. If this is a small fraction of the total interface load then
the system should show the same influence of roughness lay. If a large fraction

of the load is supported by asperity contact then the mean inlet film thickness should always be larger than that predicted by thick film theory. The lubricant carried into the work zone will be mainly trapped in the roughness valleys and thick film theory will have little relevance. Further work is obviously necessary to understand the interplay of the various influences in thin-film and mixed regime inlet zones.

3. Inlet Zone - Solid Lubricants

In hydrostatic extrusion systems where the lubrication provided by the pressurizing liquid is inadequate, it is common to use a separate solid lubricant. This can be dispersed in the pressurizing fluid but it is more common to use a soft solid coating of a wax, soap, polymer or low-strength metal on the billet. It is commonly believed that such materials act as boundary lubricants. In fact they can operate in any of the four regimes described above and just as with liquid lubricants they are unlikely to perform acceptably in the boundary regime.

It is important to distinguish between the action of hard and soft surface coatings. Hard coatings such as titanium carbide are used to provide wear resistance and reduced friction in rubbing contacts. They form an extension of the surface on which they are deposited and suffer little, if any, plastic deformation in use. Relative surface motion occurs by the coated surface sliding over the opposing surface. On the other hand, soft coatings such as soaps act as true lubricants by providing a low shear strength layer between the opposing surfaces. Only part of the applied coating may pass into the load bearing region and relative surface motion is accommodated by shear within the coating. Soft coatings are heavily deformed in use.

As with liquid lubricants, the flow in the inlet zone decides how much of a soft solid coating is carried into the work zone. While the theory of liquid lubrication is almost a century old and relatively well developed, comparatively little work has been done on the mechanics of lubrication by soft solid coatings. Early workers in the field attempted to treat solid coatings as viscous liquids with viscosities which vary with speed [17] or temperature [18]. However, this approach fails to represent the essentially plastic behavior of the commonly used coating materials at low speeds.

The upper bound method of plasticity may be used to model the flow of a rigid-plastic coating in the inlet zone of hydrostatic extrusion. A first step is to develop the equivalent of the Reynolds' equation which in the current notation for the converging flow in the inlet zone is approximately [19]

$$\frac{dp}{dx} = [\tau_d - \tau_b + 2.64(\tau_c - \tau_d)(\theta - 0.695\theta^2)]/h \qquad (12)$$

where

$$\theta = \tan^{-1}(dh/dx) \qquad (13)$$

τ_c is the shear strength of the coating and τ_b and τ_d are the shear strengths of the bond between the coating and the billet and die respectively. Eq. 12 shows that the lubricant's ability to generate the negative pressure gradient necessary for full film lubrication increases with increase in its shear strength and bonding to the billet. It also increases with decrease in the film thickness, die angle and bonding or friction with the die.

At first sight it would appear to be possible to integrate Eq. 12 in a way similar to that used with the Reynolds' equation in the liquid lubricant analysis. If such an approach is followed, an infinite value of the inlet film thickness h, is predicted which is obviously not in accord with the finite supply of lubricant coating to the inlet zone. In lubrication parlance the inlet zone is "starved" and a more complete model of the inlet flow must be adopted.

A model of the inlet zone which builds up the lubricant flow field from regions of rigid body motion separated by velocity discontinuities is illustrated in Figure 7 [19]. Both the discontinuity field and the associated velocity field or hodograph are shown. The inlet zone has been divided into two regions. In the outer back flow region part of the coating is peeled off and moves back along the die surface somewhat like the "chip" in machining. The remainder of the lubricant passes into the inner wedge region where most of the pressure build up governed by Eq. 12 occurs. From there the lubricant film is carried into the work zone. With this model the key question is what fraction of the lubricant passes into the wedge region and eventually the work zone. The upper bound method was used to find the fraction corresponding to minimum power dissipation subject to the condition that sufficient pressure must be built up to cause yielding at the inlet edge of the work zone.

The results of the upper bound analysis are shown in Figure 8 for the optimum bonding conditions ($\tau_b = \tau_c$, $\tau_d = 0$). The results are plotted using the non-dimensional film thickness H and strength ratio S defined by

$$H = h_1/h_c \qquad (14)$$

and

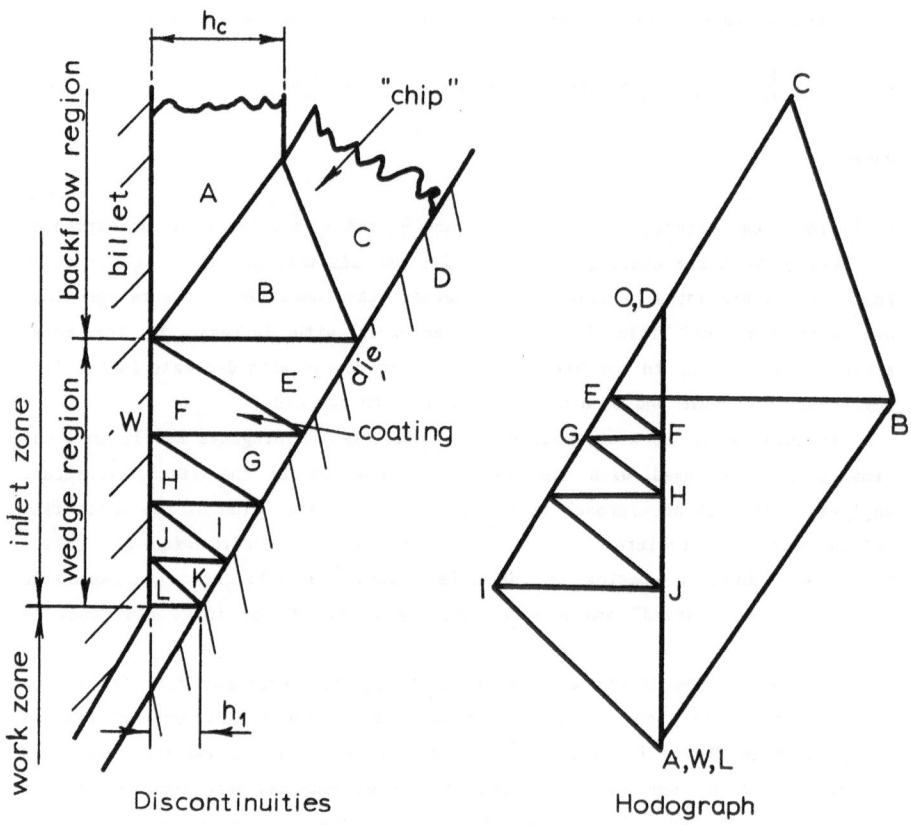

Fig. 7. Discontinuities and hodograph for inlet zone with
solid lubricant (after Wilson and Halliday [19]).

$$S = \sigma_1 / 2\tau_c \qquad (15)$$

where h_1 is the inlet film thickness, h_c the applied coating thickness, σ_1 the
billet yield strength and τ_c the lubricant shear strength. The original paper
erroneously doubled the values of S. This has been corrected in Figure 8.

The results shown in Figure 8 show that the inlet film thickness h_1 is
proportional to the applied film thickness h_c, increases with increase in
lubricant shear stength τ_c, and decreases with increase in die semi-angle θ and
billet yield strength σ_1. The sensitivity of the entrained film thickness is
much larger than that found in liquid lubricated systems. However the solution
is independent of speed and solid coatings can retain their effectiveness under
very low speed conditions where liquid lubricants must rely on the formation of

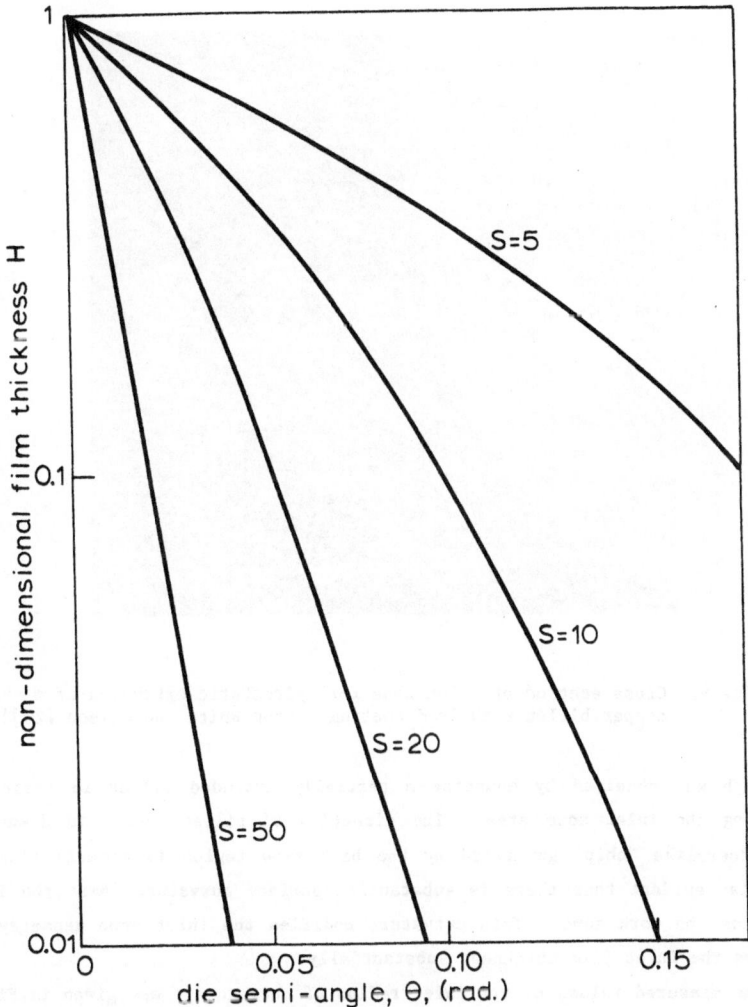

Fig. 8. Results of upper-bound analysis for inlet film thickness with optimum bonding conditions (after Wilson and Halliday [19]).

boundary films.

The simple theory presented above models many of the factors present in the actual lubrication system. Figure 9 shows a cross section of the inlet zone in the hydrostatic extrusion of copper with lead as a lubricant. The photo-

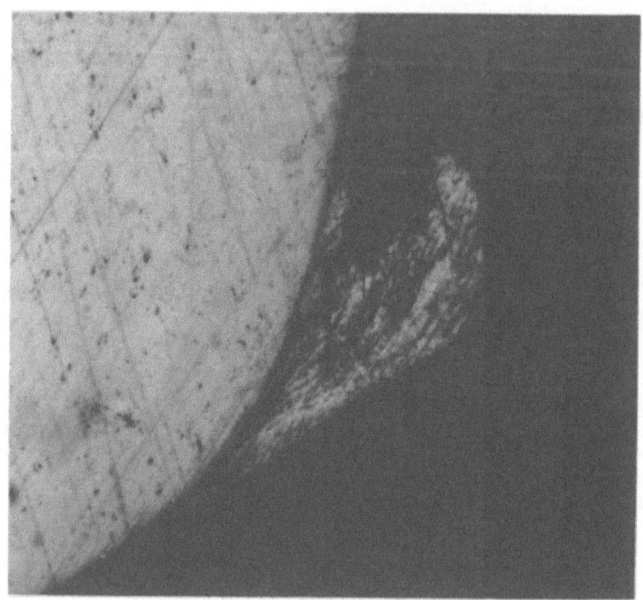

Fig. 9. Cross section of inlet zone for hydrostatic extrusion of a
copper billet with lead coating (after White and Wilson [20]).

micrograph was obtained by mounting a partially extruded billet in resin and
sectioning the inlet zone area. The direction of billet motion is downward.
The feather-like "chip" generated by the back flow region is clearly visible.
It is also evident that there is substantial surface curvature where the inlet
zone joins the work zone. This curvature modifies the inlet zone geometry and
increases the inlet film thickness substantially.

Some measured values of the inlet radius of curvature r are given in Figure
10. It seems that the curvature decreases with increasing die angle and depends
on the billet material and not the coating material. The mechanics of inlet
curvature or "rounding in" are poorly understood. There is no obvious
correlation with billet mechanical properties.

From a theoretical standpoint it is convenient to characterize inlet
curvature by a non-dimensional rounding factor R given by

$$R = r/h_c \qquad (16)$$

Repeating the upper-bound analysis with allowance for inlet curvature [20]

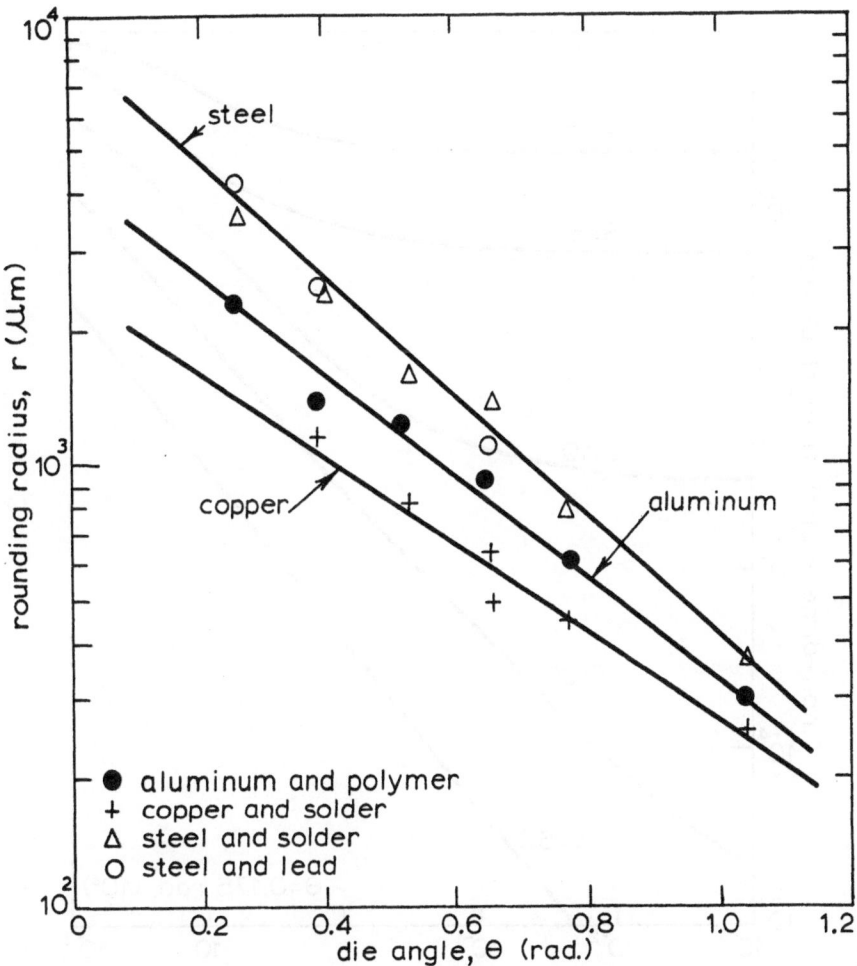

Fig. 10. Inlet radius of curvature in hydrostatic extrusion
with different billet and lubricant materials.

yields results such as those shown in Figure 11 which shows that rounding can
dramatically increase the inlet film thickness. The influence is particularly
strong with large die angles and large values of the strength ratio S.

The upper-bound theory with allowance for inlet rounding and the slight
pressure build up in the back-flow region, is quite effective in predicting
inlet film thickness in hydrostatic extrusion with soft metal coatings as
lubricants [21]. Figure 12 compares the theory with experimental measurements
for steel rod-to-rod extrusion with different die angles and lead and solder

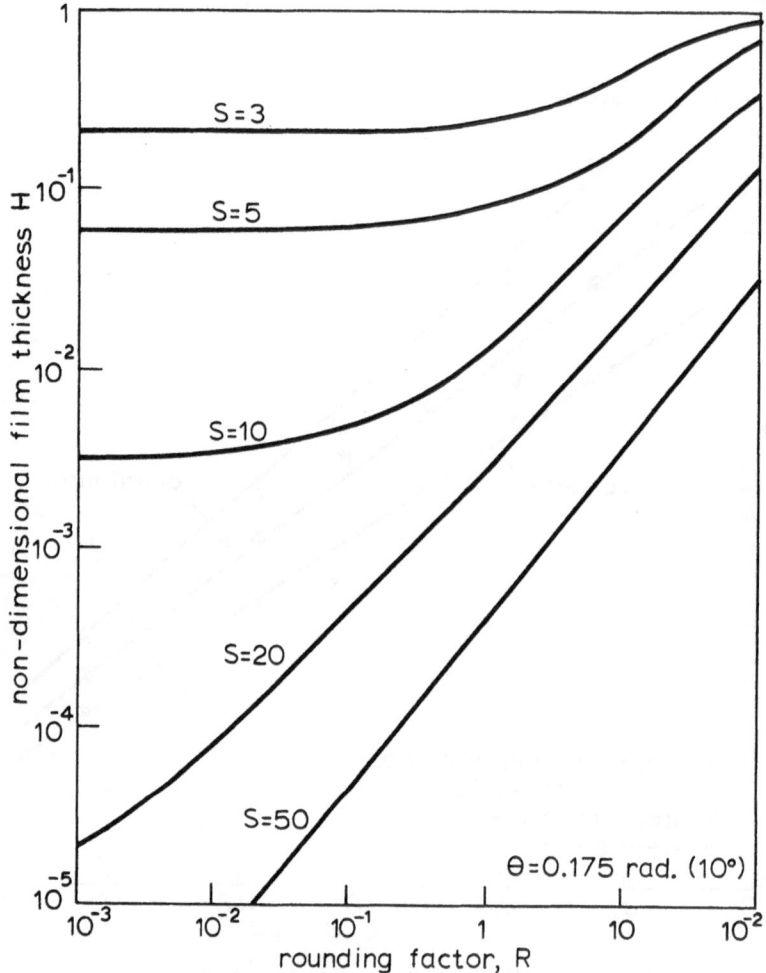

Fig. 11.   Influence of rounding on inlet film thickness
(after White and Wilson [20]).

coatings.   The solder has a hardness of about three times that of lead (S = 8
versus S = 23) and thus much thicker films are generated with solder than with
lead.   The rapid decrease of film thickness with increasing die semi-angle is
evident.   The complex shape of the curves is largely due to the influence of
rounding.   Rounding radii were determined experimentally.

Soft metal films have relatively constant shear strength.   The shear

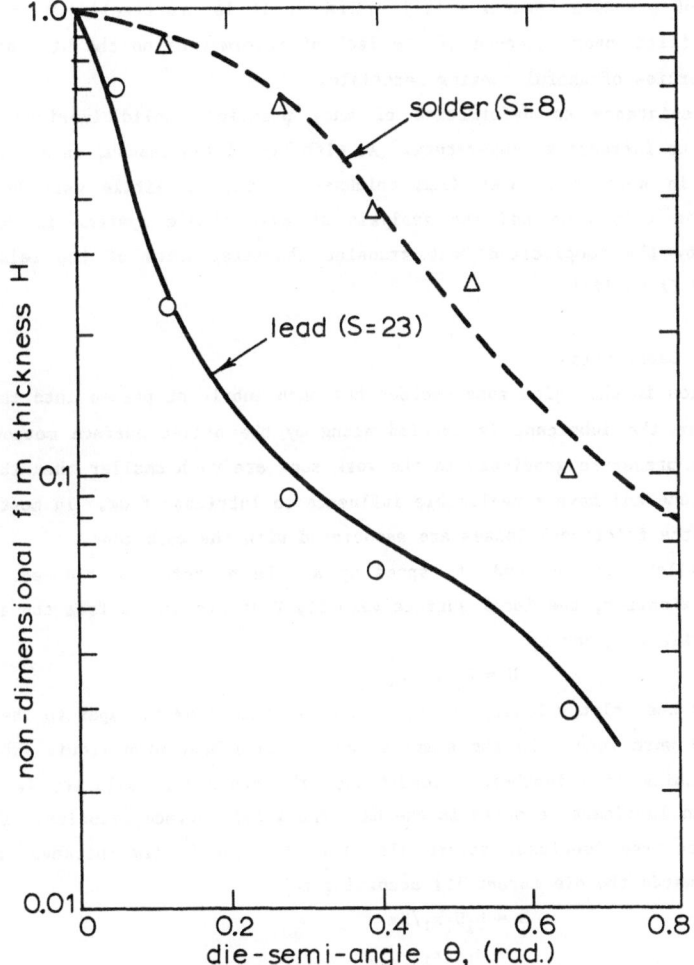

Fig. 12. Comparison of measured film thickness with theory for hydrostatic extrusion of carbon steel with lead and solder coatings (after Wilson and Johnson [21]).

strengths of many lubricants such as polymers increase significantly with increase in pressure [22]. As with liquid lubricants this increase in resistance to deformation under high pressure tends to increase the inlet film thickness [20]. It is relatively easy to modify the theory to account for the influence of pressure hardening [21]. However it is practically difficult to use the modified theory because of the lack of information on the high pressure shear properties of useful coating materials.

The resistance to deformation of many practical solid lubricants also decrease with increasing temperature. As with liquid lubricants, this will tend to result in decreased inlet film thickness. Only a little work has been published in this area and the analysis of even simple systems is rendered difficult by the complicated heat transfer characteristics of the relatively thick solid films [23].

4. Steady Lubrication

The flow in the inlet zone decides how much lubricant passes into the work zone. There the lubricant is carried along by the billet surface motion. In general the presssure gradients in the work zone are much smaller than those in the inlet zone and have a negligible influence on lubricant flow. In most cases almost all the frictional losses are associated with the work zone.

The billet surface tends to speed up as the diameter is reduced. To a first approximation, the local surface velocity U at distance x from the apex of a conical die, is given by

$$U = U_1 \, x^2/x_1^2 \tag{17}$$

where U_1 is the inlet velocity and x_1 is the distance from the apex to the inlet edge of the work zone. In the simplest case with a Newtonian liquid lubricant under the thick film isothermal conditions, the cross-film velocity is linear and the mean lubricant velocity is one half the local surface velocity. Thus in order to conserve lubricant volume flow rate the local film thickness h must decrease towards the die throat [6] according to

$$h = h_1 U_1 x_1/Ux$$
$$= h_1 x/x_1 \tag{18}$$

This distribution may be modified by thermal effects. If the billet is well insulated from the die by a thick lubricant film, the billet temperature will rise above that of the die due to the heat of plastic deformation. This tends to produce a nonlinear cross-film velocity distribution because the lubricant near the hot billet surface has a lower viscosity than that near the cold die. The result is that the mean lubricant velocity is less than half the workpiece speed and the local film thickness must increase to maintain volume flow. In extreme cases, this effect can result in the film thickness increasing towards the die throat [24]. The other main thermal effect is due to viscous heating within the film. This tends to offset the influence of plastic heating on local film thickness, especially near the inlet zone.

The flow or transport of solid films is much more complicated and less well understood than that of liquid films. Under ideal conditions with smooth surfaces solid films also tend to be transported at half the billet speed and Eq. 18 will apply. Thermal influences on the transport of solid films are apparently relatively small but the influences of surface roughness are very important [25].

Most of the frictional losses in hydrostatic extrusion generally occur in the work zone. In the thick film regime under isothermal conditions with a Newtonian liquid lubricant that can be characterized by Eq. 6, the local frictional shear stress τ_f is due to viscous shear and is given by

$$\tau_f = \mu_o e^{\gamma p} U/h \tag{19}$$

Eq. 19 may be used in conjunction with the conditions of equilibrium for the plastically deforming billet to predict the local interface pressure and friction and global variables such as extrusion pressure in the thick film regime [6]. Such an isothermal analysis neglects the influence of plastic and viscous heating which usually tend to reduce friction appreciably.

The non-dimensional plastic thermal variable K and viscous thermal variable L are appropriate measures of heating effects on lubrication. These are defined by

$$K = \alpha\sigma_1/\rho c \tag{20}$$

and

$$L = \mu_o \alpha U_1^2/4k \tag{21}$$

where α is the lubricant temperature coefficient of viscosity, σ_1 the billet yield stress, ρ the billet density, c the billet specific heat, μ_o the lubricant base viscosity, U_1 the inlet billet velocity and k the lubricant thermal conductivity. The variable L is not localized to the inlet zone and thus does not contain the correction term $e^{\gamma q}$ present in L' as defined in Eq. 11.

If the variables K and L are small, the lubrication system behaves isothermally. With large values of K and/or L heating of the lubricant film reduces its viscosity and associated friction [8]. This is evident from the results shown in Figure 13 which shows the predicted variation of non-dimensional extrusion pressure Q with area ratio for different lubrication conditions. No strain hardening is assumed and Q is defined by

$$Q = q/\sigma_1 \qquad (22)$$

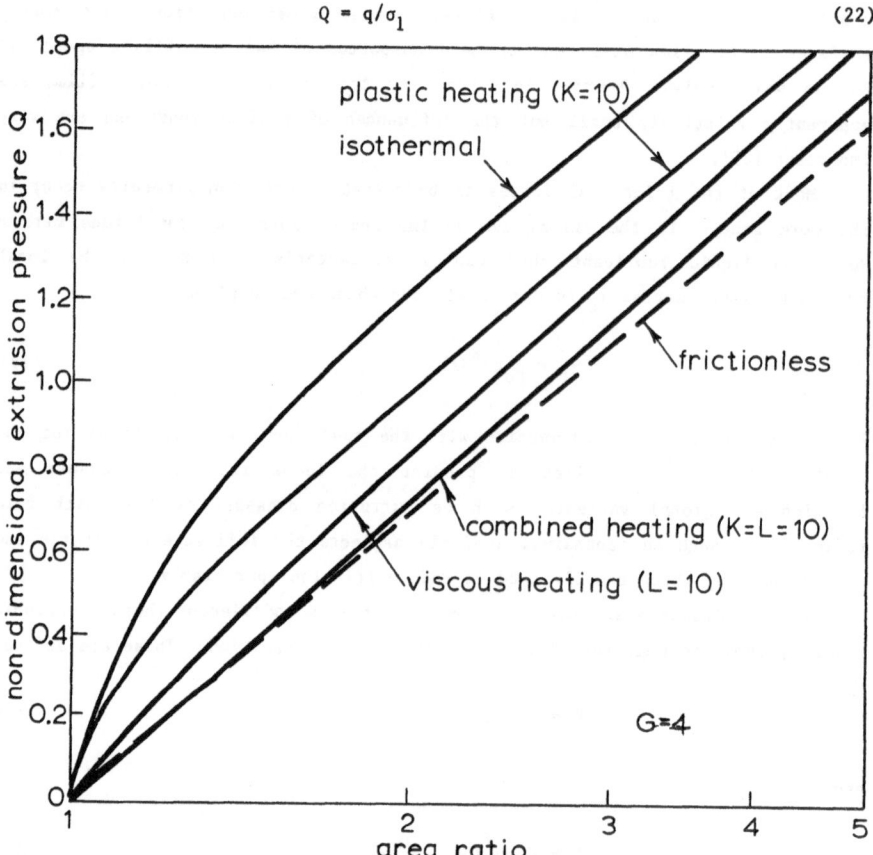

Fig. 13. Influence of lubricant film heating on extrusion pressure in hydrostatic extrusion with thick film liquid lubrication (after Wilson and Mahdavian [8]).

where q is the extrusion pressure and σ_1 is the billet yield on flow stress. When the influences of both plastic and viscous heating are combined the extrusion pressure is close to the frictionless level. If thick film lubrication conditions can be established and maintained with a liquid lubricant, friction will be negligibly small.

As shown in Figure 2a, friction levels can also be very low with a soft solid coating in the thick film regime. Thick film conditions can be achieved by using small die angles even with low shear strength coatings. However, if small die angles are unacceptable, then higher strength coatings are necessary to achieve thick film conditions. This will generally result in higher friction levels though friction is rendered less important by the use of larger die angles and lubricant softening may help keep friction to an acceptable level.

In the event that surface roughnesses are too high or the lubricant film thickness is too small, then the system will enter the mixed lubrication regime. The main features of film formation and transport will be maintained but the roughness of the workpiece will play an important role in carrying lubricant through the work zone. Friction levels will be increased due to the interaction of surface asperities. In order to control friction, careful attention must be paid to the formulation of the lubricant to ensure the provision of adequate boundary lubrication under the severe conditions of asperity clashes and new surface generation in the work zone.

After the billet has completed plastic deformation it enters the outlet zone. This region is of considerable interest to the lubrication mechanicist because it provides the environment for the lubricant pressure to drop to downstream ambient conditions [6]. From a practical standpoint it is relatively unimportant. In the full film regimes the outlet zone contributes little to frictional drag unless soft billets, high reductions and long die lands are used [26]. However under poor mixed lubrication conditions with long die lands outlet zone friction may assume important proportions.

5. Unsteady Lubrication

Under changing conditions of speed, extrusion pressure etc., unsteady lubrication conditions can exist at the billet-die interface. The inlet zone meters lubricant into the work zone at a rate governed by the current inlet zone conditions. This lubricant is then carried through the work zone by billet surface motion. Since this transport process takes some time, the local lubrication regime in the work zone does not adjust immediately to reflect the new inlet zone conditions and an unsteady lubrication state results. This is particularly important during the start of extrusion where it may lead to a peak in the pressure required for extrusion.

As a first approximation, it can be assumed that the average transport velocity of either a liquid [27] or a solid lubricant [25] is equal to the mean velocity of the billet and die surfaces. Since the die surface is stationary the lubricant film tends to be carried along at half the billet surface velocity. The result is that during the start of extrusion the developing lubricant film fails to keep up with the billet nose as shown in Figure 14. The

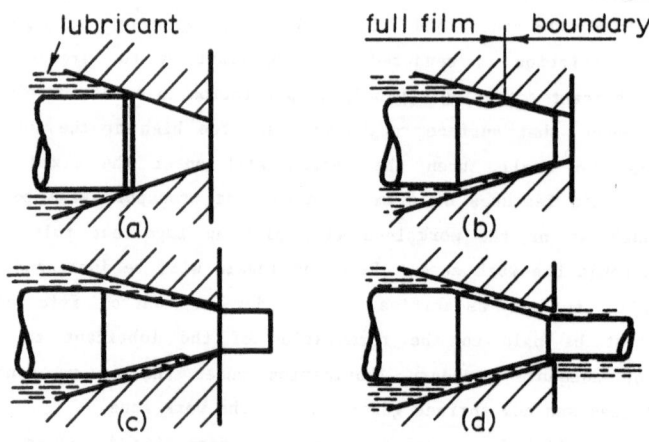

Fig. 14.   Development of lubricant film during start of extrusion
           (after Wilson [27]).

initial un-nosed billet is shown in Figure 14a. As the billet moves into the die a region which must operate in the boundary regime is formed as shown in Figure 14b. This will occur even if the conditions are generally suitable for steady state full film operation. The higher friction associated with the boundary lubricated region increases the pressure required for extrusion. The boundary region reaches its maximum extent and the pressure reaches a peak when the billet just fills the die [28] as shown in Figure 14c. If no severe surface damage occurs, the lubricant front will eventually reach the die throat and steady state conditions will ensue in the absence of further speed or pressure fluctuations. Surface damage occurring during the start up phase may increase roughness to a point where adequate lubrication is no longer possible and surface distress will proliferate with an associated increase in friction and pressure required for extrusion.

The drop in pressure required for extrusion after the pressure peak occurs may lead to the billet shooting forward [29]. In some cases the billet will be violently expelled from the system. In other cases it will come to a stop and the start up process is repeated to produce "stick-slip" motion. Generally the pressure peak is less pronounced in subsequent starts because of the presence of residual lubricant at the billet-die interface. However progressive surface damage may lead to increasing peaks.

The influence on the boundary lubricated region on the pressures required for nosing an aluminum billet [30] is illustrated in Figure 15. Three

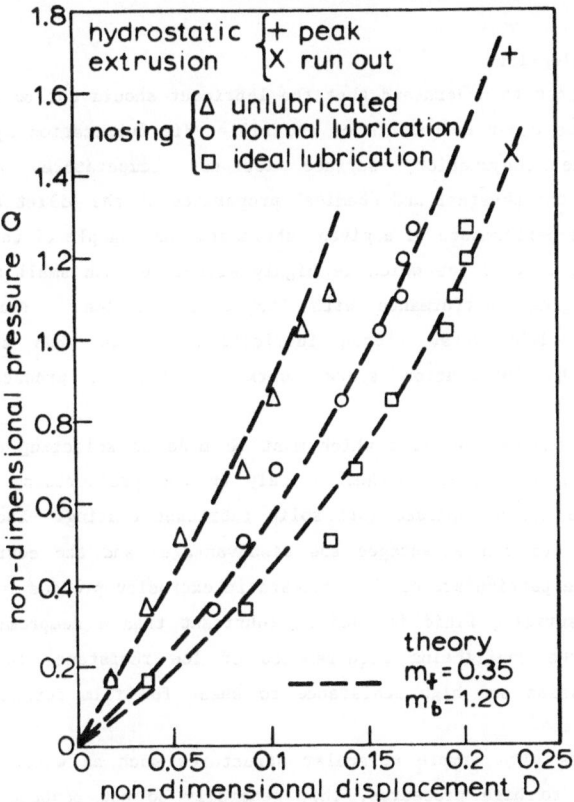

Fig. 15. Pressure required to nose aluminum alloy billets with different lubrication conditions (after Wilson et al. [30]).

conditions of lubrication were used: unlubricated, normal lubrication in which a single coating of a commercial solid lubricant was applied prior to nosing, and ideal lubrication in which the coating was reapplied periodically during the course of the nosing process. The results for the normal lubrication condition lie between the unlubricated (boundary) case and the ideal (full film) case. The peak and run-out process measured in an actual hydrostatic extrusion correspond to the normal and ideal conditions respectively. The results can be fitted by assuming friction factors $m_b = 1.2$ and $m_f = 0.35$ for the boundary and full film regimes respectively. In the normal condition the extent of the regimes was determined experimentally. Measurements of actual lubricant transport rates during nosing [29,30] show substantial deviations from the half-speed theory due to thermal and roughness influences.

6. Lubricant Selection

It is important to understand that the lubricant should not be selected in isolation from the other components of the billet-die lubrication system. The geometry of the conjunction, surface motions, temperatures and micro-topographies and the physical and chemical properties of the billet and die can all influence the performance of a given lubricant. An example of this is shown in Fig. 2, where a lubricant which is highly effective with small die angles, yields only marginal performance with larger die angles. There are no universally good lubricants. Good lubrication results when the various components of the lubrication system work together to produce superior performance.

Perhaps the biggest decision which must be made in selecting a lubricant for hydrostatic extrusion is whether to rely on the pressurizing liquid for lubrication or to use a separate soft solid lubricant coating. Each of these alternatives has its own advantages and disadvantages and the correct choice will depend on the particulars of the hydrostatic extrusion process involved.

If the pressurizing fluid is used as lubricant then a compromise must be struck between the conflicting requirements of low resistance to shear for pressure transmission and high resistance to shear for film formation in the billet-die interface.

While liquids with simple molecular structures such as water or ethanol remain liquid up to high pressures, they generally do not possess sufficient viscosity to be effective as full film lubricants. At the other extreme, mineral oils which are mixtures of complex organic molecules, have good film forming capabilities but generally "freeze", or become too viscous to transmit pressure, at pressures above about 0.4 GPa at room temperature. Although the onset of freezing may be delayed in practice by compression under adiabatic conditions. Natural fatty oils from vegetable or animal sources such as castor oil or tallow oil represent a useful compromise. They generally have lower pressure coefficients of viscosity and higher freezing pressures than mineral oils but still have sufficient viscosity for film formation. Moreover their viscosity can be reduced by dilution with alcohol for use under very high pressures and they contain free or compounded fatty acids which can form boundary films.

If a separate soft-solid coating is used as lubricant, then the problem of selecting the pressurizing liquid is greatly simplified. The liquid can be chosen on the basis of pressure transmission although it must also be capable of lubricating seals and be compatible with the coating. In particular the coating must be relatively insoluble in the liquid under the conditions in the extrusion container.

Some techniques must also be developed for the application of the coating to the billet prior to extrusion. The cost associated with applying the coating (and possibly removing it after extrusion) is the major disadvantage of using a solid coating rather than the pressurizing liquid as lubricant. Waxes, soaps, thermo-plastics and low melting point metals can be applied by dipping the billet into the molten lubricant and cooling. Metal coatings can also be applied by electro-plating. With thermosetting plastics the catalyzed precursor can be applied by dipping or spraying and the coating cured in situ. Lamellar solids such as graphite or molybdenum disulfide can be applied by burnishing or in a thermosetting binder. In situ chemical reaction must be used to create conversion coatings such as phosphates or oxalates. Whatever the coating or application method chosen, it is vital to ensure a strong bond between the coating and the billet. This generally requires great care in cleaning and surface preparation. Care is also necessary to ensure good coverage and control of coating thickness in production.

Whether a liquid or solid lubricant is used, the lubrication system should be designed to generate as thick a film as is compatible with surface roughness requirements. Thus the system should be capable of generating an inlet film thickness in steady state operation which is at least as large as the billet roughness. The die roughness should be maintained at a much lower level. Techniques for estimating inlet film thickness are described in sections 2 and 3. For given billet and lubricant materials the inlet film thickness can be adjusted by changing operating speed or die angle. In general liquid lubricants are more sensitive to speed and less sensitive to die angle than solid lubricants and this should be borne in mind in producing a prototype design. The problem is complicated in the case of solid coatings by the lack of useful mechanical property data on many coating materials.

Attention must also be paid to the unsteady lubrication phenomena at the start of extrusion described in section 5. Here the intrinsically better transport properties of solid coatings [29] give them a distinct advantage over liquids. However much can be done by billet nosing or roughening to promote lubricant transport. It is also possible to use a local coating to assist the start of an extrusion which will otherwise rely on the pressurizing fluid for lubrication.

It is also important to incorporate some boundary additives into the lubricant if it does not contain these naturally. For mild conditions fatty acids or their esters such as methyl oleate are effective. For more severe conditions the so-called "extreme pressure" agents which generally contain sulphur, phosphorus or chlorine must be adopted. For oils these additives are generally soluble organic compounds. In the case of solid coatings the method of incorporation must suit the composition. For example wax coatings can be sulphurized directly.

The performance of the lubrication system can be adjusted as a result of initial trials. For example if scoring is observed the die angle can be reduced or additional boundary additives incorporated in the lubricant. However careful consideration of the mechanics of lubrication in the earliest stages of process design is generally essential to developing a successful production process without expensive and tedious trial and error.

REFERENCES

1 D.D. Ratnagar, H.S. Cheng and J.A. Schey, J. of Lubric. Tech., 96(1974)591–594.
2 W.R.D. Wilson, J. of Lubric. Tech., 99(1977)10–14.
3 W.R.D. Wilson and J.G. Siletto, in S.C. Jain (Ed.), "Metalworking Lubrication", A.S.M.E., New York, 1980, pp.87–94.
4 W.R.D. Wilson and P. Cazeault, in T. Altan (Ed.), Proc. NAMRC II, Battelle, Columbus, 1976, pp.165–170.
5 A. Eichinger and W. Lueg, Mitt. Kaiser Willhelm Inst. fur Eisenforschung, 23(1941)21–30.
6 W.R.D. Wilson and J.A. Walowit, J. of Lubric. Tech., 93(1971)69–74.
7 R.W. Snidle, B. Parsons, and D. Dowson, in "Elastohydrodynamic Lubrication – 1971 Symposium", I. Mech. E., London, 1972, pp.169–172.
8 W.R.D. Wilson and S.M. Mahdavian, J. of Lubric. Tech., 98(1976)27–31.
9 M.J. Hillier, Int. J. of Prod. Res., 5(1966)171–181.
10 B. Avitzur, "Metal Forming Processes and Analysis", McGraw Hill, New York, 1968.

11 J.A. Walowit and J.N. Anno, "Modern Developments in Lubrication Mechanics", Applied Science Publishers, London, 1975, pp.97-108.

12 J.A. Walowit and J.N. Anno, loc.cit., pp.180-190.

13 J.A. Walowit and J.N. Anno, loc.cit, pp.20-23.

14 S.M. Bloor, D. Dowson and B. Parsons, J. Mech. Eng. Sci., 12(1970)178-190.

15 Research Committee on Lubrication, "Pressure Viscosity Report", Vol. 1 and 2, ASME, New York, 1953.

16 Y.H. Tsao and L. B. Sargent, Trans. A.S.L.E., 23(1980)70-76.

17 G.H. Tattersall, J. Mech. Eng. Sci., 3(1961)378-393.

18 E. Felder and G. Breinlinger, in T. Altan (Ed.), Proc. 4th North American Metalworking Res. Conf., Battelle, Columbus, 1976, pp.158-164.

19 W.R.D. Wilson and K. Halliday, Wear, 42(1977)135-148.

20 D.R. White and W.R.D. Wilson, Trans. A.S.L.E., 23(1980)305-314.

21 W.R.D. Wilson and J.R. Johnson, Trans. A.S.L.E., 24(1981)307-318.

22 K.D. Pae and D.R. Mears, Polymer Letters, 6(1968)269-273.

23 K.F. Kennedy, A. Wadhawan and W.R.D. Wilson, in W. Bartz (Ed.), Proc. 3rd Int. Colloq. on Metalworking Lubric., Tech. Akad. Esslingen, 1982, pp.36.1-36.7.

24 S.M.Mahdavian and W.R.D. Wilson, J. of Lubric. Tech., 98(1976)22-26.

25 S. Lak and W.R.D. Wilson, D. Dowson et al. (Eds.), Proc. 4th Leeds-Lyon Symposium on Tribology, Mechanical Engineering Publishers, London, 1978, pp.301-307.

26 P.F. Hetwetter, J.C. Uy and D.P. McCann, J. of Basic Eng., 91(1969)822-829.

27 W.R.D. Wilson, Int. J. Mech. Sci., 13(1971)17-28.

28 A.W. Dufill and P.B. Mellor, Annals of C.I.R.P., 17(1969)135-141.

29 W.R.D. Wilson and S.M. Mahdavian, Proc. NAMRC III, Carnegie Mellon University, Pittsburgh, 1975, pp.52-71.

30 W.R.D. Wilson, B.B. Aggarwal, A.B. Norelius and H. Quist, Proc. NAMRC VII, S.M.E., Dearborn, 1979, pp.129-136.

SECTION 4

DYNAMICS OF HYDROSTATIC EXTRUSION

T. MATSUSHITA

KOBE STEEL, LTD.

1. Introduction

One problem in the hydrostatic extrusion process is that control of the pressing speed of a stem does not always provide an expected extrusion speed of product, because the stem is separated from the billet by compressed pressure medium. It is essential to establish a stable extrusion condition to start the extrusion at a low breakthrough pressure and to maintain stability during operation.

Instability in hydrostatic extrusion is characterized by intermittent movement of product and fluctuation of extrusion pressure. The instability arises from many factors such as the volume of pressure medium in relation to the volume of the billet, pressing speed, surface finish and nose geometry of the billet, compressibility of pressure medium and the angle of the die. By optimizing these factors, stable hydrostatic extrusion can be governed to obtain a controlled extrusion speed and dimensional accuracy of the product. The critical design of tooling and auxiliary equipment such as a coiling machine will also be possible.

This section presents an analysis of the stick-slip motion of a billet and press ram motion by taking the compressibility of the pressure medium into account. It provides a numerical solution as a guideline to prevent instability during extrusion. Factors which cause instability are discussed and conditions for stable hydrostatic extrusion are introduced. The influence of the dynamic behaviour of a press on the stability of hydrostatic extrusion with reference to press ram motion is also discussed.

Notation

p: pressure in container

p_e: extrusion pressure

Δp_{AB}, Δp_{BC}, Δp_{CD}: excess pressure to accelerate the billet AB, BC

73

and product CD, respectively (Fig.2)

P_f: excess pressure to overcome frictional force along billet surface

A_s, A_b, A_p: cross sectional area of stem, billet and product, respectively

z, L: length of product and billet, respectively

r_b: radius of billet

h_f: clearance between container and billet

U_s, U_b: speed of stem and billet, respectively

μ: coefficient of friction

η: viscosity of pressure medium

V: volume of pressure medium

τ: shear stress

σ: flow stress

n: coefficient of work-hardening

ϵ: strain

R: extrusion ratio

α: half angle of die

ρ: density of billet

g: acceleration of gravity

t: time

v: velocity

m: mass of press ram

P_1, P_2: pressure in main ram cylinder and pullback cylinder, respectively

A_1, A_2: cross sectional area of main ram cylinder and pullback cylinder, respectively

V_1, V_2: volume of main ram cylinder and pullback cylinder, respectively

β: compressibility of working oil

λ_f: coefficient of friction of working oil along surface of pipe

ℓ, d: length and diameter of pipe connecting pullback cylinder and oil, tank, respectively

$\Sigma\zeta$: coefficient of pressure loss for working oil

γ_g: specific weight of working oil

v_p: speed of working oil flowing through pipe

μ_o, P_o, λ, k, k_1, k_2, ω_1, ω_2: constant

2. Dynamics of billet motion

2.1 Stick-slip behaviour appearing in pressure curves

Instability in hydrostatic extrusion is observed in pressure-time (or ram displacement) curves. The pressure curves are considerably dissimilar for different extrusion conditions. They are classified into various types according to the characteristics during runout [1]. In general, a breakthrough pressure peak is followed by a slightly lower runout pressure under preferable lubrication as shown in Fig.1. This is attributed to the fact that static friction is equal to, or slightly higher than, kinetic friction during runout. In some cases, the runout pressure increases because of a partial breakdown of lubricant film due to temperature rise or changes in flow stress due to temperature changes in the billet.

When the stick-slip motion of the billet occurs, pressure fluctuation is superimposed on the general pressure curves. In Type III pressure curves shown in Fig.1, a few cycles of stick-slip follow the breakthrough pressure peak and their amplitudes decrease. Here, stick-slip is not

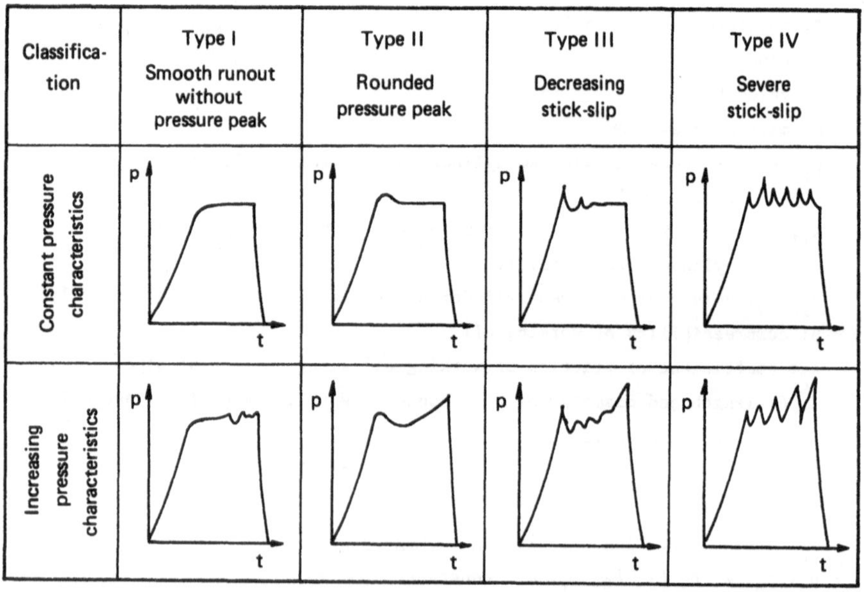

Classifica-tion	Type I Smooth runout without pressure peak	Type II Rounded pressure peak	Type III Decreasing stick-slip	Type IV Severe stick-slip
Constant pressure characteristics				
Increasing pressure characteristics				

Fig. 1. Classification of pressure curves in hydrostatic extrusion.

severe. In contrast, Type IV curves indicate that stick-slip is very
severe and continues throughout the stroke. Here, a smooth runout cannot
be obtained if extrusion is continued further. Rather the amplitudes of
stick-slip increase due to a breakdown of lubrication film.

2.2 Analysis of billet motion

In the hydrostatic extrusion process the billet is forced through a
die by a high pressure medium. The dynamic behaviour of the billet during
extrusion can be simulated by the momentum equation which is derived as a
function of compressibility of pressure medium, stem speed, billet
dimensions, volume of pressure medium, coefficient of friction, flow stress
of billet material, die angle, extrusion ratio and viscosity of pressure
medium [2].

(i) Pressure generated by volume change of pressure medium

When pressure medium is compressed from an initial volume V_o to a
current volume V, the pressure p is given by

$$p = p_o (\Delta V / V_o)^\gamma \qquad\qquad (1)$$

Here, volume change ΔV is given by

$$\Delta V = A_s U_s t - A_p z \qquad\qquad (2)$$

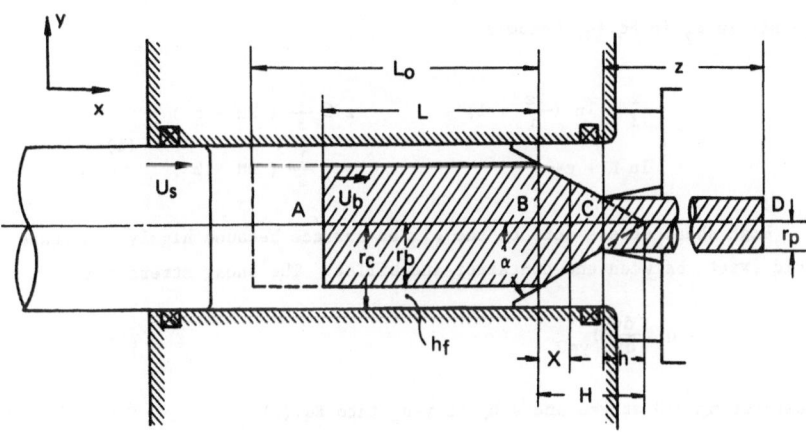

Fig. 2. Deformation pattern of billet during hydrostatic extrusion.

From Eqs. (1) and (2), the pressure in the container can be represented as

$$p = p_0 \, V_0^{-\gamma}(\, A_s U_s t - A_p z \,)^{\gamma} \tag{3}$$

(ii) Extrusion pressure

In hydrostatic extrusion, a billet is generally extruded through a conical die. Extrusion pressure for a billet having work-hardening characteristics $\sigma = \sigma_0 \, \varepsilon^n$ is

$$P_e = \frac{\sigma_0}{n+1} \, (\, \varepsilon_2^{n+1} - \varepsilon_1^{n+1} \,) + \frac{\mu \sigma_0}{\sin\alpha} \left\{ \frac{\varepsilon_2^{n+1} - \varepsilon_1^{n+1}}{n+1} \right.$$
$$\left. + \frac{\varepsilon_2^{n+2} - \varepsilon_1^{n+2}}{(n+1) \, (n+2)} \right\} + \frac{\lambda \sigma_0}{n+1} \, (\, \varepsilon_1^{n+1} - \varepsilon_2^{n+1} + \varepsilon_3^{n+1} \,) \tag{4}$$

where, $\quad \varepsilon_1 = \frac{1}{3} \, (\, \frac{\alpha}{\sin 2\alpha} - \cot\alpha \,), \quad \varepsilon_2 = \varepsilon_1 + \ln R \, (\text{ or } \varepsilon_1 + \varepsilon_z \,)$

$$\varepsilon_3 = \varepsilon_2 + \frac{1}{\sqrt{3}} \, (\, \frac{\alpha}{\sin 2\alpha} - \cot\alpha \,)$$

To reduce a breakthrough pressure at the beginning of extrusion, the billet is nosed to fit a conical entry die. A strain induced in the nosed part is expressed as a function of the length of product.

$$\varepsilon_z = \frac{2}{3} \, \ln \, (\, \frac{3z}{h} + 1 \,) \tag{5}$$

The strain ε_2 in Eq. (4) becomes

$$\varepsilon_2 = \begin{cases} \dfrac{2}{3} \, \ln \, (\dfrac{3z}{h} + 1) + \varepsilon_1 & z \leq \dfrac{1}{3} \, (\, RH - h \,) \\[3mm] \ln R + \varepsilon_1 & z > \dfrac{1}{3} \, (\, RH - h \,) \end{cases} \tag{6}$$

High shear stress acts on the billet surface because highly compressed fluid exists between the container and billet. The shear stress τ is

$$\tau = \eta \, (\frac{dv}{dy})_{y=0} \tag{7}$$

Substituting v=0 at y=0 and v=U_b at y=h_f into Eq. (7),

$$\tau = \eta \, \frac{4 U_b}{h_f} \tag{8}$$

Frictional force acting on the billet surface during extrusion is given by

$$F = \int_0^L 2\pi r_b \tau dL = \frac{8\pi\eta U_b L}{h_f} \qquad (9)$$

F is reduced to the frictional force per unit area of the cross section of the billet as follows:

$$P_f = \frac{F}{r_b} = \frac{8\eta U_b L}{r_b h_f} \qquad (10)$$

In hydrostatic extrusion, this frictional force acts as a damper because the fluid under high pressure is highly viscous to suppress rapid movement of the billet.

(iii) Momentum equation

When the stick-slip motion occurs during hydrostatic extrusion, excess pressure is required to accelerate the billet and the product. Crawley et al. produced the following equation [3].

$$\Delta P_{AB} = \frac{\rho L \ddot{z}}{gR} - \frac{\rho z \ddot{z}}{gR^2} - \frac{\rho \dot{z}^2}{gR^2}$$

$$\Delta P_{BC} = -\frac{k\rho}{g}\left(1 - \frac{1}{R}\right)\dot{z}^2 \qquad (11)$$

$$\Delta P_{CD} = \frac{\rho \dot{z}^2}{g} + \frac{\rho z \ddot{z}}{g}$$

By combining Eqs.(3), (4), (10) and (11), the momentum equation can be obtained to numerically analyse the stick-slip motion in hydrostatic extrusion as follows:

$$P_o V_o \quad (A_s U_s t - A_p z)$$

$$= \frac{\sigma_o}{n+1}(\varepsilon_2^{n+1} - \varepsilon_1^{n+1}) + \frac{\mu\sigma_o}{\sin\alpha}\left\{\frac{\varepsilon_2^{n+1} - \varepsilon_1^{n+1}}{n+1}\right.$$

$$\left. + \frac{\varepsilon_2^{n+2} - \varepsilon_1^{n+2}}{(n+1)(n+2)}\right\} + \frac{\lambda\sigma_o}{n+1}(\varepsilon_1^{n+1} - \varepsilon_2^{n+1} + \varepsilon_3^{n+1}) + \frac{8\eta U_b L}{r_b h_f}$$

$$+ \frac{\rho L \ddot{z}}{gR} + \frac{\rho}{g}\left(1 - \frac{1}{R}\right)\left(1 + \frac{1}{R}\right)\dot{z} + \frac{\rho}{g}\left(1 - \frac{1}{R}\right)z\ddot{z} \qquad (12)$$

2.3 Numerical calculation of stick-slip motion

(i) Velocity dependency of coefficient of friction

78

The fluid lubrication is dominant in hydrostatic extrusion because the viscous pressure medium is introduced into the billet/die interface by wedge action. Under the fluid lubrication condition, coefficient of friction increases with billet speed. However, boundary lubrication occurs during the slow speed of the billet, where the coefficient of friction becomes small with an increase in billet speed.

Velocity dependency of coefficient of friction is assumed for the wide range of billet speed as follows:

$$\mu = \mu_0 \, \dot{z}\omega \tag{13}$$

Two cases shown in Fig. 3 are selected to simulate the dynamics of hydrostatic extrusion by taking the friction test results [4] into account.

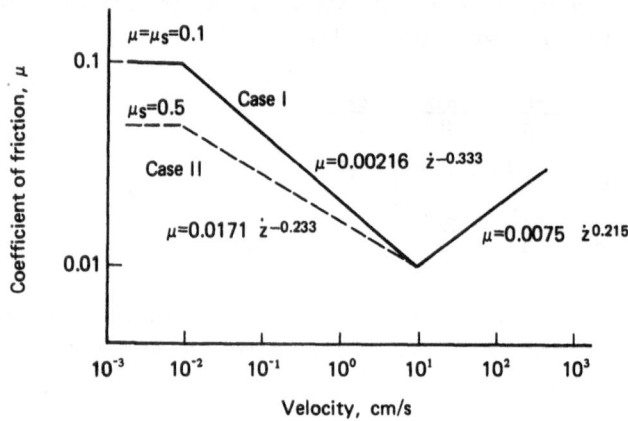

Fig. 3. Velocity dependency of coefficient of friction assumed for simulating the dynamic behaviour in hydrostatic extrusion.

(ii) Compressibility of pressure medium

The pressure medium has a high compressibility under high pressures. Value of γ in Eq.(1) was determined by measuring volume changes of castor oil which is frequently used in hydrostatic extrusion. Figure 4 shows the volume and density changes at high pressure up to 500 MPa. The value of γ is about 1.5 for the castor oil.

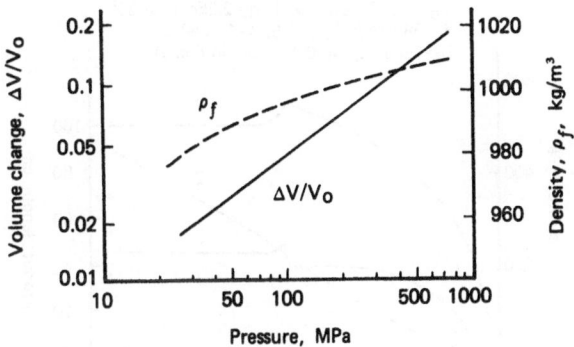

Fig. 4. Volume and density changes of castor oil under
high pressures.

(iii) Some results of numerical calculation

In this simulation, commercially pure copper and aluminium are used as examples. Their flow stress-strain relations are determined by tensile tests as follows.

$$\sigma = 420\ \varepsilon^{0.27} \qquad MN/m^2 \qquad \text{for copper}$$
$$\sigma = 132\ \varepsilon^{0.20} \qquad MN/m^2 \qquad \text{for aluminium}$$
(14)

By using Eq.(12), the effects of extrusion parameters on dynamic behaviour can be numerically determined. Figure 5 indicates the changes of pressure, product speed and product length during the extrusion. As the stem is pressed pressure gradually reaches a breakthrough point. It is then that the billet nose is extruded during pressurization and attains 90 percent of steady extrusion speed just before the breakthrough. The rounded breakthrough pressure is followed by a constant extrusion pressure. Corresponding to the pressure peak, product speed reaches its maximum and then decreases to become constant. Another simulation indicates the severe stick-slip motion throughout the extrusion of aluminium, in which only the stem speed is decreased from 75 mm/sec to 15 mm/sec.

When extruding a billet with a fully annealed nose fitted to the die face, hydrodynamic lubrication becomes dominant because the billet begins to move before the pressure reaches a breakthrough peak. This results in a low breakthrough pressure and a stable extrusion. If the billet nose is

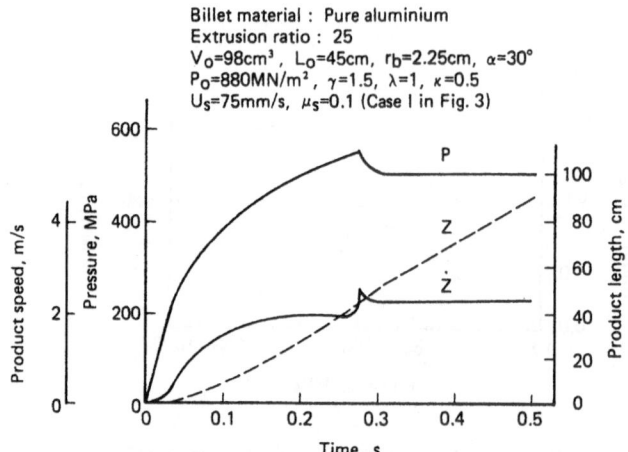

Billet material : Pure aluminium
Extrusion ratio : 25
$V_0=98cm^3$, $L_0=45cm$, $r_b=2.25cm$, $\alpha=30°$
$P_0=880MN/m^2$, $\gamma=1.5$, $\lambda=1$, $\kappa=0.5$
$U_s=75mm/s$, $\mu_s=0.1$ (Case I in Fig. 3)

Fig. 5. Pressure and product speed curves simulated for hydrostatic extrusion of pure aluminium [5].

fully work-hardened, the billet adheres to the die until the pressure is
increased to the peak as shown in Fig.6. Stable extrusion is obtained with
the good boundary lubrication indicated by Case II in Fig.3. In contrast,
with the high coefficient of friction shown by Case I in Fig. 3, severe
stick-slip occurs and the maximum product speed is calculated as 74 m/sec.
To suppress the rapid movement of the billet, a frictional force by viscous
pressure medium is effective. Reduced volume of pressure medium is also
applicable.

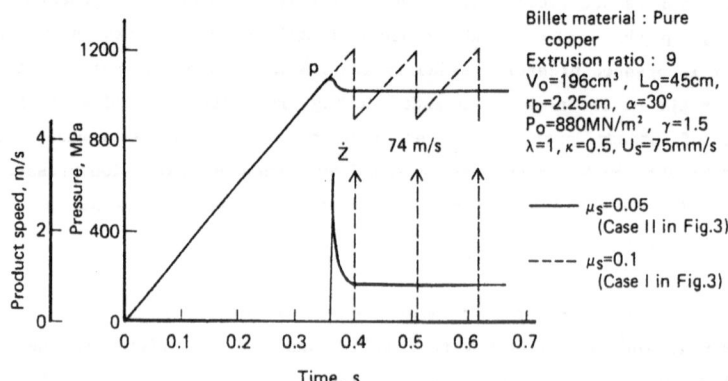

Billet material : Pure
copper
Extrusion ratio : 9
$V_0=196cm^3$, $L_0=45cm$,
$r_b=2.25cm$, $\alpha=30°$
$P_0=880MN/m^2$, $\gamma=1.5$
$\lambda=1$, $\kappa=0.5$, $U_s=75mm/s$

——— $\mu_s=0.05$
(Case II in Fig.3)

---- $\mu_s=0.1$
(Case I in Fig.3)

Fig. 6. Pressure and product speed curves simulated for
hydrostatic extrusion of pure copper [5].

2.4 Variation of product speed during extrusion

According to the calculated results, it is clear that a billet starts

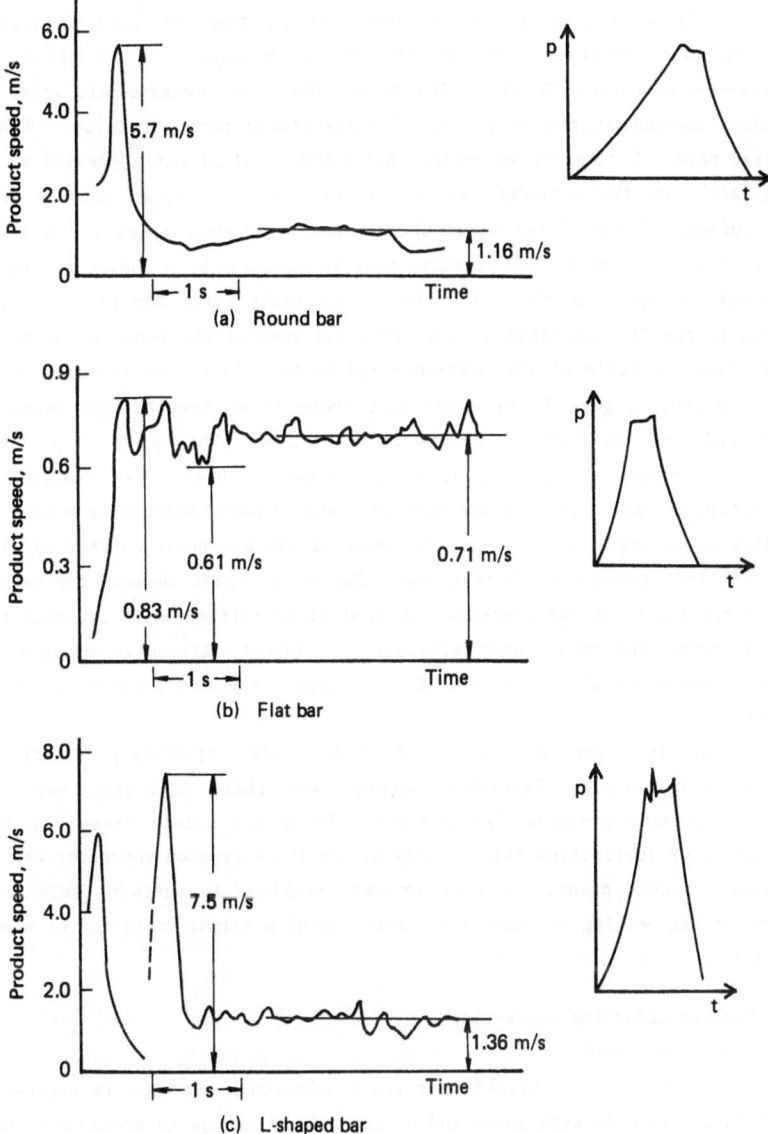

Fig. 7. Variation of product speed during hydrostatic extrusion.

to move before fluid pressure reaches the breakthrough point. Nishihara et al. examined this behaviour by measuring the displacement of the front end of the billet nose [2,6].

On the other hand, it is not always certain that the speed of product is constant whenever the smooth runout pressure is observed in pressure-time curve. To clear this doubt, the author measured the speed of product throughout the extrusion of commercially pure aluminium rod and shaped bars. During the extrusion, paint was intermittently sprayed at 17 cycles/sec on the extruded bar at the die exit and speed changes were determined. Figure 7 (a) shows the changes of product speed for a round bar. Rounded breakthrough pressure peak is followed by a runout pressure. The maximum speed is about 5.7 m/sec immediately after the breakthrough, which is rapidly decreased to 0.6 m/sec and reaches the constant speed of 1.2 m/sec. A ratio of the maximum speed to that in steady state is about five to one. Figure 7 (b) shows that there is no breakthrough pressure peak and constant runout pressure is maintained. The speed of product, however, changes irregularly from 0.83 m/sec to 0.61 m/sec during the extrusion of flat bars. In the case of L-shaped bars, stick-slip occurs as indicated in Fig.7 (c). The two peaks of product speed correspond with the two breakthrough pressure peaks. The sudden speed changes are caused by energy stored in the pressure medium which is sufficient to overcome the breakthrough but more than necessary for runout. Irregular changes of product speed are also observed. Nevertheless the runout pressure curve is smooth.

Generally, products extruded under the stick-slip condition periodically display burnished surface and their diameters vary in accordance with pressure fluctuations. These are mainly caused by the variation of lubrication film thickness. In the extrusion characterized by a smooth runout pressure curve, however, problems in surface finish and diameter are nothing to worry about even though a slight variation of speed occurs.

2.5 Factors affecting stick-slip
(i) Extrusion speed

It is generally regarded that the breakthrough pressure is decreased with an increase in stem speed and no significant change in pressure occurs beyond a certain point. Because the pressure fluctuations are a function of the ratio of breakthrough pressure to runout, reducing the breakthrough pressure is effective in governing stability in hydrostatic extrusion. When investigating the effect of extrusion speed, Low and Donaldson concluded that the stick-slip motion might be removed if the speed of billet was 27-127 mm/sec [7]. Other investigators have experienced the stable extrusion without stick-slip at the lower speed [1,6,8].

Figure 8 shows the influence of stem speed on the stick-slip motion observed in extruding an aluminium bar coated with beeswax. The ratio of billet volume to pressure medium was about 0.65 to 1. Castor oil was used as a pressure medium. At a stem speed of 0.9 and 3.3 mm/sec, severe stick-slip occurs. Amplitude of pressure fluctuation increases at a lower stem speed. The stick-slip motion continues when the pressing speed of the stem is not high enough to compensate for the sharp drop in pressure as shown in Fig.8 (b). When the stem speed is increased to be nearly equal to

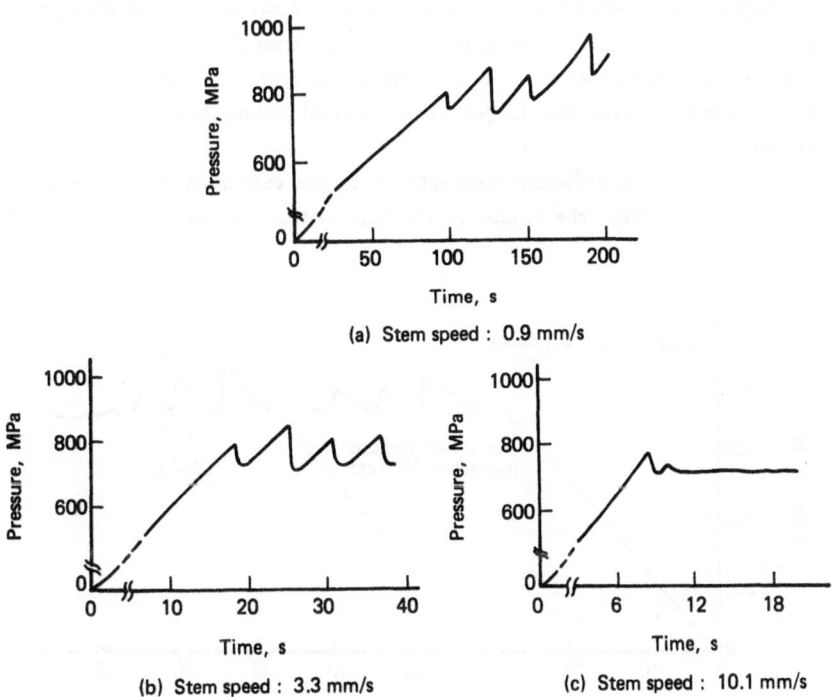

(a) Stem speed : 0.9 mm/s

(b) Stem speed : 3.3 mm/s   (c) Stem speed : 10.1 mm/s

Fig. 8.   Three types of pressure-time curves observed in hydrostatic extrusion with different stem speeds.

that of the billet during slip condition, no stick-slip is observed as shwon in Fig.8 (c) because a kinetic friction condition is maintaned.

(ii) Volume of pressure medium

In some cases, it is a problem when the speed effect is not the same for different volumes of billet and pressure medium. To avoid this confusion, a pressurization rate is convenient. The pressurization rate is delivered by differentiating Eq.(3) as follows, when a constant stem speed is assumed:

$$\frac{dp}{dt} = p_0 A_s \gamma \left(\frac{U_s}{V_0}\right)^\gamma t^{\gamma-1} \qquad (15)$$

Equation (15) indicates that a small volume of pressure medium has the same effect with an increase in stem speed.

Low et al. studied the influence of fluid volume on breakthrough pressure and stick-slip [9]. Changing the volume from 20 in.^{3} to 7 in.^{3}, the pressure was decreased by 2.2% and only a few cycles of pressure fluctuation occurred, while the larger volume caused stick-slip throughout the extrusion.

Figure 9 shows pressure fluctuations in the extrusion of an aluminium bar, in which only the volume of pressure medium was decreased compared

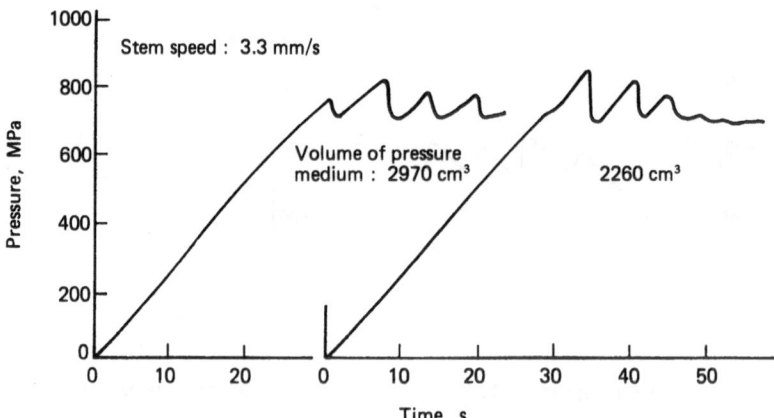

Fig. 9. Effect of volume of pressure medium on the stick-slip.

with the conditions of the experiments shown in Fig.8 (b). Reducing the volume by 24%, amplitudes of pressure fluctuation damper to a steady extrusion pressure.

(iii) Effect of lubrication

In general, the pressure medium itself provides adequate lubrication between the billet and the die. If the pressure drop is not very large, the fluid pressure is maintained enough to extrude the billet by pressing the stem with constant speed. This results in steady extrusion.

To eliminate stick-slip in hydrostatic extrusion, it is essential that the breakthrough pressure is kept equal to or slightly higher than the runout pressure. Factors affecting breakthrough pressure are billet lubricant, pressure medium, nose geometry and compatibility of die and billet materials.

Figure 10 shows the effects of billet lubricant and pressure medium on the stick-slip motion obtained in the extrusion of 7075 aluminium alloy[1]. In the case of Curve 1, high breakthrough pressure and stick-slip appear during runout. By changing the billet lubricant from 20wt% MoS_2 in castor wax to 20wt% MoS_2 in stearyl stearate, the breakthrough pressure was reduced and stick slip was eliminated.

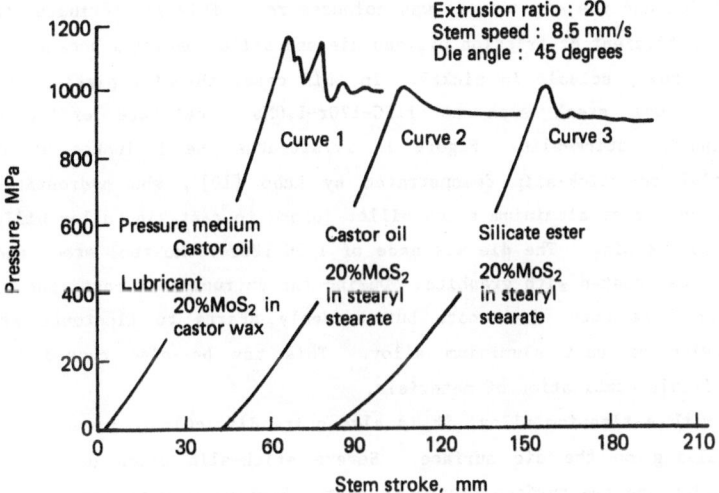

Fig. 10. Pressure curves obtained in hydrostatic extrusion of 7075 Al-alloy using different pressure medium and billet lubricant [1].

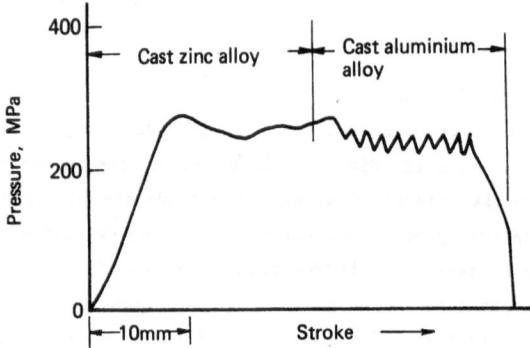

Fig. 11.   Effect of billet material on pressure-stroke curve [10].

Poor lubrication causes stick-slip motion of the billet. It was found
that compatibility of billet with the die material was important to prevent
stick-slip. For extruding a tough pitch copper billet coated with beeswax
in a cold condition, a die made of maraging steel containing 18% Ni was
used. Very severe pressure fluctuations were observed throughtout the
extrusion and the die surface was coloured red. This is attributed to the
high coefficient of friction between die and billet material because copper
can be fully soluble in nickel. In this case, chromium plating or high
carbon tool steel such as 1.5C-12Cr-1.0Mo steel are effective in
eliminating stick-slip. Figure 11 illustrates the influence of billet
material on stick-slip demonstrated by Kubo [10] , who hydrostatically
extruded a cast aluminium alloy billet joined to cast zinc alloy billet by
friction welding. The die was made of 1.5C-12Cr-1.0Mo tool steel and the
billet was coated with graphite. During the extrusion of cast zinc alloy,
the pressure curve is smooth but suddenly starts to fluctuate at the
extrusion of cast aluminium alloy. This may be also caused by the
undesirable combination of materials.

With a titanium alloy, it is also quite difficult to prevent seizing
or galling on the die surface. Severe stick-slip often occurs during
extrusion and the surface finish is poor. Similar results are obtained in
warm hydrostatic extrusion of aluminium alloys. Such a problem in
extrusion can be overcome by improving lubrication by, for instance,
coating the billet with oxide film and molybdenum disulfide for titanium

87

alloy and roughening the billet surface with small grooves for aluminium alloys (Fig.12) [11].

(iv) Billet nose geometry

Billet nose geometry has a considerable influence on the decrease of breakthrough pressure and the elimination of stick-slip. Fiorentino et al. solved the problem by changing the billet nose geometry. The standard nose was machined in a plain cone which fitted the die angle. The changed nose had a compound angle so that the upper part had a smaller angle compared with the die but the lower part fitted the die surface. By using the compound angled nose geometry, breakthrough pressure was remarkably reduced and stick-slip following breakthrough was eliminated, while in the extrusion of a billet with the standard nose high breakthrough pressure and severe stick-slip were observed [1, 2].

(v) Surface roughness of billet

Because a suitably rough surface of billet assists lubrication by

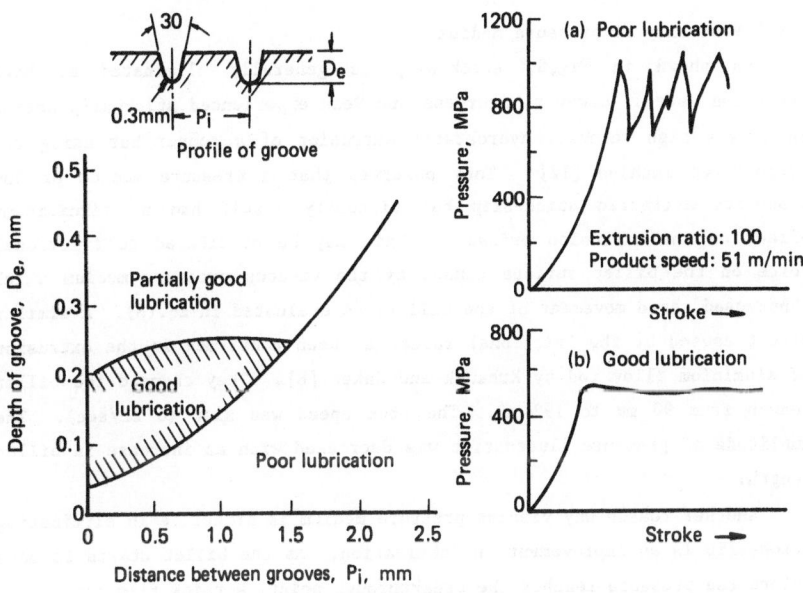

Fig. 12. Effect of groove geometry over billet surface on dynamic behaviour in different lubricating conditions [11].

trapping the coated lubricant and viscous pressure medium, surface roughness is an important factor in preventing stick-slip. Low et al. studied the effect of billet surface finish and nose geometry on extrusion pressure [9]. They found that with the sand-blasted billet, the breakthrough pressure was reduced and stick-slip became mild compared with that of a bare billet. Moreover, no stick-slip was obtained with the sand-blasted and radiused nose geometry.

Seido et al. investigated the effect of machined spiral grooves on lubricating conditions during extrusion of pure aluminium [11]. The gross area of die surface overlaid with aluminium was measured to evaluate lubrication. As shown in Fig.12, lubrication quality varies depending on the depth and distance of grooves. The most efficient lubrication is obtained when the grooves are from 0.1 to 0.2 mm deep and about 0.5 mm apart. It is also shown in Fig.12 that severe stick-slip occurs due to poor lubrication, when the depth of the grooves is either too shallow or too deep relative to the distance between them. Improved lubrication, however, results in smooth extrusion.

(vi) Viscosity of pressure medium

As shown in Fig.8, stick-slip is generally eliminated at high extrusion speed. However, Sturgess and Dean experienced stick-slip motion in a very high velocity hydrostatic extrusion of a copper bar using the Petro-Forge machine [12]. They observed that a pressure medium of low viscosity attracted stick-slip but viscosity itself had no significant effect on the extrusion pressure. This may be attributed to frictional force on the billet surface caused by the viscous pressure medium which suppressed rapid movement of the billet, as evaluated in Eq.(8). A similar effect caused by the frictional force has been confirmed in the extrusion of aluminium alloy rod by Krhanek and Jakes [8]. They changed the billet length from 90 mm to 192 mm. The stem speed was slow (6 mm/sec). The amplitude of pressure fluctuation was decreased with an increase in billet length.

Another reason why viscous pressure medium is effective in eliminating stick-slip is an improvement in lubrication. As the billet starts to move before the pressure reaches the breakthrough point, a thick film can be hydrodynamically introduced into the billet/die interface. The film becomes thicker with an increase in viscosity. Hence, the breakthrough pressure is reduced to give a smooth runout.

(vii) Temperature

Figure 13 shows temperature rises in die wall accompanied by the stick-slip motion of the billet in the extrusion of a pure aluminium bar. Temperature was measured by thermocouples inserted into holes drilled to just below the die surface. The temperature rise is remarkably rapid and high at the slip motion of the billet while it is gradual during the stable extrusion. Such a rapid change in temperature causes the billet to easily stick to the die surface due to the breakdown of lubricant film. This is a serious problem for sticky materials such as titanium and aluminium alloys.

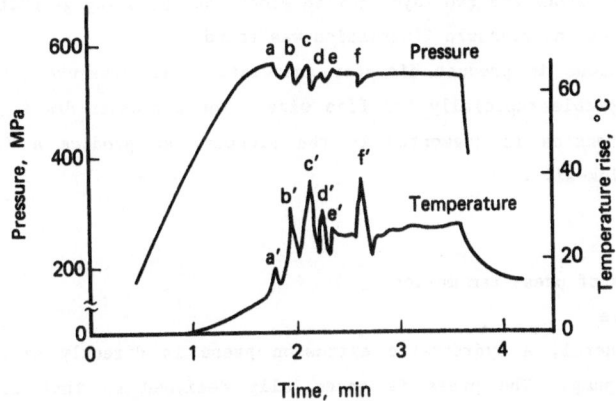

Fig. 13. Temperature change in the die wall near surface during hydrostatic extrusion with stick-slip [2].

As extrusion continues, pressure fluctuations are amplified gradually as illustrated in Type IV of Fig.1. To avoid this happening, breakthrough pressure should be reduced to near runout pressure by improving the lubrication. Pre-coating of the billet with anti-heat lubricants, correct surface roughness of billet and a viscous pressure medium are effective.

(viii) Billet dimension

Stick-slip is often inevitable in experiments in which a small billet is extruded at slow speed by using a large quantity of pressure medium. In contrast, in industrial practice, the billet is large and the volume of pressure medium is small enough to hydrostatically extrude at the correct speed. As discussed in sub-section 2.3, the motion of a large billet is insensitive to pressure fluctuation due to its large inertia. Hence, smooth extrusion without stick-slip can be achieved even if severe seizing occurs on the billet/die interface [11].

On the other hand, in the hydrostatic extrusion of wire, die size for a given reduction has a significant effect on pressure fluctuation. Lowe et al. carried out some experiments in which aluminium and copper were extruded using dies with bore diameters of 0.02, 0.07 and 0.2 in. [13]. The pressure medium was castor oil. No stick-slip occurred with the 0.2 in. die. Stick-slip, however, was always observed for the 0.02 and 0.07 in. dies. Pressure fluctuation was more severe in the extrusion of unlubricated stock and two layer coiled stock, but by using graphite grease on the billet, no pressure fluctuation was found.

Variations in product diameter associated with pressure fluctuation are unacceptable especially for fine wire. The extrusion-drawing process in which tension is augmented to the product can produce a fine wire without stick-slip.

3. Dynamics of press ram motion

3.1 Analysis

In general, a hydrostatic extrusion press is directly driven by a hydraulic pump. The press is essentially designed so that controlled delivery of the working oil provides an expected speed of the press ram. Sometimes, however, unexpected movement of the press ram occurs due to the compressibility of the working oil. This behaviour is important for hydrostatic extrusion accompanied by a high breakthrough pressure, particularly the severe stick-slip motion of the billet [14].

Figure 14 indicates schematically an oil driving system of extrusion press. The working oil is fed into the main cylinder to advance the main ram by a hydraulic pump. In pushing the stem, which is fixed to the main ram, into the container, high pressure is generated to extrude a billet through the die. The stem is returned to the initial position after extrusion by the pullback ram fixed to the main ram. For the movement of the main ram with stem and pullback ram, the momentum equation is represented as follows:

$$m \, \frac{d^2x}{dt^2} = p_1 A_1 - p_2 A_2 - p A_s \qquad (16)$$

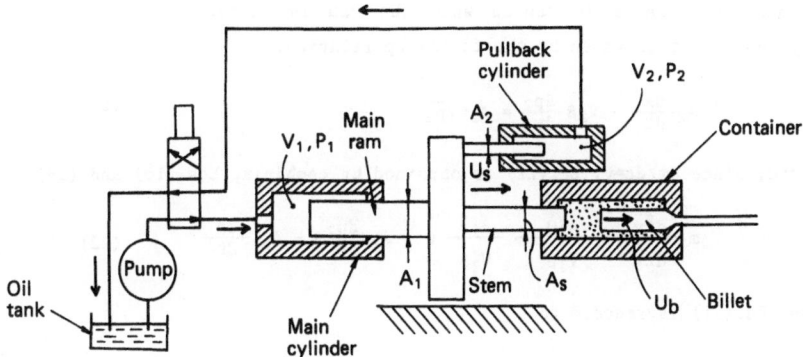

Fig. 14. An oil driving system of hydrostatic extrusion press.

Reducing Eq.(16), Eq.(17) is obtained for oil pressure in the main ram cylinder.

$$p_1 = \frac{1}{A_1} \left(m \frac{d^2x}{dt^2} + p_2 A_2 + p A_s \right) \tag{17}$$

The working oil is delivered by Q per unit time into the main ram cylinder, and an increase in pressure associated with volume change of working oil is given by

$$A_1 \frac{dx}{dt} + V_1 \beta \frac{dp_1}{dt} = Q \tag{18}$$

On the other hand, the oil in the pullback cylinder is pressurized enough to flow through a pipe to an oil tank. The pressure is represented by the following equation:

$$p_2 = \left(\frac{\lambda_f \ell}{d} + \Sigma \zeta + 1 \right) \gamma \frac{v_p}{2g} \tag{19}$$

Flow rate q of working oil out of the pullback cylinder is

$$q = \frac{\pi}{4} d^2 v_p = \phi A_2 \sqrt{p_2} \tag{20}$$

where,

$$\phi = \frac{\pi d^2}{4A_2} \sqrt{\frac{2g}{\gamma g} \cdot \frac{1}{\frac{\lambda_f \ell}{d} + \Sigma \zeta + 1}}$$

A pressure change is caused when the ram is pushed into the pullback cylinder. It is given by the following relation.

$$A_2 \frac{dx}{dt} - V_2\beta \frac{dp_2}{dt} = A_2 \phi \sqrt{p_2} \qquad (21)$$

Here, since $dx/dt = v$, Eq.(22) is obtained by combining Eqs.(16) and (18).

$$m \frac{d^2v}{dt^2} = \frac{A_1Q}{V_1\beta} - \frac{A_1^2}{V_1\beta} v - A_2 \frac{dp_2}{dt} - A_s \frac{dp}{dt} \qquad (22)$$

And Eq.(21) is reduced to

$$v = \phi\sqrt{p_2} + \frac{V_2\beta}{A_2} \cdot \frac{dp_2}{dt} \qquad (23)$$

It is generally held that a large diameter pipe used for the pullback cylinder provides a faster return speed of the ram than during extruding stroke. Hence, pressure change in the pullback cylinder is negligibly small. This results in the simplified equation as follows:

$$v = \phi\sqrt{p_2} \qquad (24)$$

Differentiating Eq.(24), Eq.(25) is obtained.

$$\frac{dp_2}{dt} = \frac{2v}{\phi^2} \frac{dv}{dt} \qquad (25)$$

Consequently, the following momentum equation is derived by substituting Eq.(25) into Eq.(22).

$$m \frac{dv^2}{dt^2} + \frac{A_1^2}{V_1\beta} v + \frac{2A_2v}{\phi^2} \frac{dv}{dt} - \frac{A_1Q}{V_1\beta} + A_s \frac{dp}{dt} = 0 \qquad (26)$$

Analysing the above equations gives the behaviour of ram motion under a constant delivery out of hydraulic pump. If the sharp drop of extrusion pressure or stick-slip occurs after breakthrough, the main ram could suddenly accelerate. Hence, the velocity of ram could not be constant.

3.2 Dynamic behaviour of press ram

When the extrusion pressure suddenly drops by Δp, the ram is rapidly accelerated. This behaviour can be analysed as follows by assuming the sharp drop of pressure. In a simple example, it is assumed that the press has no pullback cylinder and extrusion pressure is constant during steady extrusion. Then, Eq.(26) is reduced as follows:

$$\frac{d^2v}{dt^2} = \frac{A_1Q}{mV_1\beta} - \frac{A_1{}^2}{mV_1\beta} v \qquad (27)$$

Integrating Eq.(27) with the boundary conditions $v\dot=0$, $\frac{dv}{dt} = \frac{\Delta pA_s}{m}$ at $t=0$, the solution is obtained.

$$v = \frac{Q}{A_1} (1- \cos \omega_1 t + k_1\sin \omega_1 t) \qquad (28)$$

where,

$$\omega_1 = \frac{A_1}{\sqrt{mV_1\beta}} , \qquad k_1 = \frac{\Delta pA_s}{Q} \sqrt{\frac{V_1\beta}{m}}$$

Variation of the pressure is caused in the main ram cylinder as follows:

$$p_1 = \frac{1}{A_1} [\frac{mQ\omega_1}{A_1} (\sin \omega_1 t + k_1\cos \omega_1 t) + pA_s] \qquad (29)$$

Figure 15 shows the changes of pressure in the main cylinder simulated for a 4.9 MN extrusion press by using Eq.(28). The fluctuation of pressure is

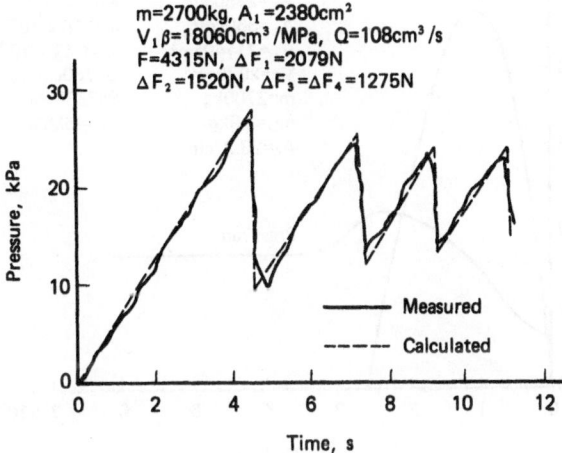

Fig. 15. Pressure fluction in main ram cylinder accompanied by stick-slip motion of press ram.

oscillatory and this indicates the occurrence of stick-slip motion of the press ram.

To avoid such unserviceable motion of the ram, the pullback cylinder is effective. According to the assumption of constant extrusion pressure during extrusion, the following equation is obtained for the press with pullback cylinder:

$$\frac{d^2v}{dt^2} + \frac{A^2_1}{mV_1\beta} \, v + \frac{2A_2}{m\phi^2} \, v \, \frac{dv}{dt} - \frac{A_1Q}{mV_1\beta} = 0 \qquad (30)$$

With the boundary conditions $V=0$, $\frac{dv}{dt} = \frac{\Delta pAs}{m}$ at $t=0$, the following solution is obtained by integrating Eq.(30).

$$v = \frac{Q}{A_1} \, [\, 1 - e^{-n't} \, (\, \cos \omega_2 t - k_2 \sin \omega_2 t \,) \,] \qquad (31)$$

where,

$$n' = \frac{A_2Q}{mA_1\phi^2} \, , \qquad \omega_2 = \frac{A_1}{\sqrt{mV_1\beta}} \, , \qquad k_2 = \frac{1}{\omega_2} \, (\, \frac{A_1\Delta pAs}{Qm} - n' \,)$$

Equation (31) indicates that the oscillated motion of the ram is damped to

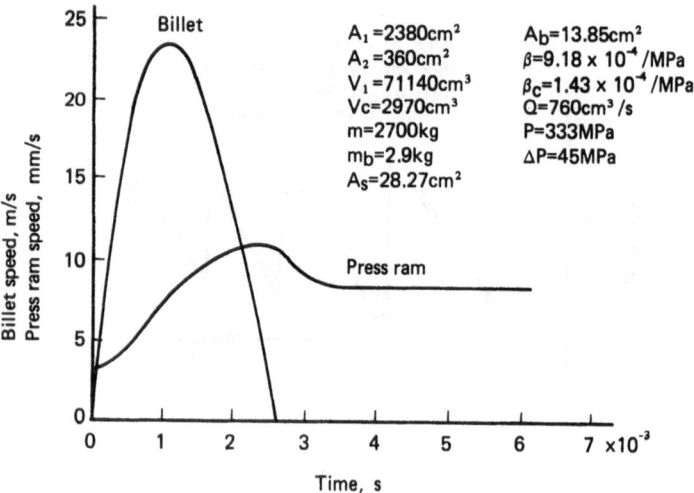

Fig. 16. Changes of press ram speed accompanied by stick-slip movement of billet [14].

be steady. This is attributed to the pullback cylinder suppressing rapid movement of the ram.

Yamaguchi numerically solved the dynamic behaviour of the press ram by taking the billet motion into account [14]. Figure 16 shows the changes in speed of billet and main ram in hydrostatic extrusion using a 4.9 MN press. While the change in billet speed is sharp, the ram is gradually accelerated and a constant speed can be obtained. The steady movement of the ram is attributed to the fact that the mass of the ram is large compared with a billet and the drop of extrusion pressure is small and insufficient to accelerate the large mass rapidly.

4. Conclusions

In hydrostatic extrusion, it is not difficult to establish stable extrusion under selected conditions. The important factors are small volume of pressure medium, preferably high velocity of stem and large mass of billet. These are easily realized in the actual use of this process. The other important factor is improvement of lubrication, particularly for reducing the breakthrough pressure. An optimization in lubrication has been achieved for some materials such as copper alloys and steels. For sticky materials, however, effective lubricants must be investigated.

With reference to press ram motion, stick-slip seldom occurs in the extrusion press with pullback cylinder even if the billet moves intermittently.

References

1   R.J.Fiorentino, B.D.Richardson, A.M.Sabroff and F.W.Boulger, Proc. CIRP. Inter. Conf. Manufacturing Tech., Sept. 25 - 28, 1967, P.941.

2   M.Nishihara, Pre-print, 2nd Czecho-Slovak Conf. High Pressure Forming, Brno, May 12 - 14, 1971.

3   J.Crawley, J.A.Pennell and A.Saunders, Conf. High pressure Engng., Sept. 11 - 15, 1967, (Paper 23).

4   S.Fukui, T.Ohi, H.Kudo, I.Takita and J.Seino, Int.J.Mech.Sci., 4(1962)297.

5   T.Matsushita, Y.Yamaguchi, M.Noguchi, M.Nishihara and T.Fujita, J.Japan Society Tech. Plasticity, 15-164(1974)709, (in Japanese).

6   M.Nishihara, T.Fujita, Y.Yamaguchi, T.Matsushita and M.Noguchi, ibid., 13-134(1972)178, (in Japanese).

96

7   A.H.Low and C.J.H.Donaldson, National Engng.Lab.Report, No.289, 1967.

8   P.Krhanek and J.Jakes, Engineering Solids under Pressure, IME, London, 1970, P.117.

9   A.H.Low, C.J.Donaldson and P.T.Wilkinson, Conf. High Pressure Engng., Sept. 11 - 15, 1967, (Paper 22).

10 K.Kubo, J.Society Mat. Sci., Japan, 20-215(1971)945, (in Japanese).

11 M.Seido, S.Mitsugi, D.Oelschlagel and A.Kobayashi, Proc. 2nd Int. Aluminium Extrusion Tech. Seminar, Nov. 15 - 17, 1977, Atlanta, Vol.1. P.132.

12 C.E.N.Sturgess and T.A.Dean, Proc. 13th MTDR Conf., Sept. 18 - 22, 1972, P.389.

13 B.W.H.Lowe and D.Goold, Conf. High Pressure Engng., Sept. 11- 15, 1967, (Paper 10).

14 M.Yamaguchi, Doctor dissertation, Univ. Tokyo, 1979.

CHAPTER 3

COLD AND WARM HYDROSTATIC EXTRUSION PROCESS

S. MITSUGI AND M. SEIDO
HITACHI CABLE, LTD.

SECTION 1

INTRODUCTION

In hydrostatic extrusion the friction between the billet and the die is reduced to the utmost limit. From the viewpoint of lubrication, therefore, hydrostatic extrusion is an ideal process of metalworking and has a number of advantages over conventional extrusion. In using hydrostatic extrusion as a forming process, however, a variety of ideas and considerations are required to make the most of its features in a series of extrusion operations including billet preparation, to say nothing of extrusion tooling.

The most important factor characterizing hydrostatic extrusion is the pressure medium. Quite a variety of materials have been studied experimentally, but it is castor oil that is mostly used at present as the standard from the point of view of viscosity, pressure dependence, lubrication, workability, economy, and so forth. In recent years, on the other hand, hot hydrostatic extrusion has been developed in which in place of castor oil the pressure medium is made of a grease-like substance which is solid or semi-solid at room temperature and is heat resistant. Great progress is being made in its application, as is described in the next chapter. The present chapter is devoted to hydrostatic extrusion with castor oil as the pressure medium, and is discussed from an industrial point of view as a metalworking process.

In the early stage of its development hydrostatic extrusion with castor oil as the pressure medium tended to employ higher and higher pressure as a cold extrusion process. Because of the limitation on the industrially usable level of pressure, however, cold extrusion has not provided sufficient processability in many cases. In order to achieve a higher degree of processing at lower pressure it is a common practice in industry to heat up the billet in the temperature range allowable from the thermal stability of castor oil, that is, up to ca. 600°C, and to lower the flow stress. It is this warm hydrostatic extrusion that is widely practised.

SECTION 2
MAIN FEATURES OF HYDROSTATIC EXTRUSION

The basic principle and operational characteristics of hydrostatic extrusion are no more than extruding the billet by pressing via a pressure medium. There are several advantages and disadvantages accompanying the process, and a full understanding of these features is necessary to effectively utilize hydrostatic extrusion as a forming process.

The process factors of hydrostatic extrusion are schematically shown in Fig. 1 [1]. The use of a pressure medium has two primary merits. First, the friction between tools (such as container, die and stem) and billet material is reduced almost to zero, making it possible to drastically lower the extrusion pressure. For a given amount of press capacity higher reductions at lower temperatures are obtained, when compared with conventional extrusion. In practice, this leads to an improved efficiency of the process along with an improvement in the quality of the extrudate. Moreover, the reduction of container and die friction contributes to the uniformity of material flow in the billet during extrusion, making hydrostatic extrusion especially suited for producing composite materials. The second merit in the use of a pressure medium is that the tooling is less contaminated by the billet material. The container, especially, is not contaminated by the billet at all: a single press can be used to extrude different kinds of metals continuously with only a slight cleaning of the die surface. This feature makes hydrostatic extrusion suitable for multiple species production.

The merits of hydrostatic extrusion due to the use of a pressure medium have been stated above, but it also has demerits. First of all, the cycle time for extrusion is long. When compared with conventional extrusion, where no pressure medium is employed, hydrostatic extrusion needs additional time for pouring the liquid pressure medium into the container and pressurizing it. Secondly, the billet needs to be provided with a nose at its leading end to prevent leakage of the pressure medium and to enhance extrudability. This makes the billet preparation more

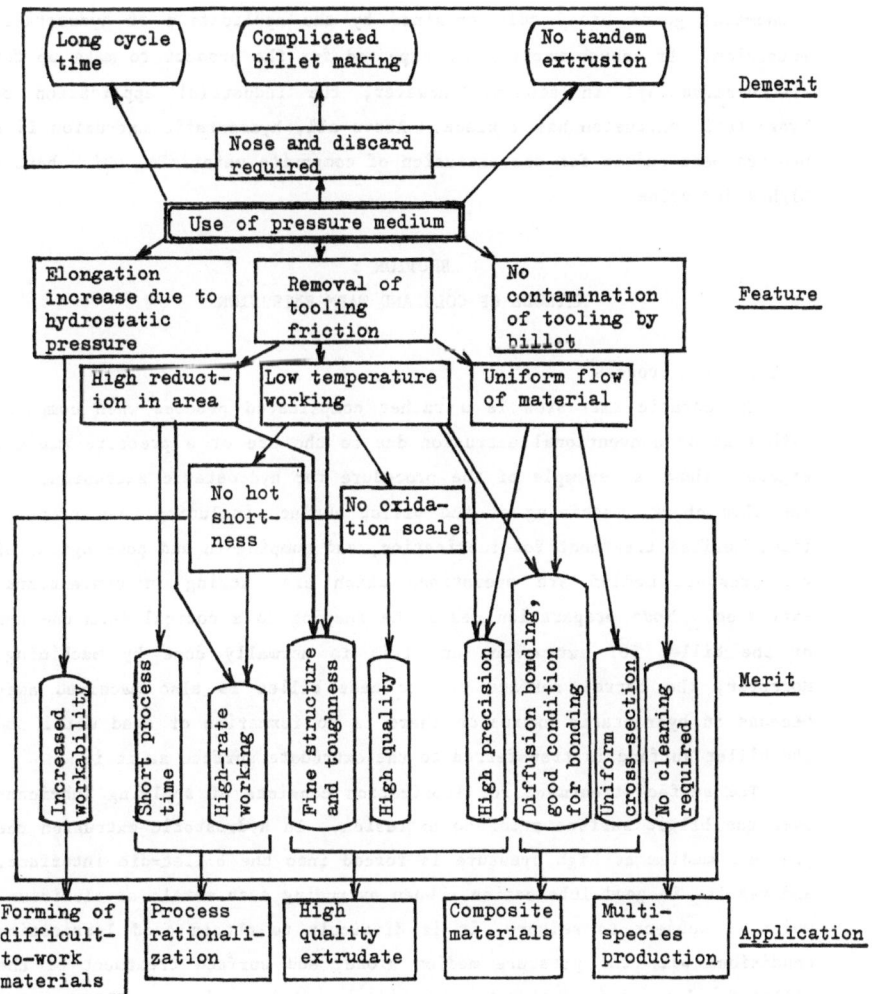

Fig. 1.   Process factors of hydrostatic extrusion [1].

involved. Thirdly, the use of a pressure medium makes a tandem extrusion difficult. Each of the demerits stated above causes a lowering of working efficiency and product yield.   Consequently, if the material can be extruded without decisive difficulties by conventional hot extrusion, economical gains are hardly obtained by the application of hydrostatic extrusion. If enough mertits are expected for the product to make up for the disadvantage in economy, however, the industrial application of hydrostatic extrusion has a place. Above all, hydrostatic extrusion is a process best suited for the extrusion of composite materials, which have a high added value.

SECTION 3

METHODS OF COLD AND WARM EXTRUSION

1. Extrusion Process

Hydrostatic extrusion is a rather complicated process when compared with that of conventional extrusion due to the use of a pressure medium. Figure 2 shows an example of the procedure for hydrostatic extrusion. In the flow chart, machining of the billet surface including nose preparation, surface treatment for lubrication, and pumping-in and pouring-out of the pressure medium are operations which are lacking in conventional extrusion. Nose preparation means the shaping to a conical form one end of the billet for extrusion, and this is normally done by machining. Moreover, the barrel surface of the cast billet is also machined away because in hydrostatic extrusion there is no formation of dead metal and the billet surface is transferred to the extrudate surface as it is.

The surface treatment for lubrication consists of applying lubricant over the billet surface prior to extrusion. In hydrostatic extrusion the pressure medium at high pressure is forced into the billet-die interface, and results in good lubrication. When extruding such metals as aluminium, which is subject to seizure, it is difficult to obtain good lubricating conditions with the pressure medium alone, and surface treatment of the billet for lubrication prior to extrusion is often required.

2. Industrial Method for Hydrostatic Extrusion

For the purpose of utilizing hydrostatic extrusion as an industrial process for metalworking a variety of procedures and apparatus have been conceived. Some of them are still in the research and development state, while some others have been put to practical use. In order to become successfully established as an industrially useful means for metal working the method should have advantages not only in its technicality but also in productivity, convenience in operation, economy in installation

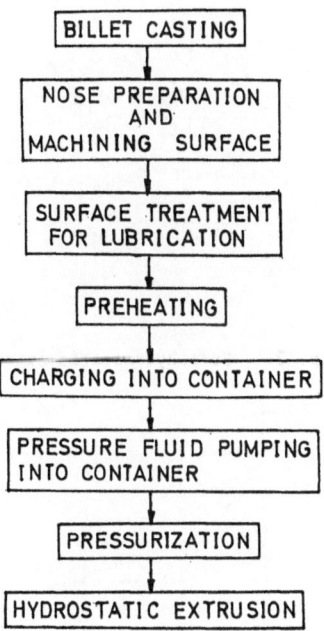

Fig. 2.   Flow chart for hydrostatic extrusion

costs, and so forth.

Figure 3 shows the standard type of hydrostatic extrusion, where the billet is extruded by the hydrostatic pressure of the pressure medium alone, without augmentation. In this sense it can be called 'Pure Hydrostatic Extrusion' or 'Orthodox Hydrostatic Extrusion'. Castor oil is mostly used as the pressure medium. It is this type of press that is mostly employed in industry, and presses of 4000 tonf and 2000 tonf capacity are currently manufactured.

In pure hydrostatic extrusion a large volume of pressure medium is used, causing an increase in cycle time for extrusion and dynamic instability such as stick-slips and so forth in some cases. In order to remove these drawbacks 'Thick Film Hydrostatic Extrusion' [2] has been proposed. Its principle is shown in Fig. 4. The amount of pressure medium between the billet and the container is reduced to minimum in so far as the hydrostatic pressure is transmitted. The lubricant applied to the billet prior

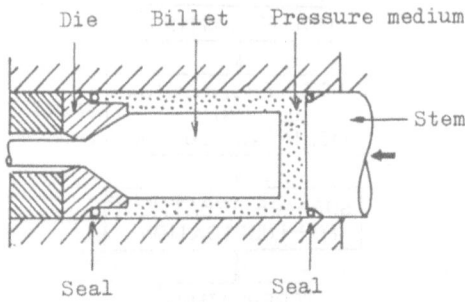

Fig. 3. Pure hydrostatic extrusion

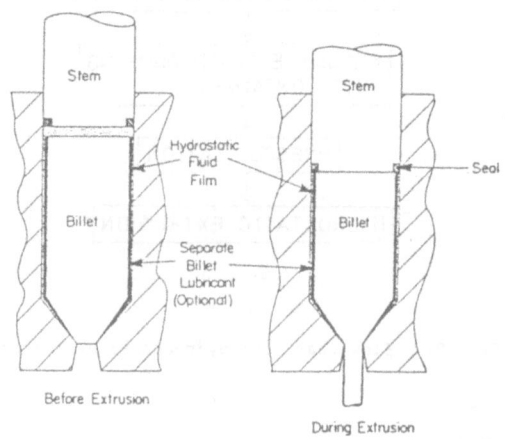

Fig. 4. Thick film hydrostatic extrusion [2].

to extrusion or to the inner surface of the container serves as the pressure medium. The idea originates in principle with 'Extrusion with Immuring Lubricant Technique' [3], proposed by Kudo and Shinozaki. The lubricating effect is further improved by the use of high-pressure seals peculiar to hydrostatic extrusion. Unlike pure hydrostatic extrusion, mentioned above, a grease-like substance, sometimes mixed with various kinds of additives, which is semi-solid at room temperature, is employed as the pressure medium. It is especially effective in hot extrusion. The 'Hydrafilm Process' [2], developed at Battelle Memorial Institute in the United States, and 'Extrusion Method with Visco-Plastic Pressure Medium' [4], developed at Kobe Steel Ltd., are in this category. As presses for this scheme those of 300 tonf and 2000 tonf capacity are currently manufactured.

In contrast to 'Pure Hydrostatic Extrusion' large extrusion ratio is realized at low hydrostatic pressure with supplementary mechanical force in 'Augmented Extrusion'. Its principle is shown in Fig. 5 [5]. According to the manner of applying the augmenting force it is classified into 'Billet Augmentation', where back pressure is applied to the billet, and 'Product Augmentation', where front tension is applied to the extrudate. An important feature of 'Augmented Extrusion' is that the phenomenon of stick-slip is prevented, which presents problems in some cases of 'Pure Hydrostatic Extrusion', and stable extrusion is obtained. Presses with 1600 tonf capacity are currently manufactured to be used for extrusion with billet augmentation.

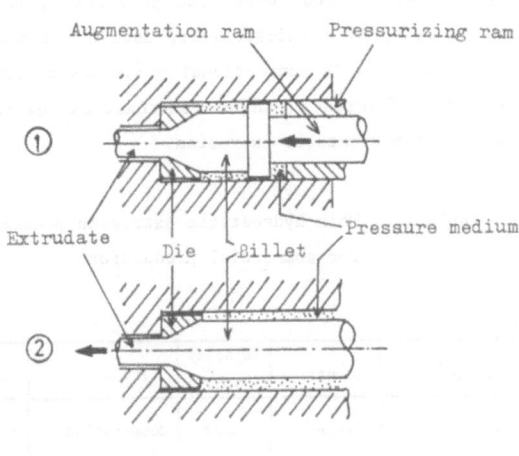

Fig. 5. Augmented extrusion [5].

In an attempt to obtain a large extrusion ratio with low hydrostatic pressure, in a similar way to 'Augmented Extrusion', Green developed the 'Hydrospin' extrusion method, in which rotary cutting is combined with hydrostatic extrusion [6]. An experimental apparatus has been built along with a pilot plant for industrial use of the method but is not in actual use.

All the methods stated above are for extruding billets. As already mentioned in Chapter 2 there are other methods for extruding wire. There have been epochmaking inventions to make the extrusion operation continuous, but they are still in the stage of experimentation and have not so far been put into practice. Consequently, the extrusion methods currently used on an industrial scale are all for extruding billets.

3. Applications

For the application of hydrostatic extrusion, various materials and shapes have been tested for extrudability and the properties of the extrudates examined. Broadly speaking, any products extrudable by conventional methods may be extruded by hydrostatic extrusion. In principle, solid or hollow rod, wire, and profiles of various metals and composites can be extruded. Industrially, however, those products which are difficult to extrude with conventional processes or are expected to be extruded with higher efficiency than conventional extrusion are the object of the application of hydrostatic extrusion.

Table 1.   Main hydrostatic extrusion presses
for commercial production.

Year	Company	Country	Capacity (tonf)	Product	Press maker
1972	A S E A	Sweden	4,000	Copper—clad aluminium	A S E A
1972	Hitachi Cable Ltd.	Japan	4000	Composites	A S E A
1972	National Standard	England	4000	Copper—clad aluminium	A S E A
1973	N. V. Lips	Holland	4000	Copper tubing	A S E A
1973	——	Japan	300	}Noble metals,	Kobe Steel Ltd.
1974	——	Japan	2000	}Special metals	Kobe Steel Ltd.
1974	Sumitomo Light Metals Ltd.	Japan	1600	Aluminium alloy tubing	Sumitomo Heavy Machineries—F & P

In Table 1 main hydrostatic extrusion presses in use for commercial production are listed. They were installed in 1970's for pure hydrostatic and augmented extrusions. Composites, noble metals, and copper and aluminium alloy tubing are extruded. Among them, composites are the product where the features of hydrostatic extrusion are utilized to the utmost extent, and wire and bus-bar of copper-clad aluminium, and superconductive wire of Nb-Ti alloy with Cu or Nb_3Sn with Cu-Sn alloy is actually marketed. In addition, according to the latest news, there are plans, both in the Soviet Union and in the United States, to install industrial hydrostatic extrusion presses for the manufacture of copper-clad titanium pipes.

SECTION 4
EXTRUSION PRESSURE

In extrusion, generally, whether a product can be manufactured or not is primarily determined by the extrusion pressure needed to form the billet into a desired cross-section. In this respect hydrostatic extrusion

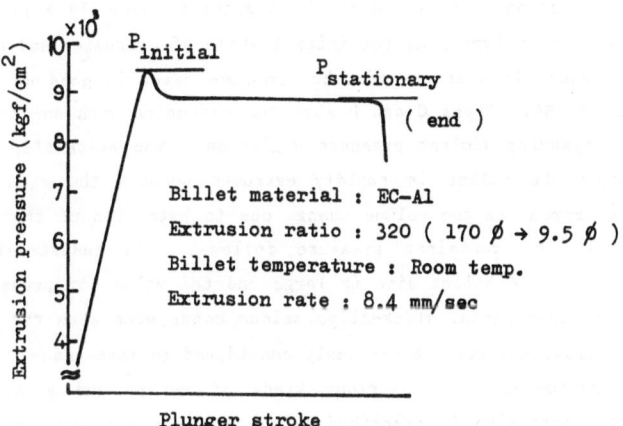

Fig. 6.   Example of pressure-stroke characteristic for
the hydrostatic extrusion.

is no exception.

A typical example of pressure-stroke characteristic of hydrostatic extrusion is shown in Fig. 6. In the figure extrusion of the billet commences at a certain pressure $P_{initial}$ and continues at an approximately constant pressure $P_{stationary}$. Of the two pressure values $P_{initial}$ mostly determines whether the extrusion is feasible or not, and for this reason it is important. It is the pressure value in the transient state at the onset of extrusion, and therefore it is variable in nature. Broadly speaking, $P_{stationary}$ is dependent on flow stress of the material, extrusion ratio, the state of lubrication, die configuration, the rate of extrusion and so forth. $P_{initial}$, on the other hand, depends on the nose geometry of the billet, the degree of fitting between billet and die, and pressurization rate to a great extent, in addition to the factors above mentioned. In actual extrusion a variety of features are provided to lower $P_{initial}$.

In actually performing hydrostatic extrusion a number of different pressure-stroke characteristics are observed as well as the typical one above described. They are shown in Fig. 7 [7], from which the state of lubrication can be estimated to some extent. In Type A in the figure it is shown that a stable extrusion is going on without pressure peak at the onset of extrusion. As stated in the beginning, normally a pressure peak is observed as in Type B at the initial stage of extrusion and subsequently the pressure is kept constant but in some cases it goes up as seen in Types B3 and B4. Types C and D show the extrusion with so-called stick-slip, accompanying violent pressure variation. The stick-slip phenomenon occurs when the billet is rapidly extruded because the stem moves too slowly to compensate the volume change due to extrusion of the billet and lowering of the container pressure follows. In industrial presses, however, where the billet size is large and the volume of pressure medium is small in comparison, stick-slips seldom occur even when the stem speed is fairly slow, and are not seriously considered in most cases.

The extrudability of various kinds of metals and alloys in cold hydrostatic extrusion is described below from the viewpoint of extrusion pressure [8].

1. Aluminium and Aluminium Alloys

The extrusion pressure vs. extrusion ratio relations in the hydrostatic extrusion of pure aluminium (A1100), corrosion-resistant aluminium

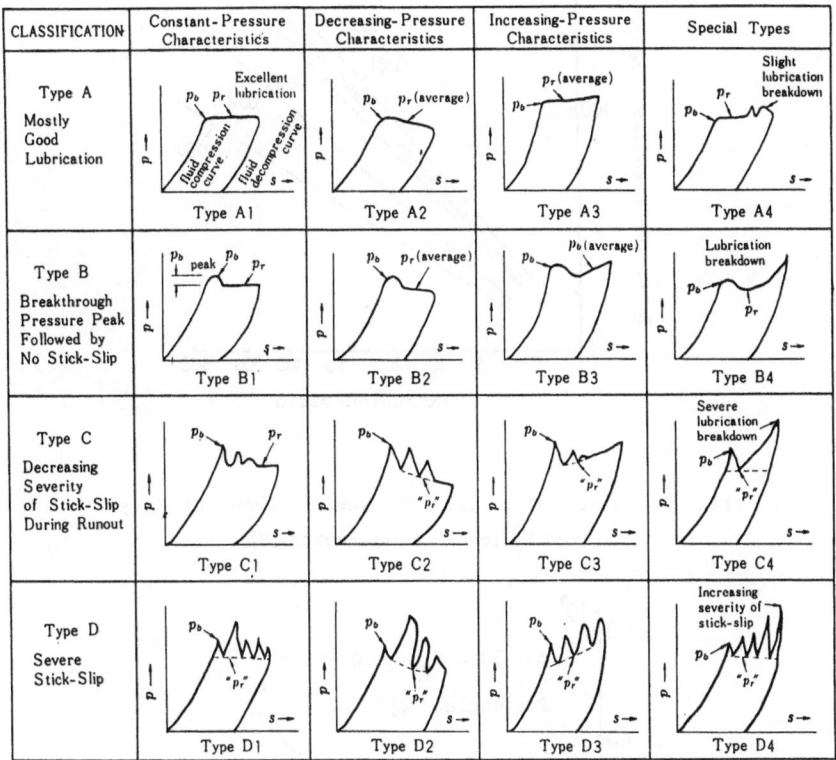

Fig. 7. Various pressure-stroke characteristics for the hydrostatic extrusion [7].

alloys (A5052, 6061, 6063), and high strength aluminium alloys (A2017, 7075) are shown in Fig. 8 [8]. As is clear from the figure, with the extrusion pressure of 15,000 kgf/cm^2, the maximum extrusion ratios are extremely high: 10,000 for pure aluminium and as high as 100 even for 7075 alloy. In the hydrostatic extrusion of 7075 alloy the limit on the extrusion speed in conventional hot extrusion is removed and sound extrusion is possible with the cast material at high speeds over 100 m/min.

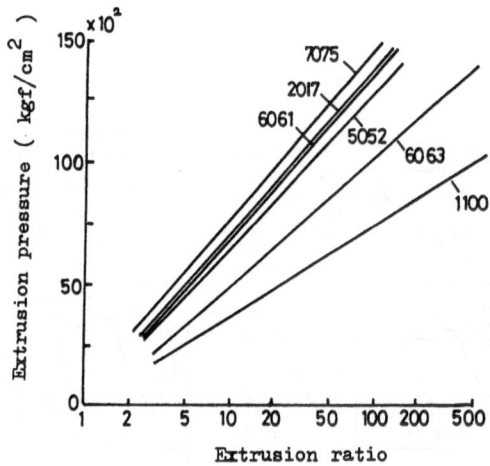

Fig. 8.   Extrusion pressure for various aluminium alloys
           in cold hydrostatic extrusion [8].

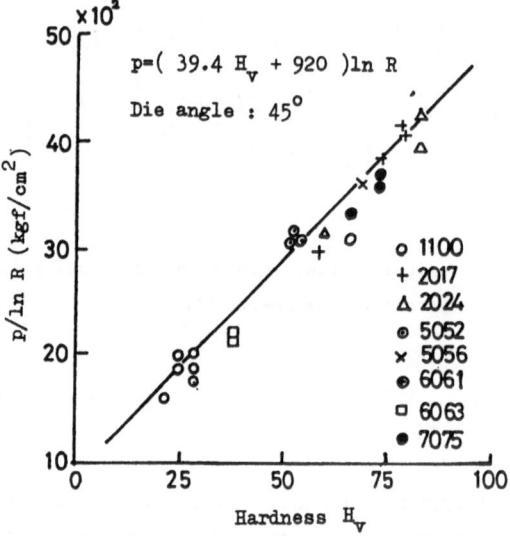

Fig. 9.   Relationship between P/ln R and billet hardness
           of aluminium alloys [8].

The effect of hardness of the billet material on the extrusion pressure in the hydrostatic extrusion of aluminium alloys of various kinds is shown in Fig. 9 [8]. The ordinate in the figure is the amount of extrusion pressure divided by ln R. If the extrusion conditions are the same, the extrusion pressure is approximately proportional to billet hardness, irrespective of the grade of aluminium alloys. From the figure, the relation between extrusion pressure, billet hardness, and extrusion ratio is given as follows.

$$p = (39.4 \ Hv + 920) \ \ln R \qquad (1)$$

where

p : extrusion pressure in kgf/cm^2

Hv : Vickers hardness of billet material

R : extrusion ratio

When compared with Pugh's equation 2 [9], obtained from quite a variety of materials, the coefficient of Hv in Eq. 1 is clearly shown to be lower, namely, a lower pressure is predicted to extrude a material with equal hardness:

$$p = (58.1 \ Hv + 620) \ \ln R \qquad (2)$$

This tendency is considered to be due to the low heat capacity of aluminium alloys, with the consequence of a large temperature rise for a given extrusion pressure, and also to the low recovery temperature.

2. Copper and Copper Alloys

The relation between extrusion pressure and extrusion ratio in the hydrostatic extrusion of Cu and Cu alloys is shown in Fig. 10 [8]. For the extrusion pressure of 15,000 kgf/cm^2, the extrusion ratio obtained is about 35 for pure copper, 15 for Cu-Ni alloy, 8 for nickel silver, and around 5 - 6 for phosphor bronze. Such copper alloys as phosphor bronze and nickel silver are normally difficult to extrude at high temperatures, and so hydrostatic extrusion, being capable of imparting high reductions at room temperature, provides an effective means of forming. The effect of hardness on the extrusion pressure is shown in Fig. 11 for Cu and Cu alloys [8]. The relation between hardness and extrusion pressure, obtained from Fig. 11 for Cu alloys, is the following:

$$p = (75.0 \ Hv + 1000) \ \ln R \qquad (3)$$

When compared with Al alloys or steels, which will be described later, Cu alloys are shown to need higher extrusion pressure at equal hardness.

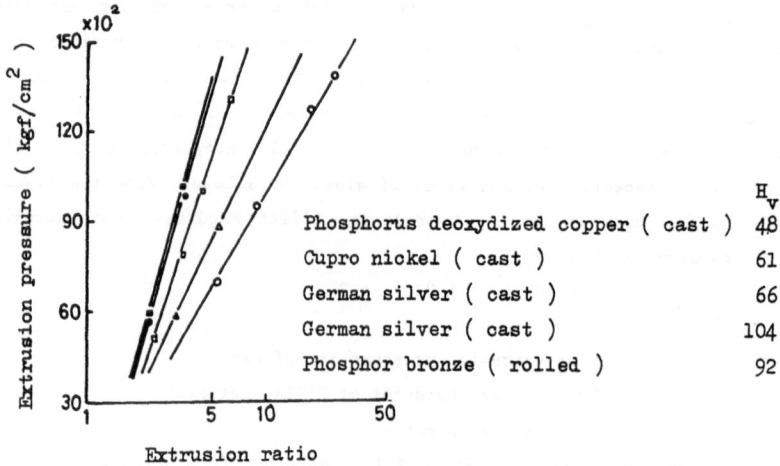

Fig. 10. Extrusion pressure for various copper alloys
in cold hydrostatic extrusion [8].

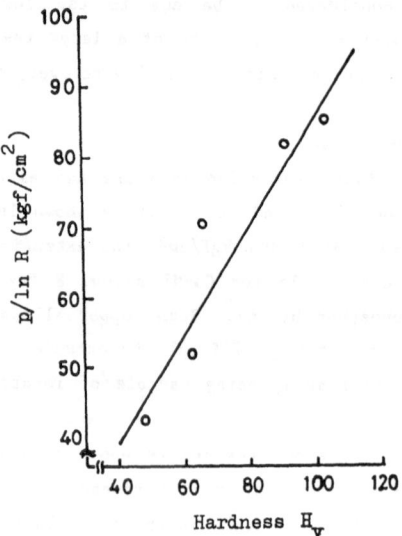

Fig. 11. Relationship between p/ln R and billet hardness
of copper alloys [8].

This is considered to be due to the higher workhardenability and recovery temperature of Cu alloys.

3. Steels

In the hydrostatic extrusion of steels large extrusion ratios cannot be expected because of their high flow stress. The relation between extrusion pressure and extrusion ratio in the hydrostatic extrusion of various kinds of steels is shown in Fig. 12 [8]. The effect of hardness on the extrusion pressure is shown in Fig. 13 [8], from which the following equation is obtained:

$$p = (56.5 \ Hv + 210) \ ln \ R \qquad\qquad (4)$$

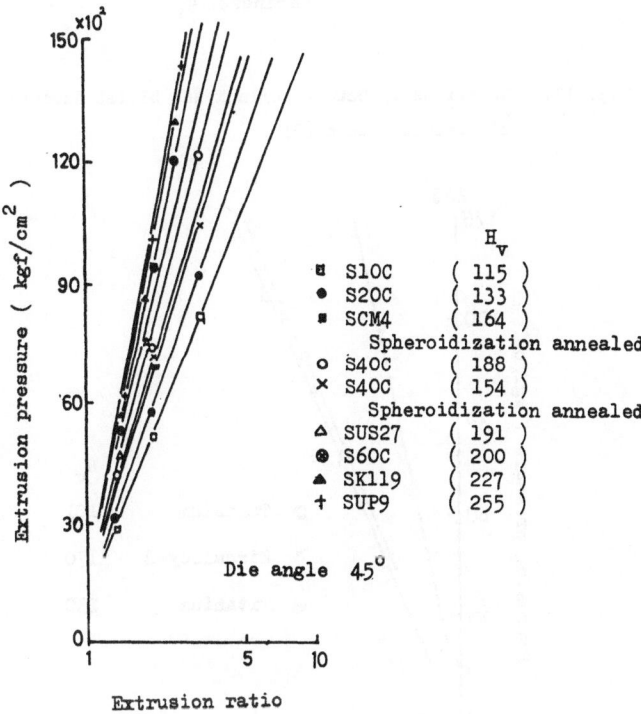

Fig. 12.   Extrusion pressure for various steels in
cold hydrostatic extrusion [8].

112

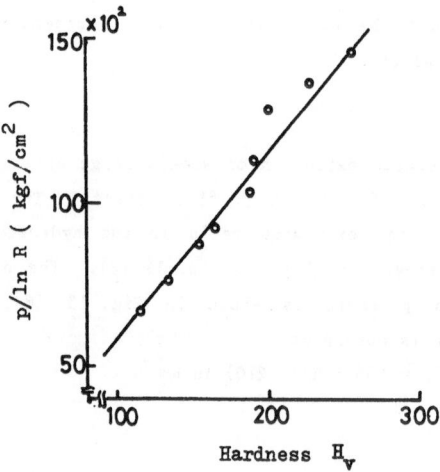

Fig. 13. Relationship between p/ln R and billet hardness of various steels [8].

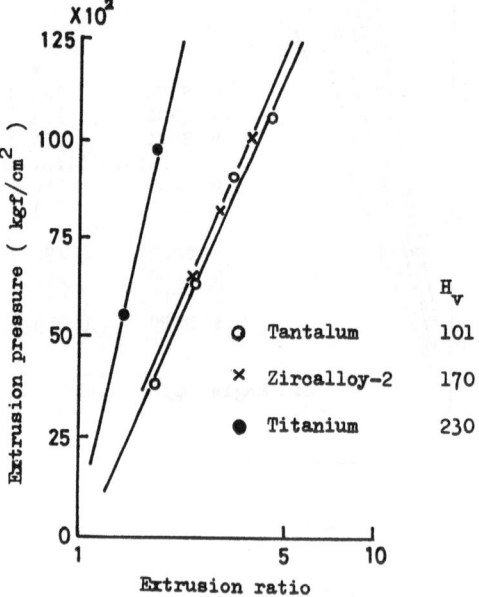

Fig. 14. Extrusion pressure for some difficult metals in cold hydrostatic extrusion [8].

4. Special Metals and Alloy

Figure 14 shows the relation between extrusion pressure and extrusion ratio in the hydrostatic extrusion of zircalloy-2, pure titanium, and pure tantalum [8]. These metals and alloy are all difficult-to-work materials and are notorious for their seizure to toolings. They are the object of effective application of hydrostatic extrusion.

5. Clad Composite Metals

The extrusion pressure of clad composite metals, which consist of core and sheath metals, can be well represented by the following equation, when both of the constituent metals uniformly deform,

$$p = a \left[(A_1 H_1 + A_2 H_2)/(A_1 + A_2) \right] \ln R \qquad (5)$$

where

A_1, A_2 : cross-sectional area of each constituent metal

H_1, H_2 : hardness of each constituent metal

R : extrusion ratio of whole billet

a : constant

For the same amount of pressure, the composites can be extruded at lower temperature with higher extrusion ratio in hydrostatic extrusion than in conventional extrusion, in a similar way to other materials. Hydrostatic extrusion is best suited for the forming of those composites which consist of metals mutually reactive.

SECTION 5

PROCESS PARAMETERS

1. Lubrication

In hydrostatic extrusion, when the relative velocity of the billet with respect to the die is higher than a certain critical value, a fluid lubrication film is formed. In the early stage of extrusion, however, a state of mixed lubrication prevails and partial metal contacts are believed to exist between the billet and the die. In the steady state of high-rate extrusion, where the state of hydrodynamic lubrication is realized, there is also a possibility of the probabilistic existence of partial contacts between the billet and the die near the die exit. The metal-to-metal contact between the die and the billet causes seizure of the billet material to the die, resulting in the degradation of the surface quality of the extrudate. Also, it brings about instability of the extrusion process, leading to a rise of extrusion pressure and many other troubles. Consequently, even in hydrostatic extrusion, it is desirable to apply lubricants to the billet surface and to reinforce the lubricating effect of the pressure medium, if the billet material is prone to seizure. The application of lubricants to the billet is also needed for those materials which are considered to have inherently good lubrication properties like copper, if the extrusion condition is adverse.

There are several methods for applying lubricants to the billet. A thin film of the lubricant can be formed on its surface through chemical reaction by soaking the billet in the prescribed solution. Solid lubricants like graphite, MoS_2, metal soap, or Teflon can be smeared onto the surface to form a film. Viscous lubricants like grease, beeswax, or highly viscous polymers are used, sometimes blended with the solid lubricants mentioned above or with some other additives. Various kinds of lubricants can be thought of as candidates to form various kinds of films. Which lubricant to use in practice can only be determined experimentally, by trial and error, taking into consideration the billet material, extrusion condition, or surface quality of the extrudate required.

Figure 15 shows the effect of billet lubrication in the hydrostatic extrusion of a spring steel, SUP9 [10]. A stable extrusion is seen with the billet with phosphate-film treatment, while with the untreated billet a seizure is seen to occur after only the tip of the billet has been extruded, resulting in the rise in extrusion pressure. Among the chemical conversion treatments, besides the phosphate film previously mentioned, there is a treatment with chromate film, which is sometimes applied to Al alloys.

When the lubricant is smeared on the billet, the state of lubrication in extrusion is greatly affected by the amount of lubricant brought into the die, in addition to the properties of the lubricant. Especially when a large billet is used, the contact area between the billet and the die is wide and the contact stroke is long. Stochastically speaking, there are many chances of die galling, and so there is need to supply without failure sufficient lubricant into the die-billet interface.

As a means of controlling the amount of lubricant supplied to the die-billet interface grooves are provided on the billet surface to pool the lubricant and their size varies [11]. An example of these grooves is shown in Fig. 16, where the groove is machined over the whole length of

x : Extrusion discontinued

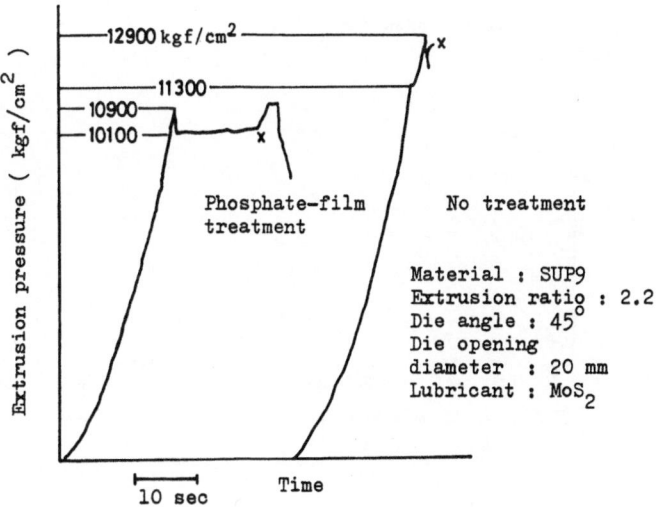

Fig. 15.   Effect of surface treatment of billet on
extrusion pressure [10].

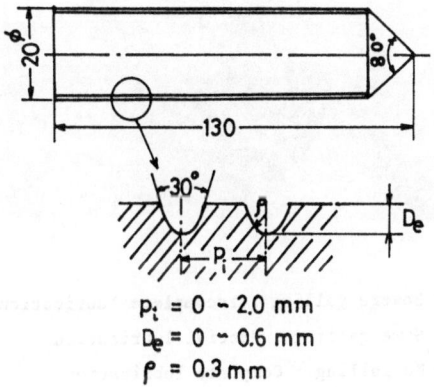

Fig. 16.   Shape of billet and type of surface grooves [11].

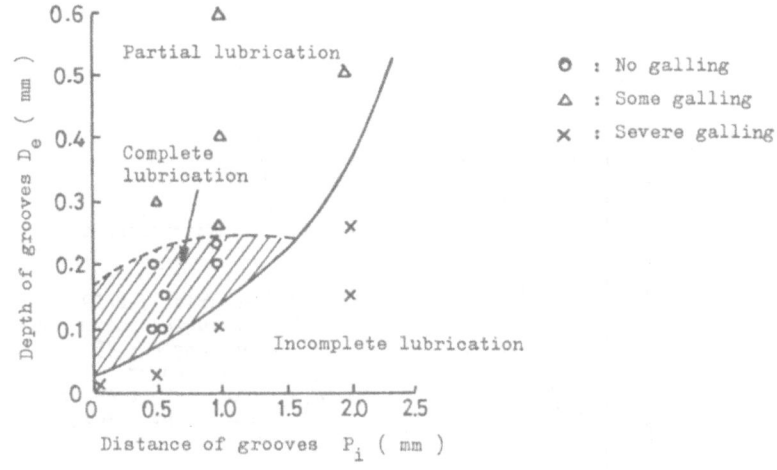

Fig. 17. Effect of depth and distance of grooves on the
         quality of lubrication during cold extrusion of
         pure aluminium at extrusion ratio of 100 [11].

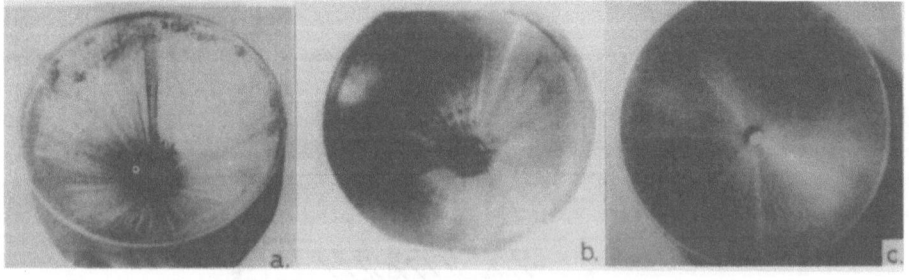

(a) Severe galling : Incomplete lubrication
(b) Some galling : Partial lubrication
(c) No galling : Complete lubrication

Fig. 18. Various degree of die galling indicating
         quality of the lubrication condition [11].

the billet to form a spiral and its depth as well as pitch is varied. The effect of the groove on lubrication is shown in Fig. 17, taking the extrusion of pure aluminium as an example. A blend of highly viscous polymer with graphite powder is employed as the lubricant, and the quality of lubrication is judged from the degree of galling of Al to the die. Examples of die galling are shown in Fig. 18. The amount of Al adhering to the die surface is reduced by improving the quality of lubrication, and a die surface without any galling is the result of perfect lubrication. From Fig. 17 it can be seen that depending upon the depth and pitch of the groove different qualities of lubrication are obtained, and the best extrusion is realized when the groove depth is 0.1 - 0.2 mm and its pitch is around 0.5 mm. The increase in groove pitch results in the degradation of the quality of lubrication, while too shallow or too deep grooves are

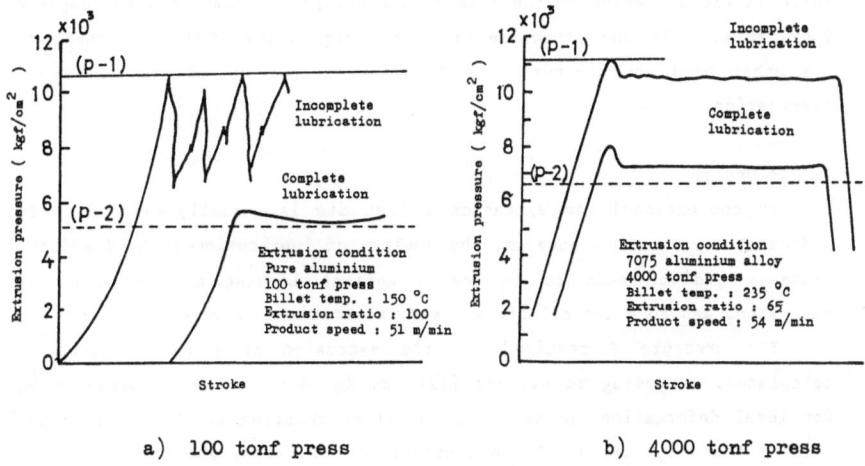

a) 100 tonf press   b) 4000 tonf press

Fig. 19   Pressure-stroke characteristics for extrusion
            with complete and incomplete lubrication [11].

            (P-1) Pressure for extrusion with galling
                   throughout the whole die area
            (P-2) Calculated pressure for extrusion
                   without die friction

not recommended. It goes without saying that the optimum dimensions of the groove are dependent on the billet material and the extrusion ratios.

Figure 19 shows a pressure-stroke diagram when Al and Al alloy are extruded under controlled conditions of lubrication through groove forming. The extrusion pressure is shown to be drastically reduced by the improvement in lubrication. In the figure the extrusion pressure P - 2, shown by broken lines, is the pressure needed for a frictionless extrusion, calculated on the assumption that the maximum extrusion pressure with incomplete lubrication is the pressure needed for an extrusion with galling over the entire area of the die. The closeness of the experimental values of extrusion pressure with complete lubrication to the calculated value shows that the friction along the die is extremely low. It is also shown in Fig. 19 that in extrusion with the small-scale 100 tonf press a violent stick-slip occurs with incomplete lubrication, while little variation in the extrusion pressure is observed with complete lubrication. In the extrusion with the large-scale 4000 tonf press, on the other hand, the extrusion is stable, irrespective of the quality of lubrication.

2. Die Design

In conventional ram extrusion a flat die is normally employed. In hydrostatic extrusion, however, the quality of lubrication is good and the pressure medium needs to be sealed against leaking at the onset of extrusion, and so a conical die is used, as mentioned earlier.

The pressure P required for the extrusion of a round billet is calculated, according to Avitzur [12], by Eq. 6 as a sum of pressure P_d for ideal deformation, pressure P_s to effect shearing at the entrance and exit of the die, pressure P_f to overcome die friction, and pressure P_c to overcome container friction:

$$P = P_d + P_s + P_f + P_c \tag{6}$$

where

$$P_d = Y \cdot f(\alpha) \cdot \ln R, \qquad f(\alpha) \doteq 1 \tag{7}$$

$$P_s = (2/\sqrt{3}) Y [(\alpha/\sin\alpha) - \cot\alpha] \tag{8}$$

$$P_f = (m_2 \cot\alpha \ln R + m_3 1/d_f)(Y/\sqrt{3}) \tag{9}$$

$$P_c = (4L/D) m_1 Y/\sqrt{3} \tag{10}$$

119

Y : flow stress of billet material, R : extrusion ratio,
d_f: die opening diameter, L : billet length, α : die semicone angle,
l : length of die bearing, D : outer diameter of billet,
m_1, m_2, m_3 : coefficient of friction of container wall, die surface, and
die bearing, respectively.

Among the four pressures, those which are dependent on die configuration
are P_s and P_f. The former becomes smaller with the die angle, while the
latter greatly varies with the quality of lubrication on the die surface
and approaches zero if an appropriate method of lubrication is employed in
hydrostatic extrusion. If the billet material is adhering to the die
surface during extrusion, as is the case with direct or indirect
extrusion, it assumes a large value. The relation between extrusion
pressure and die angle, calculated from Eqs. 6 - 9, is diagrammatically
shown in Fig. 20 [11]. It is calculated for two kinds of extrusion ratio
for complete and incomplete lubrication between the billet and the die.
When there is no friction, the pressure P_s decreases with the decrease in
die angle, and the total extrusion pressure gradually decreases. When
seizure occurs due to incomplete lubrication, on the other hand, the

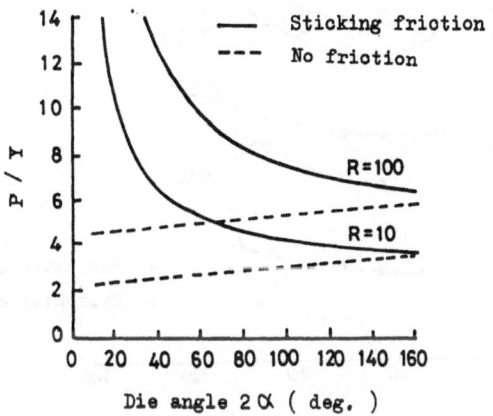

Fig. 20.   Relation between extrusion pressure and die
angle calculated for complete and incomplete
lubrication [11].

extrusion pressure remarkably increases with the decrease in die angle. For this reason a flat die is normally used in ram extrusion.

Figure 21 shows the result of experiments on the effect of die angle on the extrusion pressure [13]. For any amount of reduction hydrostatic extrusion marks lower values of extrusion pressure than ram extrusion. Unlike the curves of no die friction in Fig. 20, the existence of the so-called optimum die angle, where the extrusion pressure becomes minimum, is shown in hydrostatic extrusion (Fig. 21). This means, in actual hydrostatic extrusion, the die friction is not completely zero. The optimum die angle in hydrostatic extrusion, however, is smaller than that in ram extrusion.

Since there exists an optimum die angle, as described above, it is desirable to choose the die angle near the optimum one from the viewpoint

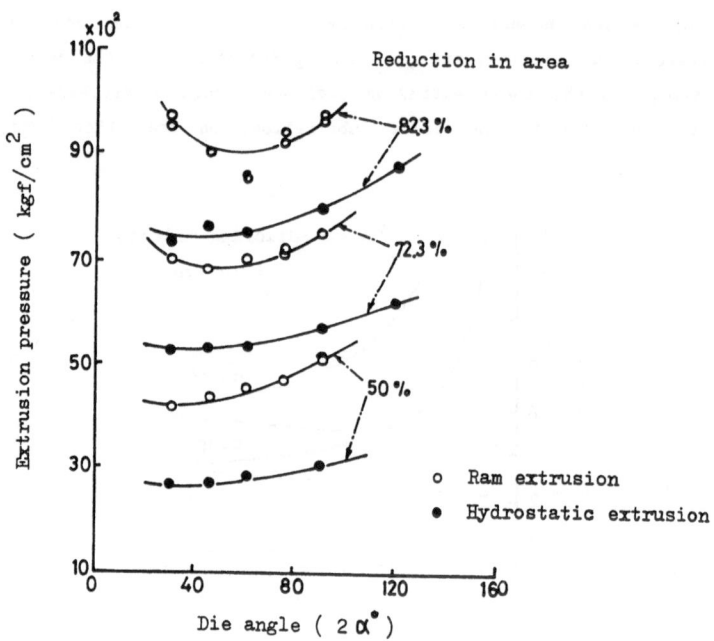

Fig. 21. Extrusion pressure for various die angle in
hydrostatic and conventional ram extrusion [13].

of minimizing the extrusion pressure in hydrostatic extrusion. For a small die angle, however, a small nose-angle of the billet is required, meaning more work in machining the billet. Also, the amount of discard increases, resulting in a decrease of material yield in extrusion. Consequently, in industrial applications, larger die angles are preferred so long as the extrusion operation does not suffer, in spite of a pressure increase to a more or lesser degree. For the hydrostatic extrusion of seizure-free material like copper, dies of more than 100° whole angle are sometimes used.

In addition to the simple conical die described above, curved or multiple-cone [14, 15] die profiles are devised, as shown in Fig. 22, in order to further improve the quality of lubrication for some billet materials under some extrusion conditions. In the extrusion of profiles, if the simple conical die is used, the die exit is not in the same plane and an imbalance occurs in lubrication and metal flow, causing extrusion defects. Consequently, in these cases, a die with a conical shape in the entrance and curved or plane surface in the exit is sometimes used.

As for the material for the hydrostatic extrusion die, commercial stock of cold or hot work tool steels can be utilized. In hydrostatic extrusion it is generally believed that the die is long-lived and stable because of the surrounding pressure of the pressure medium. In actuality, however, radial cracks sometimes initiate from the die bearing area, as shown in Fig. 23 [16], after a relatively small number of extrusion cycles, even if the materials above mentioned are employed. To prevent

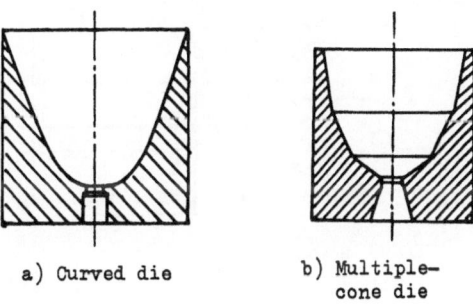

a) Curved die   b) Multiple-
                         cone die

Fig. 22.   Modified die profile [14, 15].

this type of failure, multilayer shrink-fit dies or compound dies, as shown in Fig. 24 [17], are used. By employing these dies the number of extrusion cycles is increased before crack initiation occurs. Moreover, when the compound die is employed, cracking occurs only in the die insert, which is replaceable, and therefore it is economical.

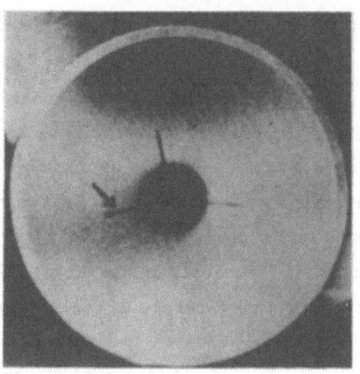

Fig. 23. Example of die cracking [16]

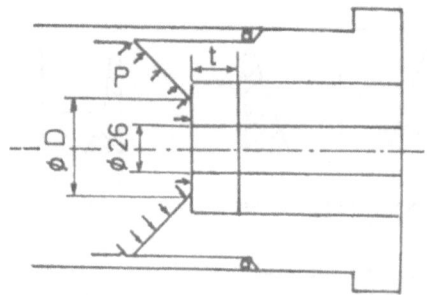

Fig. 24. Example of compound die [17].

3. Billet Nose Geometry

In conventional hot extrusion the billet is used as cut, that is, the
end surface is square to the billet axis without any additional machining.
In hydrostatic extrusion, on the other hand, a conical die is normally
used. Accordingly, in order to seal the pressure medium in the billet-die
interface at the onset of extrusion and also to minimize the pressure rise
and commence the extrusion smoothly, the billet end is machined to have a
nose. When a simple conical die is used, a conical nose is provided with
the nose angle equal to the die angle or 2 - 3 degrees smaller in order to
reduce the threshold pressure of extrusion. From the viewpoint of billet
setting to the die the equal angle is preferable, but as a pressure seal a
line seal is more reliable than a surface seal and for this purpose the
smaller nose angle is preferred.

In Fig. 25 the relation between nose length and break-through
pressure in extrusion is shown [13]. In the figure the ratio c of the
contact area between the nose and die to that of the complete nose is
taken as the axis of abscissa, c = 1 corresponding to the complete nose.
The break-through pressure assumes its minimum when the nosing is complete

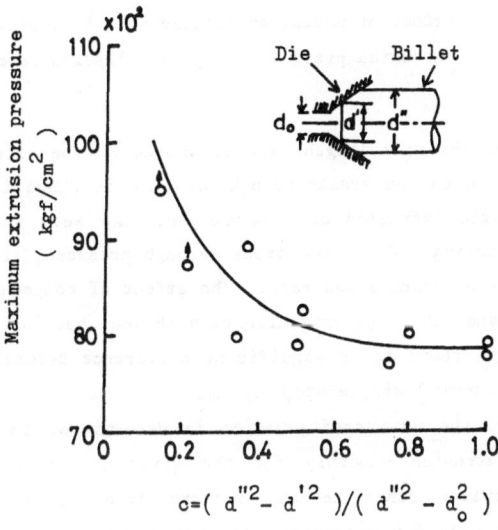

$$c=(\ d''^2 - d'^2\ )/(\ d''^2 - d_o^2\ )$$

Fig. 25. Effect of contact area between die and billet
on initial extrusion pressure [13].

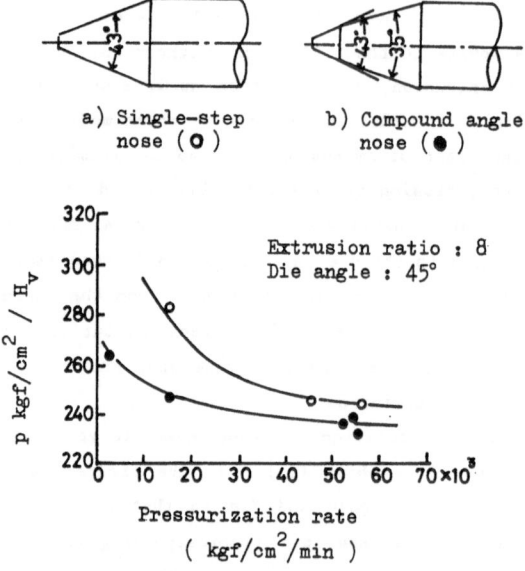

Fig. 26.   Effect of nosing on relationship between initial
extrusion pressure and pressurization rate [18].

and increases as the nose length becomes shorter. The effect of billet-
nose configuration on the break-through pressure is shown in Fig. 26 [18]
in the hydrostatic extrusion of pure copper. As seen in the figure a
compound angle nosing reduces the break-through pressure, its effect being
dependent on the pressurization rate. The effect of compound angle nosing
is significant when the pressurization rate is low, but insignificant when
the rate is high; there is no significant difference between the compound
angle nosing and normal single-step nosing.

Summing up, the nose configuration is determined, in principle, to
commence the extrusion smoothly for the given die configuration.   In
industrial extrusion other economy factors like the yield of billet
material or the cost for machining the nose are taken into consideration
along with the factors mentioned earlier. For these reasons, in actual
extrusion, those modified types of nosing, shown in Fig. 27, are also
employed in addition to the simple conical nose.

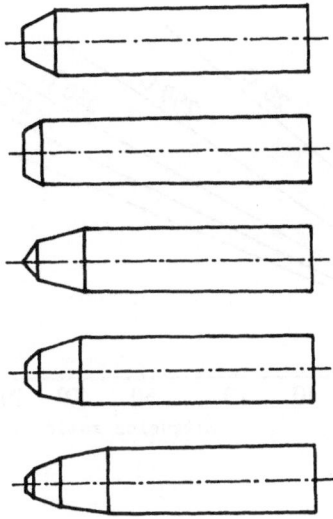

Fig. 27.   Various shapes of billet.

4. Extrusion Temperature

As described in Section 4, in hydrostatic extrusion a relatively high extrusion ratio can be obtained at room temperature. When compared with conventional hot extrusion, however, an extremely high extrusion pressure is often required to obtain equal or higher levels of reduction in cold hydrostatic extrusion. In laboratories it may be possible to perform extrusion with such a high pressure, but in industry it is questionable to handle extremely high pressure when the equipment and manipulations required are taken into consideration. Consequently, there is strong tendency in recent years to go to warm or hot hydrostatic extrusion where high reductions are obtained with low pressures.

Figure 28 shows the effect of extrusion temperature on extrusion pressure in the hydrostatic extrusion of pure copper tubing [19]. For the same level of extrusion pressure the extrusion ratio obtainable is seen to dramatically increase as the extrusion temperature rises. Table 2 lists the extrusion ratios that are obtained in the warm hydrostatic extrusion

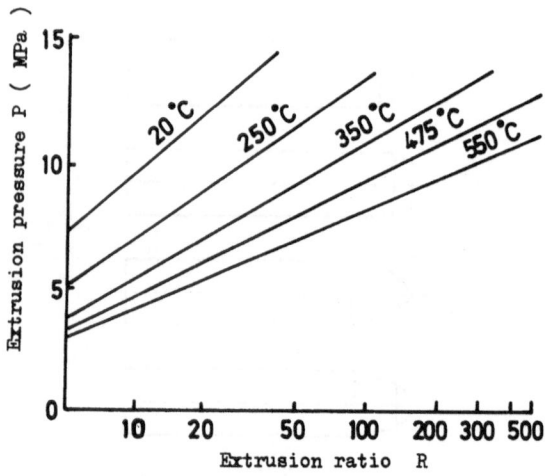

Fig. 28.   Effect of billet temperature on extrusion
pressure for DHP-copper tube [19].

Table 2   Extrusion ratios in industrial warm hydrostatic
extrusion of tubes of copper and its alloys [19].

Material	Billet diameter (mm)	Product (mm)	Extrusion ratio (R)	Billet temperature (°C)
Cu	158	⌀ 15 x 1t	450 : 1	550
CuZn37	155	⌀ 18 x 1.4t	250 : 1	600
CuZn20Al	155	⌀ 29 x 1.5t	140 : 1	600
CuZn28Al	155	⌀ 28 x 1.4t	160 : 1	600

of tubes made of copper and its alloys [19].   At Lips N.V. in Holland,
10,000 - 12,000 tons of copper tubing are produced annually by warm
hydrostatic extrusion at around 550°C in a three-shift operation [19, 20].
This temperature is almost the maximum practicable in the industrial

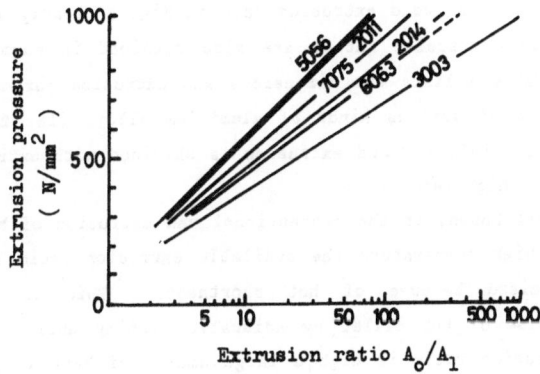

Fig. 29.   Extrusion pressure vs. extrusion ratio in warm hydrostatic extrusion of aluminium alloys [21].

Table 3   Examples for hydrostatic extrusion of various
aluminium alloy tube and rod [16, 22, 25].

Alloy No. by AA	Extrusion product		Extrusion speed (m/min)	Extrusion ratio (R)	Billet diameter (mm)	Reference
	Shape	Dimension (mm)				
2017	Tube	∅ 23.5 x 1.25t	100~200	113	∅ 115	[16]
7075	"	"	"	"	"	"
2014	"	∅ 16 x 1t	–	375	∅ 155	[22]
3003	"	∅ 8.5 x 0.5t	–	1413	"	"
2011	Wire	∅ 7	225	522	∅ 160	[25]
2024	"	∅ 13	160	151	"	"
5052	"	∅ 10	"	256	"	"

application of hydrostatic extrusion using a liquid form of pressure
medium such as castor oil.

Similarly, Fig. 29 shows the relation between the extrusion pressure
and the extrusion ratio in the hydrostatic extrusion of various kinds of
Al alloys at 200°C (except 2011 alloy, which is extruded at 150°C) [21].

A comparison with the cold extrusion data in Fig. 8 clearly shows that in Al alloys large extrusion ratios are also obtained in warm hydrostatic extrusion. Table 3 lists the dimensions and extrusion parameters of the extrudates made of various kinds of aluminium alloys [16, 22, 25]. In each of the materials a sound extrusion is obtained with extrusion ratios over 100 and at high rate.

As is well known, in the conventional ram extrusion of high-strength Al alloys at high temperature the available extrusion ratio and rate are extremely limited because of hot shortness. This is due to the temperature rise of the billet by adiabatic heating during deformation. When the extrusion ratio is high, a large amount of heat is generated by deformation, and it is important to control the hot shortness by taking the appropriate means even in hydrostatic extrusion.

The temperature rise ΔT in the extruding billet is calculated by the following:

$$\Delta T = K P / J c \rho \tag{11}$$

K : correction factor for the heat loss by conduction to tooling and billet

P : extrusion pressure

J : heat equivalent of work

c : specific heat of billet material

ρ : density of billet material

From Eq. 11 the extrudate temperature T_e at die exit is given by the following

$$T_e = T_b + \Delta T \tag{12}$$

where T_b is the billet temperature before extrusion. In order to avoid hot shortness it is desirable to keep the extrudate temperature T_e as low as possible by choosing the parameters in Eqs. 11 and 12 in an appropriate manner. In particular, the lower the billet temperature T_b, the more effectively the extrudate temperature T_e is lowered apart from the increase in P, and consequently in ΔT due to the increase in flow stress, as long as a good quality of lubrication is maintained.

In Fig. 30 the appearance of 5056 alloy rod, produced by warm hydrostatic extrusion with a 4000 tonf press [11] is shown. The figure shows (a) good surface quality and (b) surface crack due to hot shortness. Figure 31 shows the result of experiments on the initiation of hot shortness in hydrostatic extrusion of 7075 and 5056 alloys [11]. The

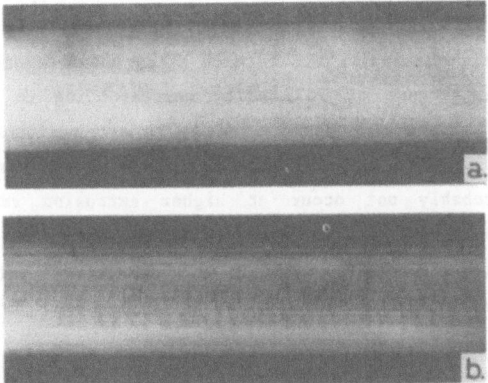

Fig. 30. Appearance of hydrostatically extruded 5056
aluminium alloy rod [11].
a) Good surface quality
b) Surface crack due to hot shortness

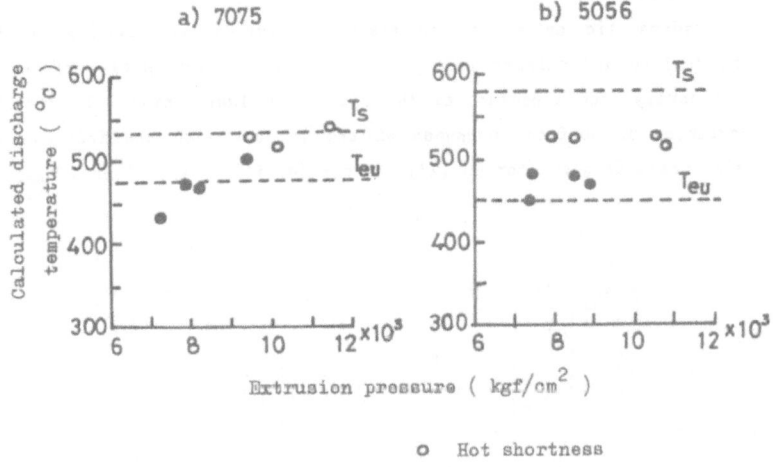

Fig. 31. Calculated discharge temperature and occurence
behaviour of hot shortness for high strength
aluminium alloys extruded by 4000 tonf press [11].
(T_s, T_{eu}: Solidus and eutectic temperatures respectively)

extrudate temperature T_e is plotted against the extrusion pressure in the figure, where two T_e's are shown for the same pressure, corresponding to different billet temperatures T_b's. In each alloy hot shortness occurs at high T_e's, the temperature of 7075 alloy corresponding to its solidus temperature. The experiments shown in Fig. 31 are performed with the extrusion rate of 50 - 100 m/min. Under a state of good lubrication hot shortness will probably not occur at higher extrusion rates. When compared with conventional hot extrusion, where the extrusion rate is 1 - 2 m/min for high-strength aluminium, hydrostatic extrusion is distinctively advantageous: the limit on the extrusion rate is practically nonexistent.

Needless to say, what is discussed above applies not only to high-strength Al alloys but also to any other material prone to hot shortness.

SECTION 6
PROPERTIES OF PRODUCT

1. Surface Quality

In hydrostatic extrusion the state of hydrodynamic lubrication is easy to realize and seizure rarely occurs, but the extrudate surface is not necessarily smooth because of the thick-film lubrication. In Fig. 32 the variation of surface roughness of the product with the increase in extrusion ratio is shown for Cu [23]. A turned billet is employed, R_{max}

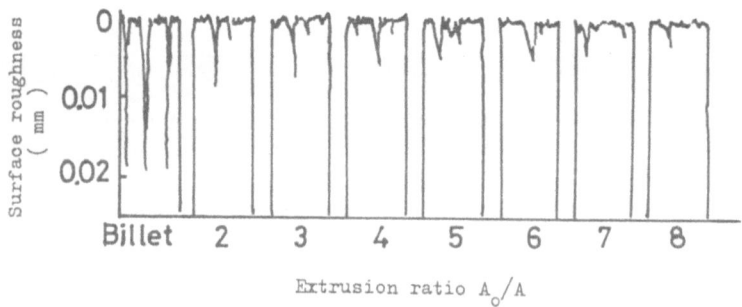

Fig. 32.   Surface roughness of hydrostatically
extruded copper wire [23].

being around 20 μm, and the surface roughness is seen to decrease with the
increase in extrusion ratio. When the billet material is soft, a product
with orange peel surface is obtained. When the billet material is hard,
the product surface is about as glossy as that obtained by conventional
extrusion to a greater or lesser degree.

In order to improve the surface quality it is necessary to control
the billet surface and the lubricant to be smeared on it, the viscosity of
the pressure medium and the pressure to be used, in addition to die
configuration, surface roughness, and material property, extrusion
temperature and so forth. By a combination of appropriate conditions it
is possible to obtain an extremely smooth surface with a roughness of less
than 1 μm. In industrial mass production, however, too strict conditions
are required to control the surface quality, if hydrostatic extrusion is
employed as the last step of production, and so it is of advantage to add
a finishing process after hydrostatic extrusion [16].

2. Dimension

The dimension of the extrudate is less than that of the die opening
in hydrostatic extrusion, the difference in diameter sometimes amounting
to 1 mm in the large-size extrudate with rectangular cross-section. The
causes of the dimensional difference are manifold: contraction of die due
to high pressure, thermal expansion of billet and die, elastic recovery of
extrudate, lubricant films, material flow and so forth.

Figure 33 shows the variation of extrudate diameter in the
longitudinal direction and also that of corresponding extrusion pressure
when Al-Mg alloy (A5056) is cold-extruded to 50 mm round rod [24]. The
outer diameter of the extrudate assumes its maximum at its tip, gradually
decreases as the extrusion proceeds, and finally settles down to a
constant value in the steady state range of extrusion. The outer diameter
in this range is close to that of the die opening less the amount of
contraction due to the deformation of the die due to pressure and the
elastic recovery and thermal deformation of the billet material.

In tube extrusion, one of the most important dimensional factors is
eccentricity. A number of factors in relation to materials, tools, and
machines are considered in conventional extrusion, and the same is true
with hydrostatic extrusion. As a phenomenon peculiar to horizontal
hydrostatic extrusion there is the effect of tilting of the billet. In
hydrostatic extrusion the billet is extruded while floating in the

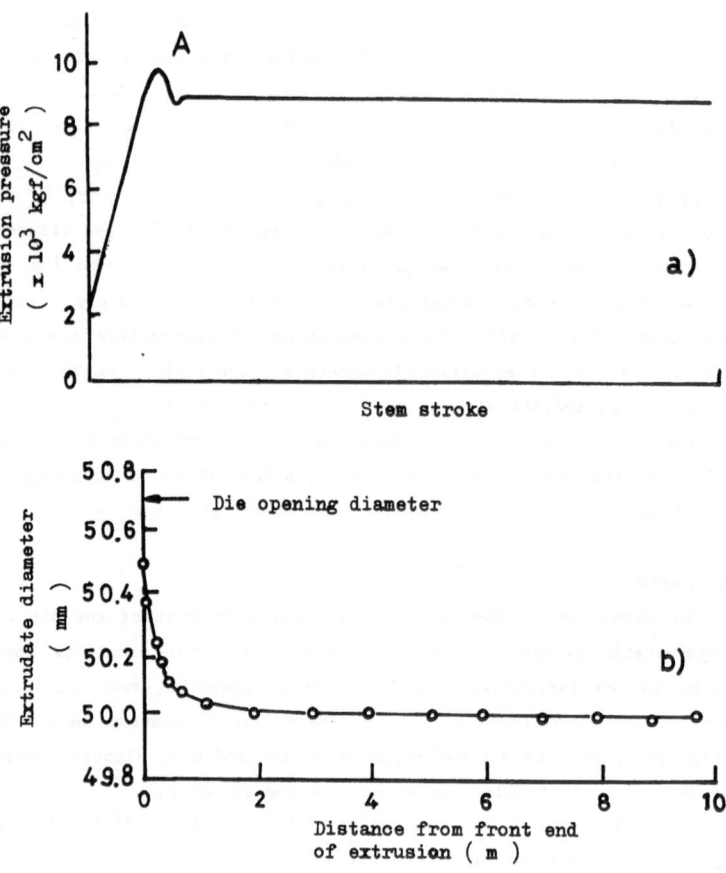

Fig. 33.   Diameter variation along the longitudinal direction for
            the hydrostatically cold extruded 5056 aluminium alloys
            by 4000 tonf press [24].

            a) Pressure–stroke characteristics
            b) Diameter change of extruded product

pressure medium and therefore it is ncessary to maintain the alignment of
the billet axis, unlike in conventional extrusion. If the centre position
deviates, the billet is extruded with its axis tilted, resulting in worse
eccentricity.   The   effect   of   billet   inclination   on   the   product

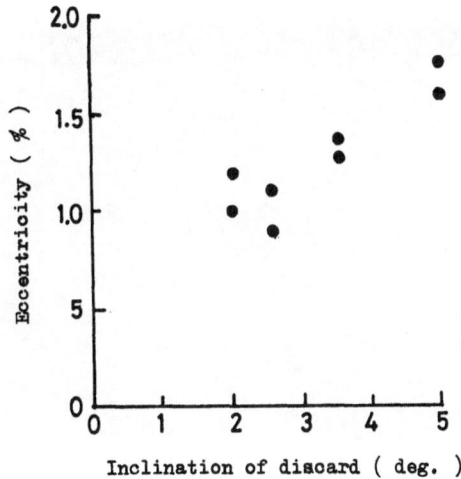

Fig. 34.   Effect of billet inclination on eccentricity
of high strength aluminium alloy tube at the
tail end of extrusion [16].

eccentricity at the tail end of the abnormal extrusion above mentioned is
shown in Fig. 34 [16].   As causes inducing the inclination the following
are considered: (a) nonuniform lubrication in the billet-die interface,
and (b) deviation in billet centreing.

3. Microstructure

With hydrostatic extrusion lower extrusion temperatures can be
employed than in conventional hot extrusion, and it is one of its main
features that fine microstructures are obtained.   Figure 35 shows the
microstructure of a cross-section of 2011 alloy, hydrostatically extruded
at 200°C with the rate of 225 m/min from 160 mm to 7 mm round rod [25].
An extremely fine recrystallized structure is obtained without peripheral
coarse grains, which appear in hot extrusion.   Figure 36 shows the
microstructure of a cross-section of oxygen-free copper, hydrostatically
extruded at 400°C from 160 mm to 16 mm round rod [26].   When compared with

134

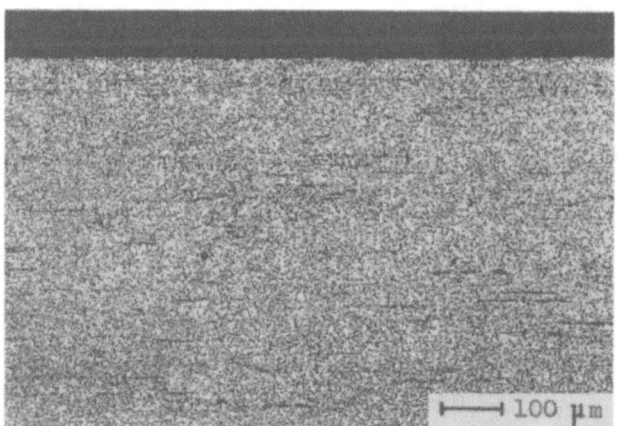

Fig. 35.   Microstructure of hydrostatically warm
extruded 2011 aluminium alloy [25].

a) hydrostatically   b) conventional
warm extruded   hot extruded

Fig. 36.   Microstructures of hydrostatically warm and
conventional hot extruded oxygen free copper
[26].   (Magnification X100)

a similar structure obtained by hot ram extrusion at about 800°C, it is clearly seen that the former has a finer structure.

4. Mechanical Properties

Figure 37 shows the relation between the mechanical properties and extrusion ratios of copper rods hydrostatically extruded at room temperature [27]. In the range of small extrusion ratios the tensile strength increases, while the elongation and reduction of area decrease with the increase in the extrusion ratio. In the range of large extrusion ratios, on the other hand, there is a tendency that the strength decreases while the elongation increases. A similar tendency is seen in Al wire also, as shown in Fig. 38 [27]. As the rate of extrusion increases the softening of the extrudate is enhanced.

The softening phenomenon of the extrudate with high rates of extrusion described above is due to the recovery and recrystallization of the billet material by the temperature rise caused by the heat of deformation during extrusion. Figure 39 shows the variation of hardness in the longitudinal direction of the discard [27]. At the extrusion ratio

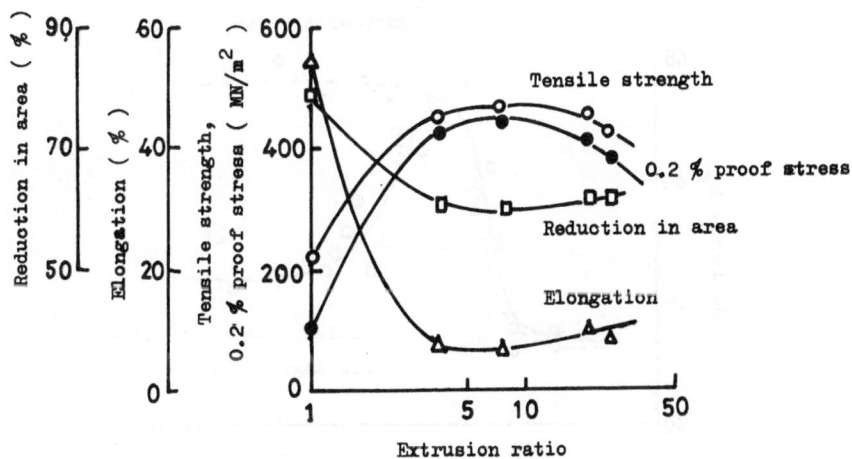

Fig. 37. Mechanical properties of copper rods hydrostatically extruded at various extrusion ratio [27].

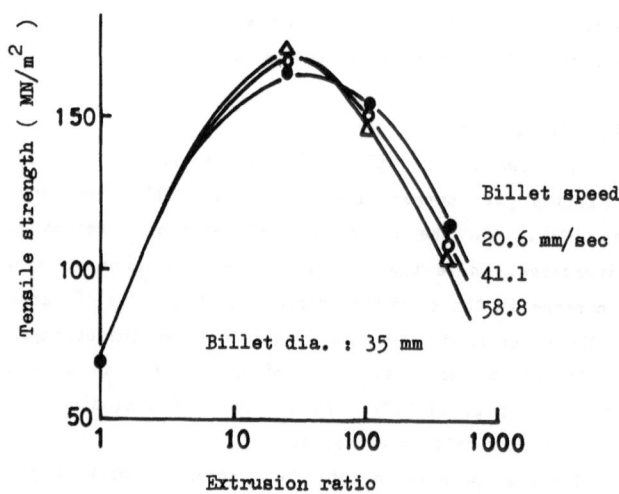

Fig. 38.   Strength of commercially pure aluminium wire
            hydrostatically extruded at various speeds
            in cold condition [27].

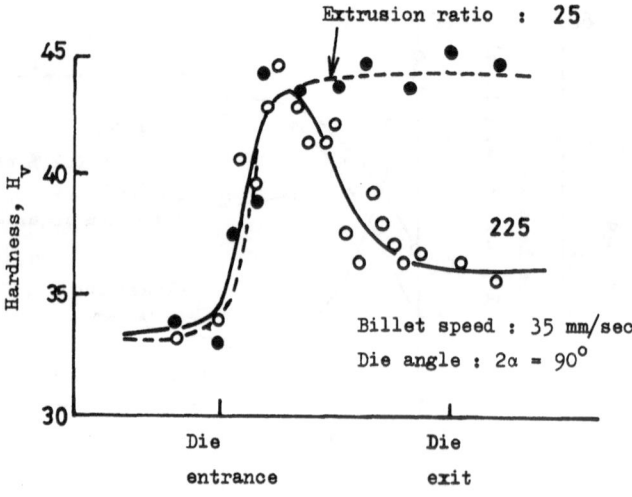

Fig. 39.   Variations of hardness in deformation of
            commercially pure aluminium [27].

of 25 the billet monotonously workhardens in the deformation zone, while at the extrusion ratio of 225 the hardness assumes its maximum in the middle of deformation zone and subsequently decreases.

REFERENCE

1. M. Seido and S. Mitsugi, J. Japan Soc. Tech. Plasticity, 21 (1980) 942.

2. R.J. Fiorentino, G.E. Meyer and T.G. Bryrer, Metallurgia and Metal Form., 39 (1972) 200.

3. H. Kudo and Y. Shinozaki, J. Japan Soc. Mech. Eng., 69 (1966) 601.

4. M. Nishihara, M. Noguchi, T. Matsushita and Y. Yamaguchi, Proc. of 18th Intern. Machine Tool Design and Research Conf., (1977) 91.

5. B. Lengyel and J. M. Alexander, Wire Ind., 39 (1972) 978.

6. D. Green, J. Inst. Metals, 99 (1971) 76.

7. R.J. Fiorentino, B.D. Richardson, A.M. Sabroff and F.W. Boulger, Proc. C I R P Intern. Conf. Manufacturing Tech., (1967) 941.

8. M. Nishihara, T. Fujita, Y. Yamaguchi, T. Matsushita and M. Noguchi, Research and Development, Kobe Steel Engineering Reports, 22-3 (1972) 31.

9. H.Ll. D. Pugh and A.H. Low, J. Inst. Metals, 93 (1964/65) 201.

10. M. Nishihara, J. Japan Soc. Tech. Plasticity, 13 (1972) 565.

11. M. Seido, S. Mitsugi, D. Oelschlägel and A. Kobayashi, Proc. of 2nd Intern. Aluminium Extrusion Seminar, (1977) 133.

12. B. Avitzur, Metal Forming, Processes and Analysis, McGraw-Hill, New York, 1968, p.153.

13. M. Nishihara, Y. Yamaguchi, T. Matsushita and S. Kawamoto, J. Japan Soc. Tech. Plasticity, 10 (1969) 149.

14. H. Kodama, A. Kobayashi and K. Tsunokawa, Japanese Patent Publication No. 37325 (1980).

15. Y. Yamaguchi, M. Noguchi, H. Ide and K. Nishioka, Japanese Patent Publication No. 24712 (1979).

16. T. Ieda, J. High Pressure Inst. Japan, 17 (1979-1) 46.

17. T. Ieda and Y. Tanaka, J. Japan Soc. Tech. Plasticity, 23 (1982) 965.

18. M. Nishihara, T. Fujita, Y. Yamaguchi, T. Matsushita and M. Noguchi, J. Japan Soc. Tech. Plasticity, 13 (1972) 178.

19. R. Hogland, S. Friborg and D. Ermel, Metall, 31 (1977) 515.

20. H. Pötschke, Metall, 31 (1977) 145.

21. R. Hogland, S. Friborg and D. Ermel, Aluminium, 55 (1979) 223.

138

22. R. Hogland, S. Friborg and D. Ermel, Metall, 33 (1979) 827.

23. S. Johnsson, Wire World Inter. 11 Jan./Feb. (1969) 23.

24. M. Seido, Investigation on Hydrostatic Extrusion of Clad Composite Metals, Ph D Thesis, Tokyo Institute of Technology, 1984, p.76.

25. M. Seido and S. Mitsugi, J. Japan Inst. Light Metals, 32 (1982) 100.

26. T. Ohta, K. Tsunokawa, A. Kobayashi and S. Mitsugi, J. High Pressure Inst. Japan, 14 (1976) 69.

27. T. Matsushita, Y. Yamaguchi and M. Noguchi, 109th A I M E Meeting, Feb.24-28, (1980) TMS Paper Selection A80-30.

CHAPTER 4

HOT HYDROSTATIC EXTRUSION PROCESS

M. NISHIHARA
KOBE STEEL, LTD.

SECTION 1

INTRODUCTION

Hydrostatic extrusion was proposed as a new lubricated extrusion process about 20 years ago, although the idea was patented by J. Robertson in 1893. From that time on, the process has been applied to composite materials, special alloys, and exotic or precious metals, with the results of diverse research and development indicating its eventual utilization in industry.

High extrusion ratio, high extrusion speed and superior quality of products are the advantages of hydrostatic extrusion over conventional extrusion. In spite of these advantages the application of this method is limited so long as the operating temperature is restricted.

For the purpose of obtaining higher extrusion ratios at higher temperatures, experimental studies were undertaken. Sauve [1] carried out extensive work on the hot hydrostatic extrusion of carbon steel, stainless steel and other materials, in which solid non-glass pads were used as a lubricant and pressure medium. The pad is in the form of flat cake which is rapidly transformed into a Newtonian viscous liquid on contact with a hot billet. In this C.E.A. (Centre d'Etudes Nuclaires) process, the lubricant is maintained under pressure by means of a metal seal which hermetically closes the container in front of the billet before extrusion and another seal which closes the container behind the billet. Takahashi et al. [2] reported that high-speed tool steel and stainless steel could be hydrostatically extruded at 1200°C by using molten salts or low-melting-point oxide mixtures as a pressure medium. The design of production equipment for hydrostatic extrusion at elevated temperature and the experimental results on copper (99.8% deoxidized high phosphorus) and 63/37 brass at temperatures up to 350°C were reported by Larker and Nilsson of ASEA in Sweden [3]. Fiorentino et al. proposed the hydrafilm extrusion process, which is a modification of the real hydrostatic extrusion, and carried out the extrusion of high-speed tool steel at a temperature of 1400°F with an extrusion ratio 2.6 under a ram pressure of 125,000 psi using this method [4][5].

Kostava et al. [6] studied the hot hydrostatic extrusion of metals and alloys over a temperature range from 250 to 1600 °C. The pressure medium was a mixture of tar (or asphalt) with flaked graphite in a ratio of 1 : 1 by volume, which is used as a lubricant in hot stamping. The melting point of this medium is approximately 200 - 250°C. Thermal conductivity was low and lubrication quality was good.

Hot hydrostatic extrusion using gas was tried by Konyaev in the Soviet Union, although the details have not yet been published [7].

Nishihara et al. [8] have developed a process of hot hydrostatic extrusion using a visco-plastic pressure medium, which is usable up to 1250°C.

Since 1975 a 40 MN press in a warm hydrostatic extrusion plant in Holland has been extruding semi-finished copper tubes from cast billet with an extrusion ratio of 500 at 550°C. Similar presses are in use producing copper-clad aluminium busbar at ASEA in Sweden and at Hitachi Cable in Japan. As the temperature range of the process increased, hydrostatic extrusion has been applied to common products, for example, shapes of aluminium alloy and tubes of copper alloy. But there is a limit to the operating temperature in the process when using a liquid pressure medium such as castor oil. The maximum temperature with castor oil is 550°C.

As the flow stress of metals and alloys decreases at high temperatures, it is possible to extrude these materials at a high extrusion ratio with a lower pressure. Consequently, a relatively small force suffices for a hot hydrostatic extrusion press in comparison with a cold extrusion press. As the extrusion pressure is lowered, the life of the container is extended and the structure of the extrusion press is simplified.

SECTION 2

METHOD OF HOT HYDROSTATIC EXTRUSION

1. Process of hot hydrostatic extrusion

In cold and warm hydrostatic extrusion, castor oil and other liquids
have been extensively used as a pressure medium. But in hot hydrostatic
extrusion, the temperature of the billet is so high that these liquids can
not be used. Even if these liquids are used, the time required to charge
and discharge the liquid is too long to avoid an unfavourable drop of
billet temperature and to reduce the cycle time of extrusion.

Nishihara et al. [8] proposed and developed a process in which
visco-plastic material instead of liquid is used as pressure medium. The
sequence of this process is illustrated in Fig. 1. As the visco-plastic
pressure medium is of such a nature that it does not flow by itself, a
piece can be introduced into a container without using a charging pump.
Consequently, the construction of extrusion press is as simple as a
conventional direct extrusion press. A preheated billet is charged into
the container by moving the stem of the press. A piece of pressure medium
is immediately supplied into the container. Air in the container is purged
out mostly by billet loading, so the pressure medium fills the gap between
billet and container. The stem of the press is forced into the container
to generate the pressure for extruding the billet. In this sequence, the
extrusion cycle time is shorter than 2 minutes, so that the temperature
drop of billet is minimized and the temperature rise of tools is
suppressed. If appropriate material is used as pressure medium, this
process is applicable also to cold and warm hydrostatic extrusion.

2. Pressure medium

The pressure medium is an essential component in hydrostatic
extrusion. In cold and warm hydrostatic extrusion, vegetable oils such as
castor oil are commonly used, because they have a lower pressure
coefficient of viscosity than mineral oils. In cold conditions the
viscosity of castor oil is approximately 10^3Pa.s at a pressure of 10^3 MPa.

142

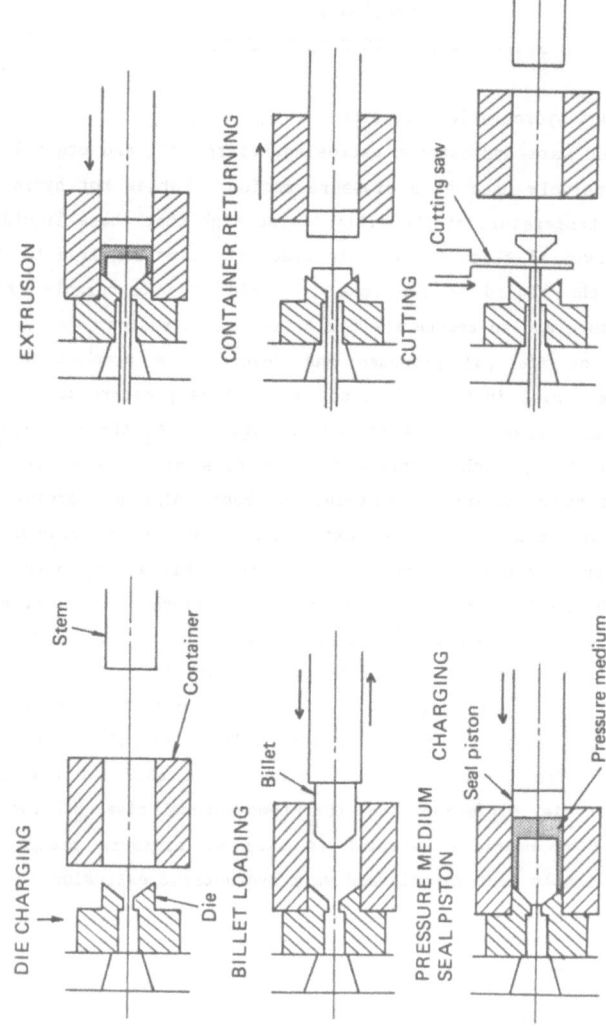

Fig. 1. Operating sequence of hot hydrostatic extrusion using visco-plastic pressure medium.

The disadvantages of using a liquid pressure medium are (1) unstable phenomenon (stick-slip) due to high compressibility of liquid, (2) long cycle time due to the charging and discharging time of the liquid medium, (3) difficulty of operation in hot conditions due to the ignition and burning of the liquid.

In the hot hydrostatic extrusion process developed by Nishihara, visco-plastic material is used instead of the viscous liquid used in cold and warm hydrostatic extrusion. Visco-plastic material does not flow until the shear stress attains a certain critical value called yield stress. On the other hand, Newtonian and non-Newtonian viscous liquids cannot withstand shear stress, however small (Fig .2).

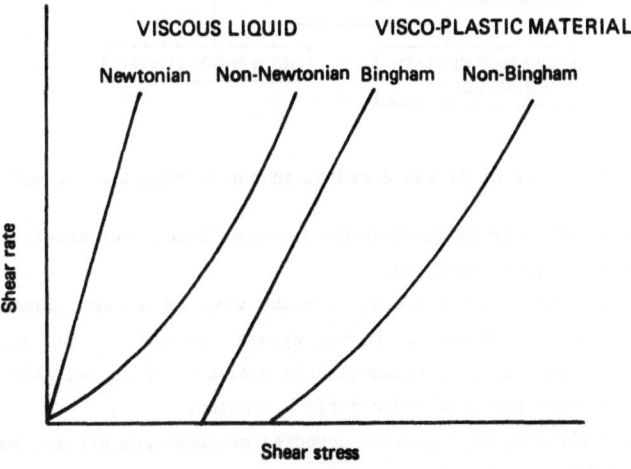

Fig. 2. Flow of viscous liquid and visco-plastic material.

A piece of visco-plastic pressure medium can be handled as a soft solid. Consequently a pumping system is not needed in hot hydrostatic extrusion using visco-plastic pressure medium. Charging of a piece of pressure medium into the container before or after the billet loading is enough to extrude a billet.

The pressure medium in hot hydrostatic extrusion has two roles. The first is to provide good lubrication between billet and tools. The second is to insulate a hot billet from the container wall and tools. These roles and the resulting efficacies are summarized in Fig. 3.

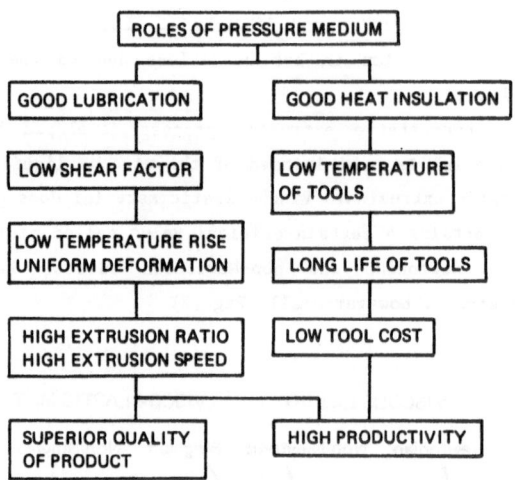

Fig. 3. Roles of pressure medium in hot hydrostatic extrusion.

The visco-plastic pressure media for use in hot hydrostatic extrusion are classified into three groups:

(a) wax or fat including beeswax, carnauba wax, montan wax, lanolin, lard and complex wax in which complexing agents are added. The mixture of beeswax and castor oil is a visco-plastic medium which is suitable for warm extrusion and even for cold hydrostatic extrusion.

(b) grease : (1) soap type greases composed of petroleum oil and soaps such as fatty acids, soaps of natrium, calcium or lithium. A mixture of petroleum oil, natrium soap and graphite is a good medium. (2) non-soap type greases in which silicas, metal oxides and the like are mixed. The mixture of petroleum oil and bentonite is a heat-resistant medium (up to 1200°C).

(c) high polymers including polyethylene, polyisobutylene and so forth. The visco-plastic property of these high polymers depends on their molecular weight and the quality of additives.

Vegetable oil such as castor oil starts to ignite at 200 - 300°C, so it is hardly usable above 500°C even if it is cut off from air. Some metal oxides, salt or glass can be used as pressure medium at high temperatures, but when they adhere to extruded products they are hard to remove, as experienced in Sejournet process. In the Soviet Union, gas was used as pressure medium at 1200°C, but at high pressure and high temperature it is extremely dangerous and unpractical.

A visco-plastic pressure medium containing some additives can be used at 1250°C.

The stable operation of hot hydrostatic extrusion depends primarily on the proper preparation of pressure medium by considering the property of billet material and the extrusion pressure and temperature.

SECTION 3

THERMAL PROBLEMS IN HOT HYDROSTATIC EXTRUSION

1. Temperature rise of product

There is a drop in the temperature of the heated billet which is charged into container. In hot hydrostatic extrusion using a visco-plastic pressure medium, a heated billet is surrounded by the medium which is a good heat-insulator because of its poor fluidity. Consequently, the temperature drop of the billet in this process is smaller compared to a process using a liquid medium. The temperature drop of a larger billet is naturally smaller than that of a smaller one.

When a billet is deformed through the die during extrusion, a large part of the deformation energy is converted into heat, resulting in the temperature rise in the extruded product. Temperature rise is also caused by friction between billet and tools, which is dependent on lubricating conditions. For determining the optimal condition of extrusion, estimation of temperature distribution in billet and extruded product is particularly important, which is commonly practised by finite element method [8].

Figure 4 is an example of calculated temperature distribution of a copper alloy billet. The frictional shear factor m=1.0 refers to unlubricated direct extrusion, m=0.1 to hot hydrostatic extrusion. As shown in Fig. 4, the temperature rise of 150°C in hydrostatic extrusion is lower than 225°C in direct extrusion. In unlubricated extrusion, the heat generated by friction along the die surface causes the local temperature rise and also the consequent rise of average temperature. On the other hand, in hot hydrostatic extrusion the unfavourable rise of surface temperature and average temperature of billet is fairly low. For obtaining

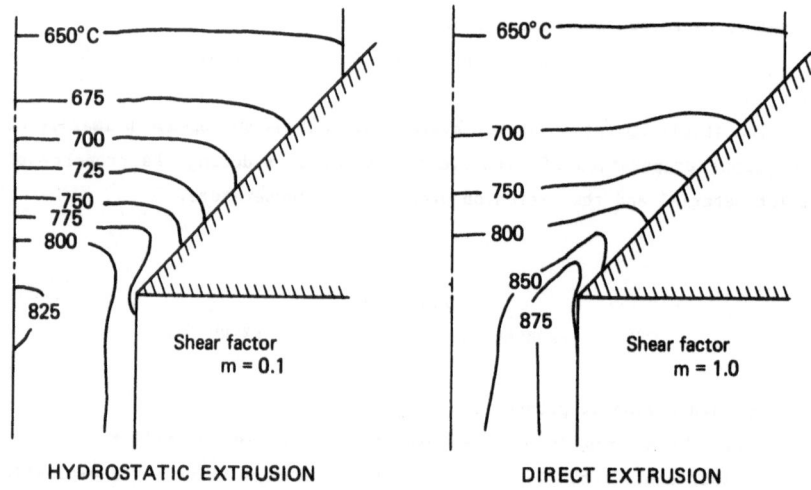

Fig. 4. Temperature distribution of billet in deformation zone. Copper
alloy (billet temp. 650°C, flow stress 286 MPa, extrusion ratio
7.4).

sound products, this unfavourable rise of temperature must be suppressed as
much as possible.

Supposing the deformation energy is adiabatically converted into heat,
average temperature T_p of product at die exit is obtained

$$T_p = T_b + kp/(J\rho C) \qquad (1)$$

where T_b is initial temperature of billet, k a constant, p extrusion
pressure, J mechanical equivalent of heat, ρ density of billet, C specific
heat of billet. The constant k is equivalent to a ratio of average
temperature rise to adiabatic temperature rise. The calculated values of k
are between 0.5 and 0.8 for a wide range of die temperatures, billet size,
extrusion speed and extrusion ratio, which agree with those by
measurements.

Aluminium and aluminium alloys are commonly formed by a non-
lubricating process such as direct or indirect extrusion in the temperature
range between 300 and 500°C. Pure aluminium and Al-Mg alloys are extruded
at a relatively high speed. If high-strength aluminium alloys such as 2024
and 7075 are extruded at such a high speed, the products will have many

cracks. These cracks are caused primarily by local temperature rise due to friction between billet and tools and also by non-uniform flow of metal in a flat die (bridge die or port-hole die).

In order to prevent these cracks and coarsening of crystal grains by the excessive temperature rise, the extrusion speed (product speed) of 2024 aluminium alloy in conventional direct extrusion is limited to 0.03 m/s, and in indirect extrusion to 0.1 m/s at the highest. In hot hydrostatic extrusion, however it is possible to obtain sound products at a much higher extrusion speed. 2024 aluminium alloy is successfully extruded without defects at the extrusion speed of 5 m/s with the extrusion ratio up to 200. This is 50 to 150 times greater than in conventional processes. Judging from experiments on the hydrostatic extrusion of these high-strength aluminium alloys at various temperatures and pressures, temperatures from 220 to 260°C and an extrusion pressure lower than 1000 MPa are the optimal conditions of extrusion. Products of these materials by hot hydrostatic extrusion using a stepped conical die have smooth surface similar to drawn products and are free from peripheral coarse grain. A fine-grained structure is obtained along the whole length of product.

In extruding a material which becomes brittle at elevated temperatures, there is an upper limit to the temperature at which a sound product without cracks can be obtained. Cu-7Sn-5Zn-0.8Al (copper alloy) is an example of this kind of material. Sound products of this alloy cannot be extruded above 770°C even by hydrostatic extrusion. Because the temperature of the product is the sum of the initial billet temperature and the temperature rise during extrusion, the billet temperature must be determined when considering the excess rise of temperature in extrusion.

2. Die and mandrel for high temperature use

For extruding aluminium alloys in hot conditions, a stepped conical die with an angle of approach at die inlet of $2\alpha < 60°$ is capable of preventing seizures (galling) and the formation of dead metal. Tungsten carbide is a suitable die material.

There is no redundant strain in metal flow with a sigmoidal die [9]. But local seizures are observed at the surface of a sigmoidal die when it is used in the hot hydrostatic extrusion of aluminium alloys. On the other hand, no seizure is observed at the surface of a three-stepped conical die which approximates to a sigmoidal die. This fact suggests that pressure medium and lubricant are stagnant at the corner of the stepped die surface and they are effective in maintaining a lubricating film between die and billet.

For extruding aluminium alloy without seizures, a three-stepped conical die having a die angle which satisfies the following relations, is recommended (Fig. 5).

$$
\begin{aligned}
\theta_1 &< 60° \\
\theta_2 - \theta_1 &< 90° \\
\theta_3 - \theta_2 &< 90° \\
\theta_1 < \theta_2 &< \theta_3
\end{aligned}
\tag{2}
$$

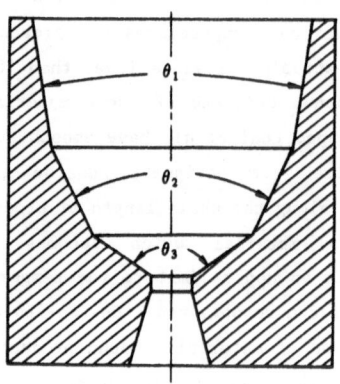

Fig. 5. Three-stepped conical die.

The flow stress of a material drops markedly above a certain temperature, for example, 400°C for copper, 700°C for aluminium brass and 800°C for cupronickel. For this reason, hot hydrostatic extrusion of these materials above such temperatures is performed at a high extrusion ratio. Die approach and die bearing, however, are liable to be deformed by extrusion at high temperatures. For reducing the production cost, a built-up die is useful instead of a monoblock die (Fig. 6). The built-up die is composed of die (approach), die-ring, die-ring holder and die seat. For hot hydrostatic extrusion of copper alloys, the die-ring is made of tungsten carbide or super alloy, and other parts are made of alloy tool steel or super alloy.

As pure copper is non-sticky material, a simple conical die is usable for hot extrusion. A two-stepped conical die (e.g. 100°/135°) is suitable for copper alloy, which is hard and sticky. For harder materials, such as phosphor bronze, a die with smaller angle is recommended.

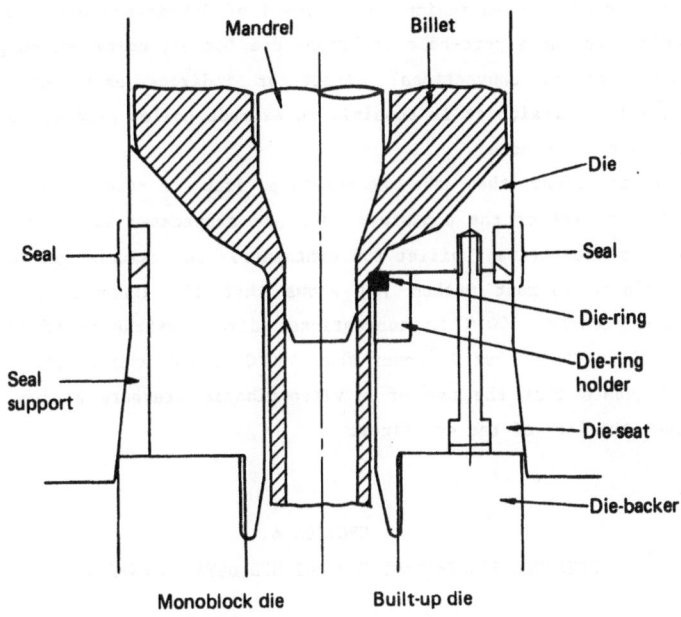

Fig. 6. Die for hot hydrostatic extrusion.

For extruding tubes and hollow shapes, a fixed mandrel or travelling mandrel similar to those employed in conventional extrusion is used. A tipped mandrel consists of mandrel tip and mandrel bar, which is a type of fixed mandrel used with a double-acting extrusion press. The mandrel tip is in direct contact with the billet at high temperature even in hydrostatic extrusion. Materials for the mandrel tip need to be heat-resistant, tough and non-sticky. Low coefficient of thermal expansion and high thermal conductivity are also required. Tool steels for high temperature use, sintered hard alloys and super alloys are suitable for these requirements. A mandrel bar made of tool steel or maraging steel is suitable for use in hot hydrostatic extrusion. To prevent an excessive rise of temperature a mandrel for high temperature use is commonly cooled by water running through its bore. The mandrel tip wears and deforms gradually during extrusion cycles, and cracks on its surface and sometimes breakage of the threaded connection of tip and bar are observed. Because the life of a mandrel is the controlling factor in the cost of tube production, a mandrel for high temperature use must be designed carefully.

As hydrostatic extrusion is a method of lubricated extrusion, it is impossible to use a port-hole or bridge die for extruding hollow products, as practised in conventional direct or indirect extrusion. But in hydrostatic extrusion it is possible to extrude hollow products by using a regular die and mandrel.

It is clear that the temperature drop of the billet and the temperature rise of the container wall are suppressed by the presence of pressure medium between billet and container. The calculation obtained by using finite element method [8] shows that the inner surface of the container reaches 700°C in conventional direct extrusion of the billet preheated to 800°C, but is lower than 400°C in hot hydrostatic extrusion. This suggests that the use of a visco-plastic pressure medium helps to lengthen the life of the container.

SECTION 4
PRESSURE REQUIREMENT FOR HOT HYDROSTATIC EXTRUSION

Extrusion pressure depends on extrusion ratio, extrusion temperature (billet temperature) and extrusion speed (stem speed). For higher extrusion ratio, higher extrusion speed and lower extrusion temperature, higher pressure is required to extrude a billet (Fig. 7).

Aluminium and aluminium alloys

As the flow stress of aluminium and aluminium alloys decreases markedly above a temperature of 200°C, it is possible to extrude these materials with a high extrusion ratio at elevated temperatures. Figure 8 shows the relation between extrusion pressure and extrusion ratio for the extrusion of pure aluminium and various aluminium alloys at a temperature of 245°C. Hot hydrostatic extrusion of aluminium alloy, especially of high-strength aluminium alloys, is liable to generate cracks and coarse crystal grains in the products. To prevent these cracks and coarse grains,

151

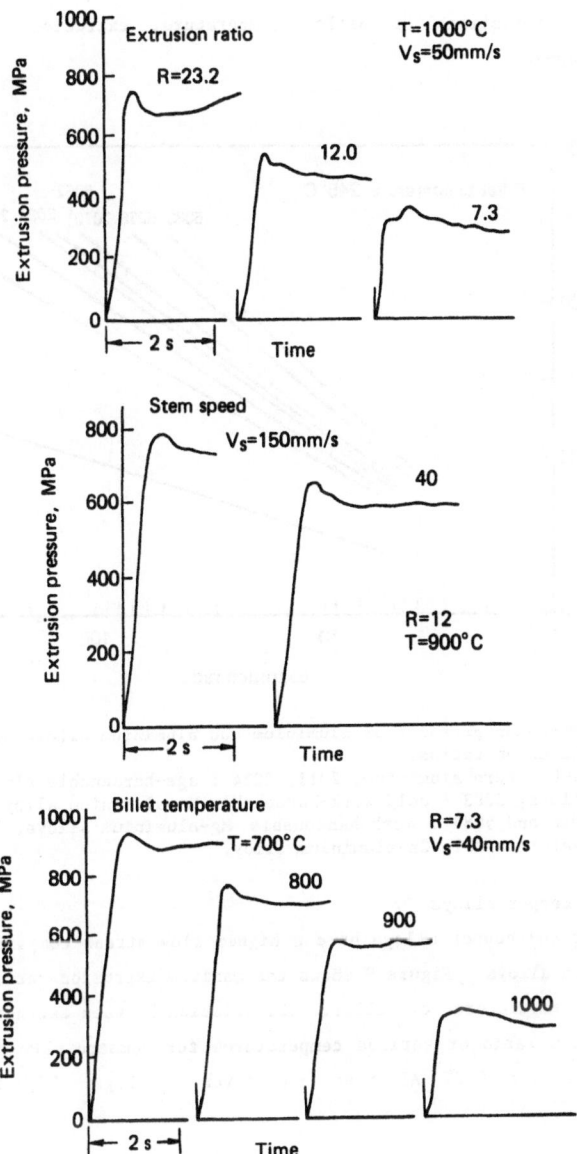

Fig. 7. Effects of extrusion ratio, extrusion speed (stem speed) and
billet temperature on extrusion pressure (0.4% carbon steel).

the temperature of the product during extrusion should be kept lower than a critical temperature which is specific to each alloy. This means that there are limitations of billet temperature, extrusion pressure and extrusion speed.

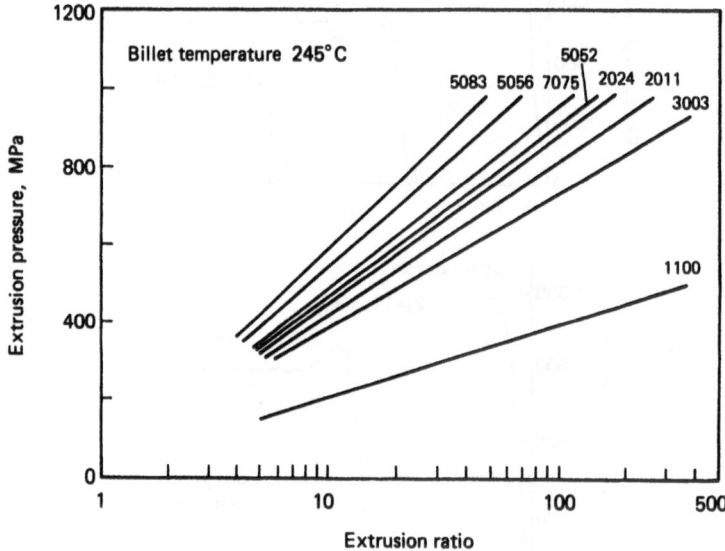

Fig. 8. Extrusion pressure of aluminium and aluminium alloys at various extrusion ratios.
1100 : pure aluminium, 2011, 2024 : age-hardenable aluminium alloys, 3003 : cold-work-hardenable Mn-aluminium alloy, 5052, 5056 and 5083 : work-handenable Mg-aluminium alloys, 7075 : heat-treatable Zn-aluminium alloy.

Copper and copper alloys

Copper and copper alloys have a higher flow stress compared with that of aluminium alloys. Figure 9 shows the maximum extrusion ratio obtainable at various temperatures of billet. The relation between extrusion pressure and extrusion ratio at various temperatures for commercially pure copper, 65/35 brass and 0.2% Al-brass are given in Figs. 10, 11 and 12, respectively.

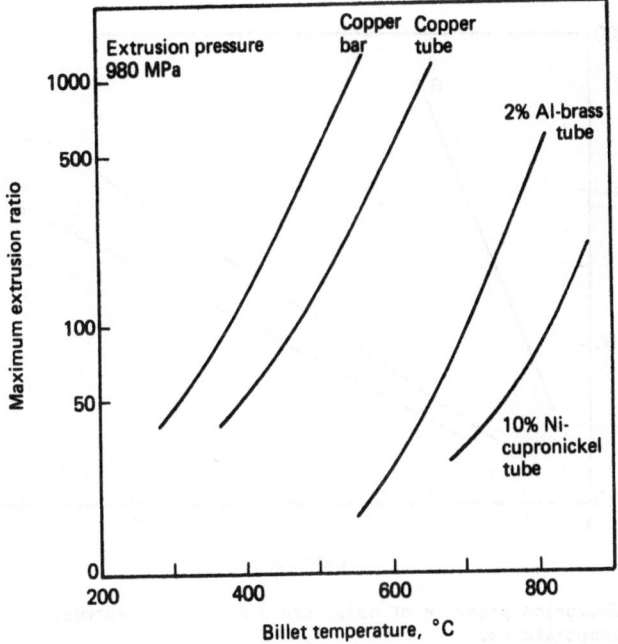

Fig. 9. Maximum extrusion ratio of copper and copper alloys
at elevated temperatures.

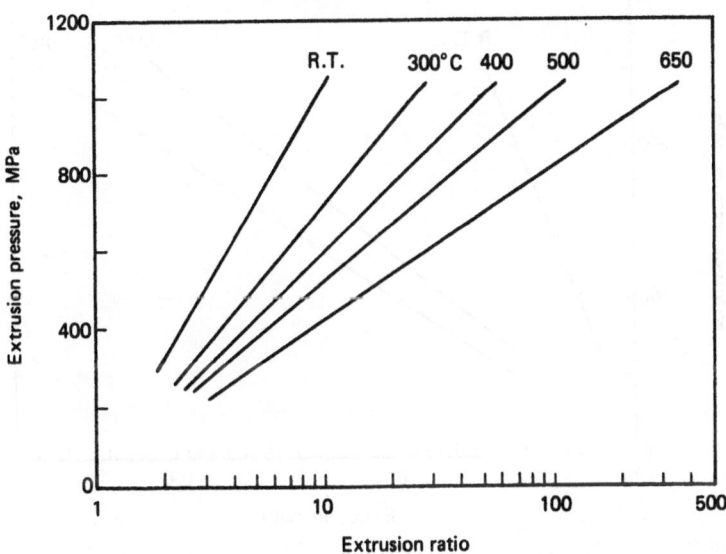

Fig. 10. Extrusion pressure of pure copper at various extrusion ratios and
temperatures.

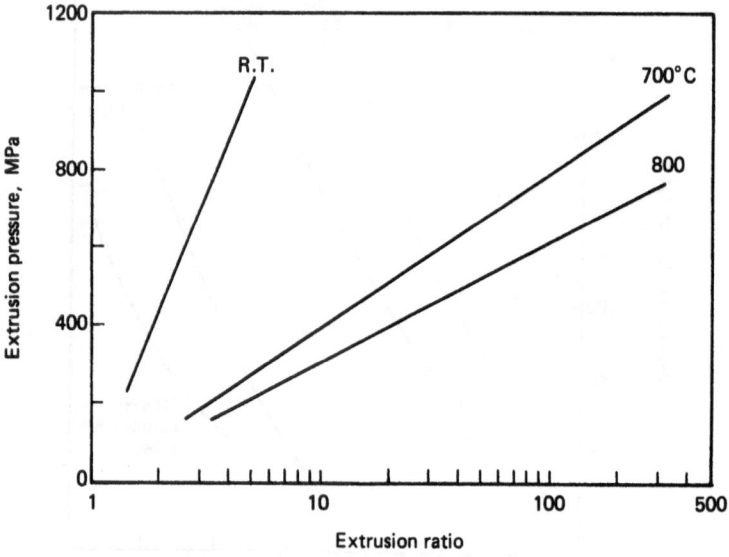

Fig. 11. Extrusion pressure of 65/35 brass at various extrusion ratios and temperatures.

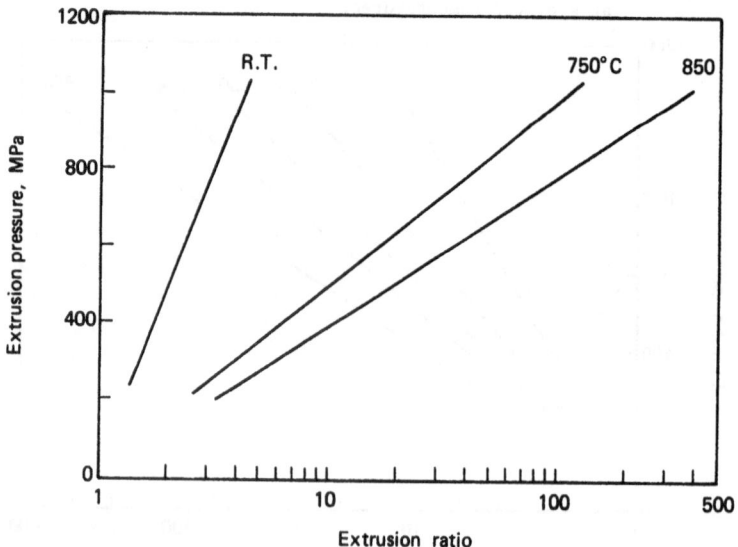

Fig. 12. Extrusion pressure of 0.2% Al-brass at various extrusion ratios and temperatures.

Carbon steel and alloy steel

In Figs. 13, 14 and 15, the relation between extrusion pressure and extrusion ratio at various temperatures for 0.4% carbon steel bar, low alloy steel (Cr-Mo steel, SCM415) tube and stainless steel (SUS304) bar are respectively given. For extrusion of carbon steel and low alloy steel, heat-resistant grease is used as a pressure medium. The billets of these materials heated in an induction furnace at various temperatures were extruded through a 90° conical die made of tool steel at the stem speed of 50 - 250 mm/s. The bar of stainless steel was extruded using a mixture of silicone and glass powder as the medium at the stem speed of 100 - 250 mm/s.

Extrusion pressure p is approximately expressed by

$$p = K\bar{\sigma}_y \ln R \tag{3}$$

where K is a constant depending on the lubricating condition in deformation zone, $\bar{\sigma}_y$ the flow stress of billet material and R extrusion ratio. The flow stress is a function of temperature, strain and strain rate. The mean strain rate in hydrostatic extrusion is of the order of 10 to 100 s^{-1}. Within this range of strain rate, the increase of flow stress with that of strain rate is fairly small. Experimental results of hot hydrostatic extrusion of 0.4% carbon steel show that extrusion pressure in parallel with flow stress decreases with the increase of temperature up to 1000°C. K for break-through pressure is between 1.4 and 1.7, and for run-out pressure between 1.2 and 1.4. These values suggest that a good lubricating condition is preserved at temperatures up to 1100°C. The experiment shows that extrusion pressure in hot hydrostatic extrusion is 20% lower than that in conventional glass-lubricating direct extrusion. This difference is mainly due to the elimination of friction at the billet/container interface in hydrostatic extrusion.

Extrusion pressure p_d in direct extrusion is given by

$$p_d = p \left(1 + \frac{4\mu L}{D} \right) \tag{4}$$

where p is basic extrusion pressure, μ coefficient of friction at billet container interface, L billet length and D diameter of container bore. In the optimal condition for glass-lubricating extrusion, μ is assumed to be 0.02. Basic extrusion pressure p in (4) corresponds to the extrusion

156

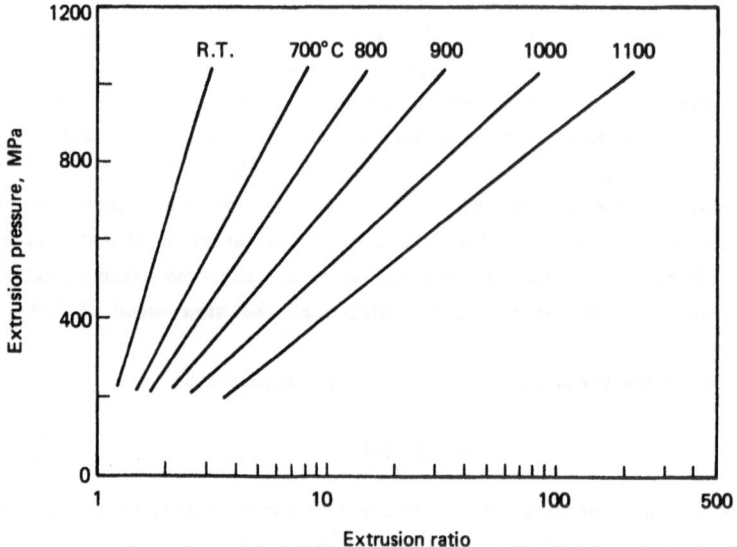

Fig. 13.   Extrusion pressure of 0.4% carbon steel at various extrusion
           ratios and temperatures.

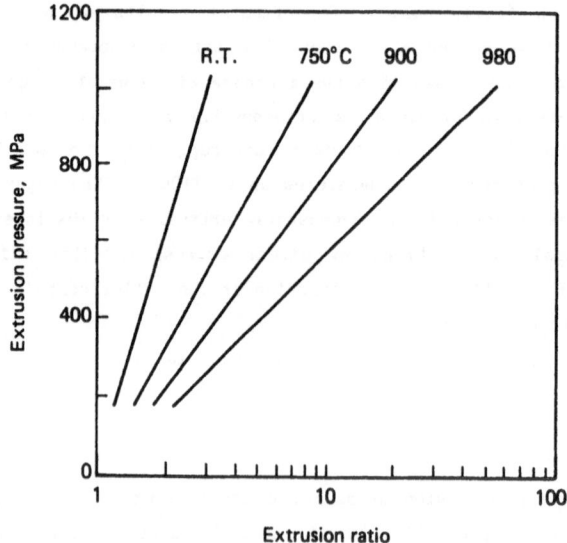

Fig. 14.   Extrusion pressure of low alloy steel (1Cr-0.2Mo steel, SCM415)
           at various extrusion ratios and temperatures.

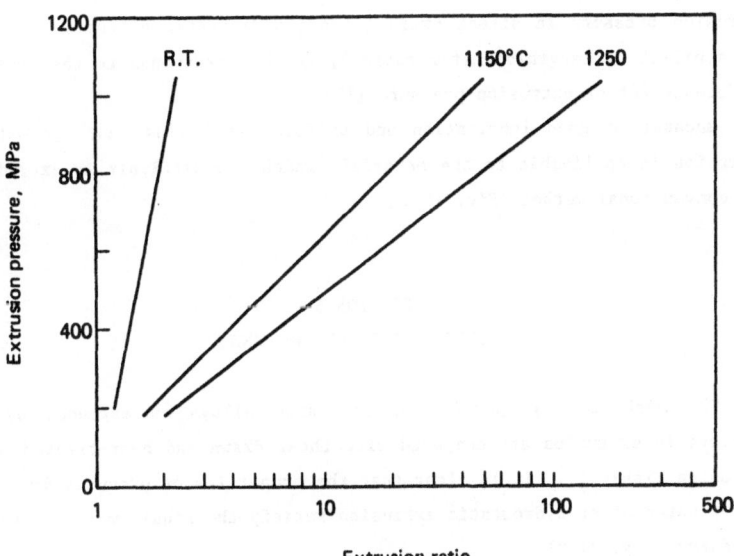

Fig. 15. Extrusion pressure of stainless steel (SUS 304) at various extrusion ratios and temperatures.

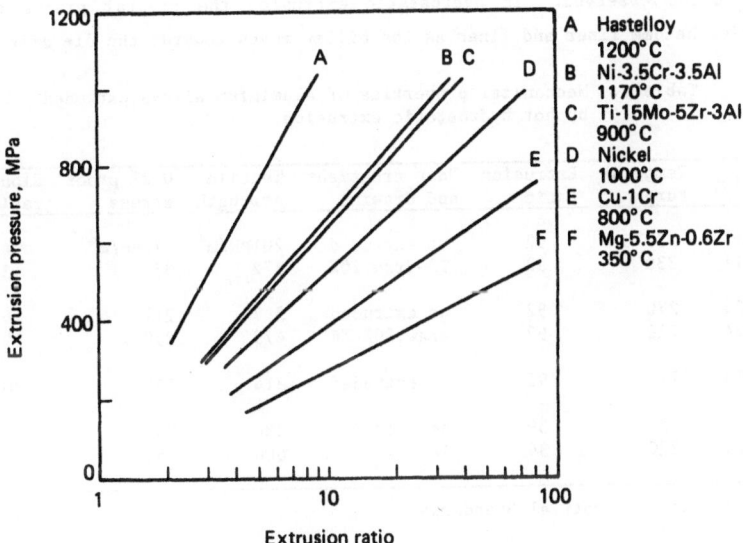

Fig. 16. Extrusion pressure of difficult materials.

pressure p in hydrostatic extrusion given by (3). The difference of
extrusion pressure in direct extrusion and hydrostatic extrusion is 0.24p
for a billet of length/diameter ratio 3, which corresponds to the observed
difference 20% of extrusion pressure [10].

Because of good lubrication and uniform metal flow, hot hydrostatic
extrusion is applicable to the materials which are difficult to extrude by
the conventional method (Fig. 16).

SECTION 5
QUALITY OF EXTRUDED PRODUCTS

The mechanical properties of aluminium alloys as extruded by hot
hydrostatic extrusion are compared with those drawn and heat-treated after
extrusion (Table 1). It is clear that the properties of products drawn and
heat-treated after hydrostatic extrusion satisfy the requirements of various
standards (e.g. JIS*).

Metal flow in hot hydrostatic extrusion is smooth and uniform because
of good lubrication in the deformation zone. In unlubricated direct
extrusion, the formation of dead metal and the entrapment of oxide scale
are often observed. In hydrostatic extrusion the crystal grains of a
billet become finer and finer as the billet moves towards the die exit.

Table 1. Mechanical properties of aluminium alloys extruded
by hot hydrostatic extrusion.

Alloy	Tempera-ture	Extrusion ratio	Heat treatment and drawing	Tensile strength	0.2% proof stress	Elon-gation
2011	330°C	92	as extruded	$201MN/m^2$	$77MN/m^2$	24%
2011	330	92	T3-draw.20%	372	335	13
2024	270	92	as extruded	367	211	16
2024	270	92	draw.20%-T4	479	298	19
5183	275	92	as extruded	314	162	30
7075	220	39	as extruded	386	253	14
7075	220	39	T6	600	555	12

* Japanese Industrial Standards

In Table 2 the mechanical properties and grain sizes of copper and copper alloys are given. The properties of products obtained by one or two passes of drawing and annealing after extrusion are as good as those obtained by several passes after conventional direct extrusion.

The surface roughness of copper alloy tube is less than 10 microns and is improved to less than 5 microns on both inside and outside surfaces by one pass of drawing. The deviation of wall-thickness of the tube is found to be within 10 percent. This small percentage of deviation is probably attributable to the fact that a billet is completely free from upsetting and self-centering of the mandrel is maintained during extrusion.

Table 2. Mechanical properties of copper and copper alloys extruded by hot hydrostatic extrusion.

Material	Process	Tensile strength	0.2% proof stress	Elongation	Grain size
copper	A	$210MN/m^2$	$58MN/m^2$	47%	35 micron
copper	B	238	53	49	30
copper	C	246	60	47	25
2% Al-brass	D	428	185	60	20
10% Ni-cupro-nickel	E	449	207	72	15
	F	365	156	46	20
	G	353	157	45	28

A : Hot hydrostatic extrusion (temperature 450°C, extrusion ratio 75)
B : Hot hydrostatic extrusion - drawing (50.3%) - annealing (630°C)
C : Direct extrusion (temperature 750°C, extrusion ratio 20) - tube reducing - multi-pass drawing - annealing (630°C)
D : Hot hydrostatic extrusion (temperature 770°C, extrusion ratio 100) - drawing (41.6%) - annealing (620°C)
E : Hot hydrostatic extrusion (temperature 900°C, extrusion ratio 34)
F : Hot hydrostatic extrusion - drawing (38%) - annealing (650°C)
G : Direct extrusion (temperature 1000°C, extrusion ratio 22) - tube reducing - multi-pass drawing - annealing (650°C)

The mechanical properties of 0.4% carbon steel extruded at various temperatures are shown in Fig. 17 [10]. At the lower extrusion ratio (R = 7.3 and 12), the maximum tensile strength is obtained at 1000°C. At the higher extrusion ratio (R = 19.5 and 23.2), tensile strength becomes high with extrusion temperature. Ductility of the product (reduction of area in Fig. 17) depends on extrusion ratio, but not much on extrusion temperature.

In other words, a bar of 0.4% carbon steel with high strength and high ductility is obtained by hot hydrostatic extrusion at high extrusion temperature and high extrusion ratio. Figure 18 illustrates the surface roughness of this bar which varies with extrusion ratio and temperature. The minimum roughness is observed at the extrusion ratio 23.2 at 1000°C, which is less than the initial roughness (7 microns) of the billet and is comparable to that of bars obtained by cold hydrostatic extrusion.

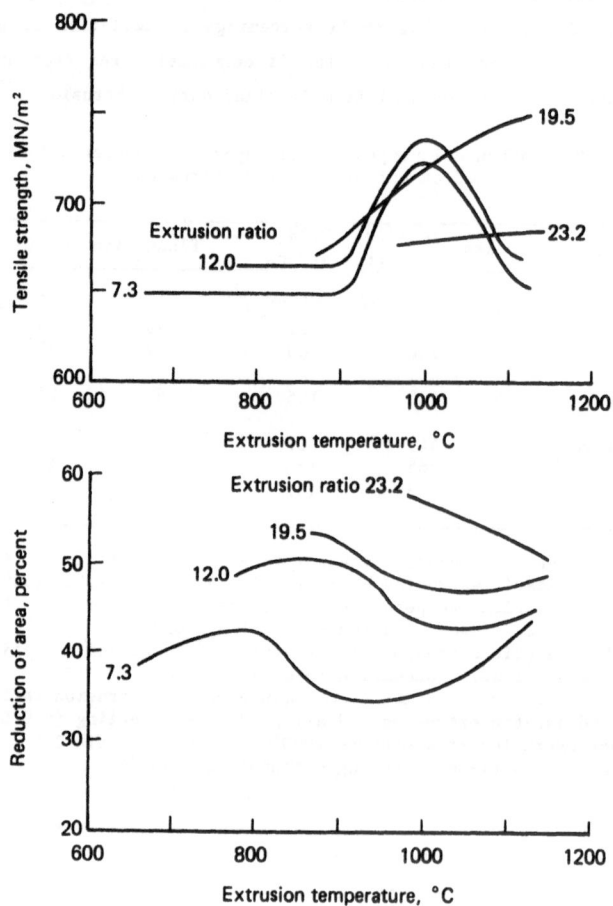

Fig. 17. Mechanical properties of 0.4% carbon steel extruded at various temperatures.

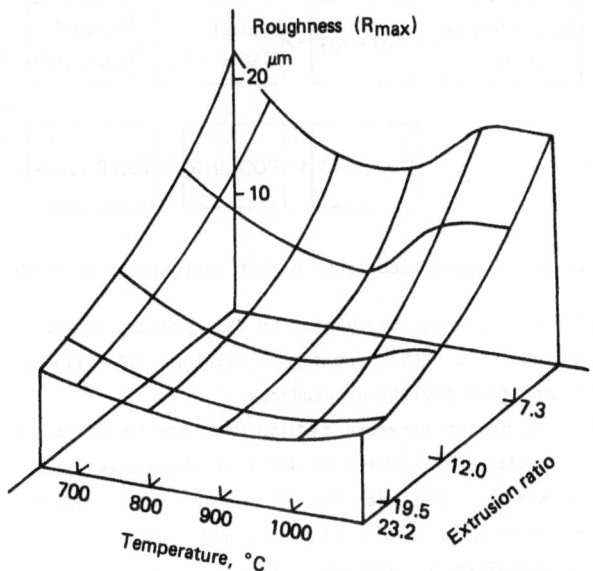

Fig. 18. Effects of extrusion ratio and temperature on surface roughness of product (0.4% carbon steel).

SECTION 6
PRODUCTION PROCESS USING HOT HYDROSTATIC EXTRUSION

The advantages of hot hydrostatic extrusion in the production of bars, tubes and shapes of various materials are as follows: minimization of processing facilities; reduction of working members; minimum stagnation of half-finished products; reduction of production hours; saving of energy; and reduction of production cost.

The production route of copper alloy tube by hot hydrostatic extrusion is shown in Fig. 19. As the high extrusion ratio, for example R = 150 at 800°C for aluminium brass, is obtainable by this process, one or two passes of drawing after extrusion is sufficient to obtain the final product. On the other hand, the extrusion ratio by conventional direct extrusion is 20 at the highest. Several cycles of drawing and annealing after extrusion are needed for the final product.

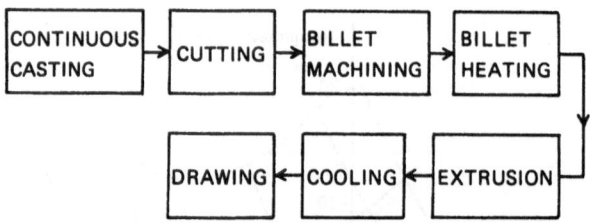

Fig. 19. Production route by hot hydrostatic extrusion.

In order to carry out production by this process without hindrance, it is essential to find the optimal conditions of extrusion for high productivity and good quality of products.

Use of the proper pressure medium is a key to successful extrusion, because the quality of extruded products is dependent on the property of the pressure medium. The quantity of pressure medium to be charged into the container should be calculated carefully. The increase in the inner diameter of container by extrusion pressure, the thermal expansion of billet and the compressibility of pressure medium must all be taken into account.

The shape of billet is one of the important factors of smooth extrusion. The length of billet nose L_n is

$$L_n = (D_b^2 - D_n^2)/(D_b^2 - D_f^2) \qquad (5)$$

where D_b is billet diameter, D_f product diameter, D_n the diameter of billet nose at the top. If D_f is equal to D_n, L_n in (5) becomes 1 where the billet has a complete nose giving the minimum pressure of extrusion. The angle of billet nose must be 2 – 10° smaller than the die angle (2α) to prevent close contact of the billet nose with the die which will cause the temperature drop of the billet nose and the consequent increase of extrusion pressure. There is an optimal surface-roughness of billet for obtaining the best lubricating condition between billet and die. In the extrusion of copper alloy (extrusion ratio 100, billet temperature 800°C), the product with surface-roughness R_{max} 15μ is obtained from a billet with surface-roughness R_{max} 10 – 20 μ. The heating of the billet is also important in obtaining products of good quality. Taper heating by using an induction furnace is sometimes needed to prevent temperature drop at the rear end of a long billet.

Another key to successful extrusion is the maintenance of tools (die and mandrel) in good condition. Tools must be checked as often as possible for deformation and cracks.

Judging from experience in plant operations the hot hydrostatic extrusion process using a visco-plastic pressure medium is suitable for the extrusion of (1) materials of poor workability, such as high-strength aluminium alloys, titanium and titanium alloys, and copper alloys, (2) thin tubes and shapes having close tolerances, e.g. condenser tubes, and (3) composite materials, such as titanium-clad copper bar and super-conducting cable.

REFERENCES

1 C.Sauve, Lubrication problems in the extrusion process, Journal of Institute of Metals, 93 (1964-65) 553-559.

2 S.Takahashi, N.Asami, T.Ishimori, A.Tamai, M.Nakatsukasa and H.Kono, Hydrostatic extrusion at high temperature, in Engineering Solids under Pressure, I.M.E., London, 1971, pp. 166-177.

3 H.T.Larker and J.O.H.Nilsson, Hydrostatic extrusion at elevated temperatures, in Engineering Solids under Pressure, I.M.E., London, 1971, pp. 161-165.

4 R.J.Fiorentino, G.E.Meyer and T.G.Byrer, The thick-film hydrostatic extrusion process, Technical Paper MF71-103, Society of Manufacturing Engineers, 1971.

5 R.J.Fiorentino, Comparison of cold, warm and hot extrusion by conventional and hydrostatic methods, Technical Paper MF72-159, Society of Manufacturing Engineers, 1972.

6 A.A.Kostava et al., High-temperature hydrostatic extrusion of metals and alloys, Transl. from Russian, BISI 18104, Kuzmechno-Shtampovochnoe Proizvodstvo, (1977) 9-11.

7 Yu.S.Konyaev, Some perspective elaborations in the field of hydraulic extrusion, in High Pressure Engineering, I.M.E., London, 1977, pp. 191-192.

8 M.Nishihara, M.Noguchi, T.Matsushita and Y.Yamaguchi, Hot hydrostatic extrusion of nonferrous metals, Proc. 18th Int. MTDR Conf., (1977) 91-96.

9 M.L. Devenpeck and O.Richmond, Strip-drawing experiments with a sigmoidal die profile, Journal of Engineering for Industry (Transactions of the ASME), November (1965) 425-428.

10 M.Nishihara, T.Matsushita, Y.Yamaguchi, T.Yamazaki and M.Noguchi, Hot hydrostatic extrusion of steel, Proc. 20th Int. MTDR Conf., (1979) 87-82.

CHAPTER 5

PLANT EQUIPMENT FOR HYDROSTATIC EXTRUSION

SECTION 1

EXTRUSION PRESS AND AUXILIARY EQUIPMENT

M. YAMAGUCHI

KOBE STEEL, LTD.

1. Introduction

Research into various kinds of hydrostatic extrusion equipment has been carried out since the hydrostatic extrusion process was invented, and in line with advances in the extrusion process itself the equipment has also developed.

The hydrostatic extrusion process is different in principle from conventional ram-type extrusion, and therefore, in the early years of research hydrostatic extrusion presses with very different structures from those of the ram-type extrusion process were designed and manufactured. In particular, the cold extrusion process using a pressure medium at an extremely high pressure and of low viscosity needed special structures of a container and seals to endure high pressure and to prevent the pressure medium from leaking. In addition, various structures to prevent a stick-slip phenomenon were examined, since it was considered to be a serious problem.

However, as the cold extrusion process has come into operation and the hot extrusion process has developed, the above problems have been resolved, thus allowing hydrostatic extrusion presses to be similar in structure to conventional ram-type extrusion presses.

The ram-type extrusion presses have progressed towards better practical use ever since the 18th century, and accordingly hydrostatic extrusion presses with a similar structure are now technically and economically feasible in industry and seem to make a favorable impression on users.

This paper describes the basic structures, movements and characteristics of the various hydrostatic extrusion presses.

2. Basic Type of Hydrostatic Extrusion Press

Hydrostatic extrusion equipment is used for extruding a work (billet) through a die by applying the pressure medium at high pressure, and basically consists of a container, a die and a high-pressure generating device, as shown in Fig. 1. According to the combination and arrangement of these components, the equipment can be divided into various types, which are explained as follows.

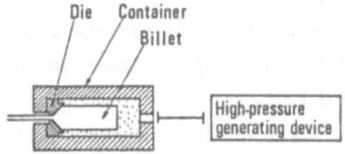

Fig. 1. Basic components of hydrostatic
extrusion equipment.

2.1 Pipe-Connection Type and the Direct-Connection Type

According to the type of connection between the high-pressure generating device and the container, the equipment can be divided into two types, that with a pipe-connection and that with a direct-connection.

(a) Pipe-Connection Type

In this type of structure the high-pressure generating device and the container are separated, and both are connected with piping.

An example of using an intensifier as a high-pressure generating device is shown in Fig. 2.

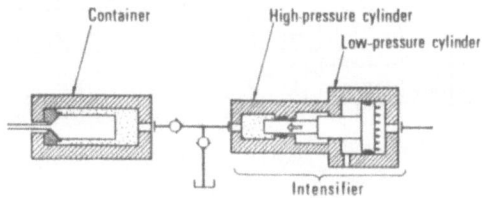

Fig. 2. Pipe-connection type.

If the internal capacity of the container is larger than the high-pressure cylinder volume of the intensifier, this should be reciprocated several times until extrusion is completed. In such an operation, extrusion is interrupted every time the intensifier reaches the stroke end. Even if an intensifier of the multi-cylinder type is used, it is difficult to maintain a constant rate of discharge and impossible to obtain defect-free extrusion products. Therefore, the one-stroke volume of an intensifier should be larger than that of the pressure medium required for one-cycle extrusion.

When the container capacity is large, the intensifier capacity has to be large correspondingly, causing problems with the intensifier, both in terms of structure and cost.

For this reason, the pipe-connection type is limited to small-scale industrial and laboratory use only.

(b) Direct-Connection Type

This type is basically constructed so that the container and the intensifier are connected directly without piping.

Since the high-pressure cylinder volume of the intensifier discharging the quantity of pressure medium required for one-cycle extrusion is roughly equal to that of the container, it is structurally simple to connect both the devices directly, compared with connecting both with high-pressure piping, and the former is easier to maintain, in for example the measures needed to prevent leakage of liquid. Furthermore,

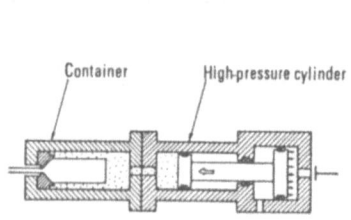
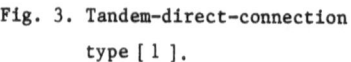

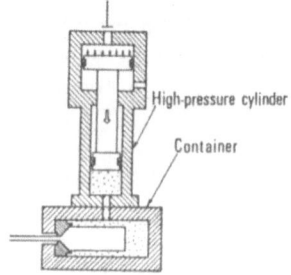

Fig. 3. Tandem-direct-connection type [1].

Fig. 4. Cross-direct-connection type [2].

at a pressure of about 1 GPa the viscosity of the pressure medium increases, thus causing the flow resistance of the liquid to increase in
the piping.

These demerits can be eliminated by connecting the container directly to the intensifier without piping. For such an arrangement using a
direct-connection, there are two possibilities, the tandem type, shown in
Fig. 3, and the cross type, shown in Fig. 4.

This arrangement is known as the press type, because the equipment
laid out with a direct-connection is very similar in structure to conventional presses, while the equipment is called a hydrostatic extrusion
press.

A hydrostatic extrusion press, laid out in tandem or cross type and
provided with a container whose dimensions are planned independently of
the diameter of the high-pressure cylinder, is advantageous when extruding very slim billets, thus enabling the manufacture of a press for
extruding billets with a length-diameter ratio of 40 to 50.

With reference to the tandem type using a direct-connection, in
cases where it does not matter that both the container and the high-
pressure cylinder diameters are equal, a straight unit type, as shown in
Fig. 5, which incorporates both into single unit is available. This type
can eliminate the high-pressure cylinder, to which ultra-high pressure is
applied, shorten the total length of the press, and further facilitate
both the supply of billets and the pressure medium. In terms of practical application, the presses being used are almost all of this type.

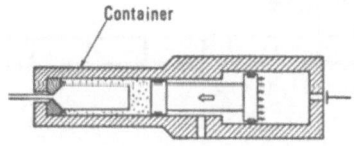

Fig. 5. Straight unit type.

168

2.2 Vertical Type and Horizontal Type

As described above, hydrostatic extrusion presses in practical use
are almost all of the straight unit type, and according to the direction
in which the press ram moves, are classified into the vertical type, with
a ram moving vertically, and the horizontal type, with a ram moving
horizontally.

Though the vertical type press can save installation area, products
are extruded downwards and when these are long, measures should be taken
to divert them into a horizontal direction by providing a guide trough
below the press. For this reason, when extruding shaped or large-
section products, which are subject to adverse effects if they become
bent, this type is not suitable.

The horizontal press can take out the extruded products, even if
they are long, in a straight line, since they are extruded horizontally.
The presses in industrial use are mostly of this type which facilitates
access to every part of the press, as well as being excellent for maintenance work. Fig. 6 shows an example of the horizontal press.

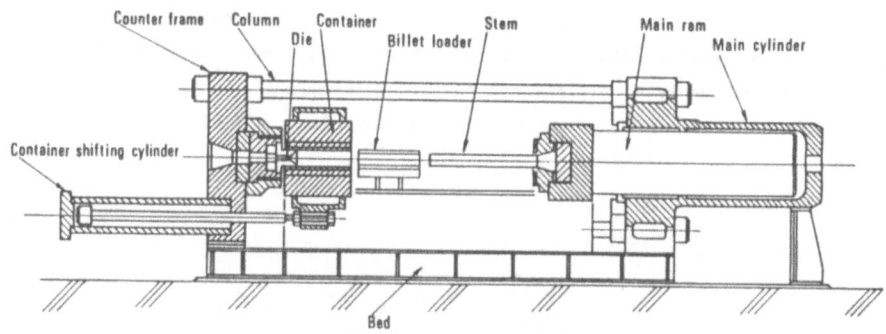

Fig. 6. Horizontal type hydrostatic extrusion press
(single-acting type).

2.3 Single-Acting Type and Double-Acting Type

The press shown in Fig. 6, having one cylinder for driving, is a single-acting press.

This type enables the extrusion of various solid products according to the shape of the die hole, and further enables the extrusion of hollow products by using a special mandrel. However, when a special mandrel is fixed to the container or the stem, the billet length is restricted, as will be discussed later, and efficiency in handling the mandrel is also reduced. Therefore, for an extrusion press which mainly manufactures hollow products, a press type to allow the mandrel to move independently of the stem is used, as shown in Fig. 7.

For this arrangement, the press is provided with a cylinder which drives the mandrel, independently of the cylinder which drives the stem, and this is called a double-acting press.

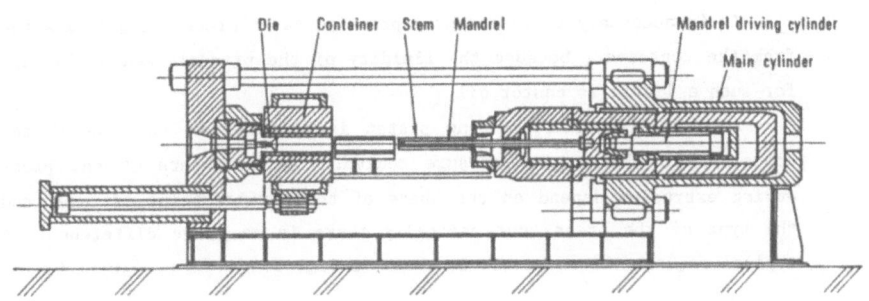

Fig. 7. Double-acting type hydrostatic extrusion press.

3. Movement of Hydrostatic Extrusion Press

The purpose of this press is to obtain a product by extruding a billet through a die, but before and after the extrusion work many subsidiary processes are required. In addition, the movements of the press necessary for completing a one-cycle operation depend on the kind of pressure medium, the shape of the product, etc. Therefore, for example, a press which extrudes solid products, though it may have the basic structure shown in Fig. 5, is provided with subsidiary devices so that these subsidiary movements are performed smoothly and efficiently.

To understand the structure of a press, it is important to have a knowledge of its movements. A liquid such as castor oil is generally used as a pressure medium, when the billet temperature is 500°C or less, while a special visco-plastic material is used when the billet temperature is over 500°C. The press movements for a one-cycle operation differ according to the kind of the pressure medium, as described below.

3.1 Movement of Press using Liquid-Type Pressure Medium

It is necessary to construct a press so as to prevent liquid leakage from the container, because the fluidity of the pressure medium is high for such a liquid as castor oil.

A pressure-medium charging platen is provided at one end of the container to charge the pressure medium. The movements of the press during extrusion depend on the shape of the product being extruded and the type of the press, but basically there is no large difference. A typical sequence of movements is shown in Fig. 8.

3.2 Movement of Press using Visco-Plastic-Type Pressure Medium

Visco-plastic material can be handled as a lump, because its fluidity is very low at room temperature. Therefore, when applying the piston type of high-pressure seal, as will be explained later, the pressure medium can be charged simultaneously with the seal piston, thus eliminating the need for any special equipment, such as a pressure-medium charging platen.

An outline of the sequence of movements for solid product extrusion using a seal piston is shown in Fig. 9.

171

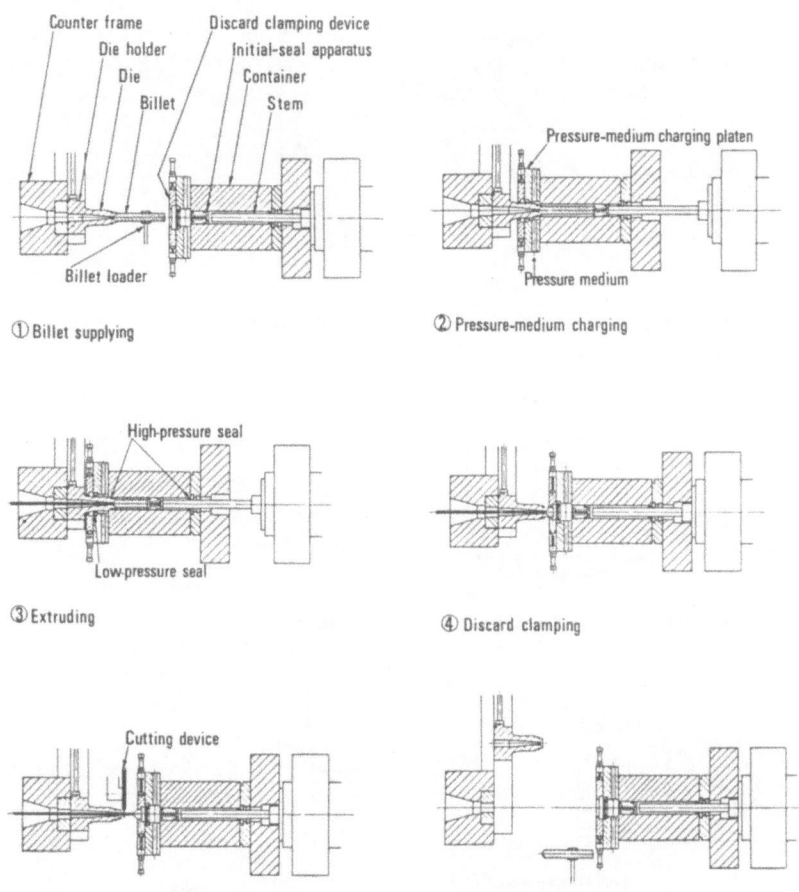

Fig. 8. Movement sequence of the press using
a liquid-type pressure medium.

172

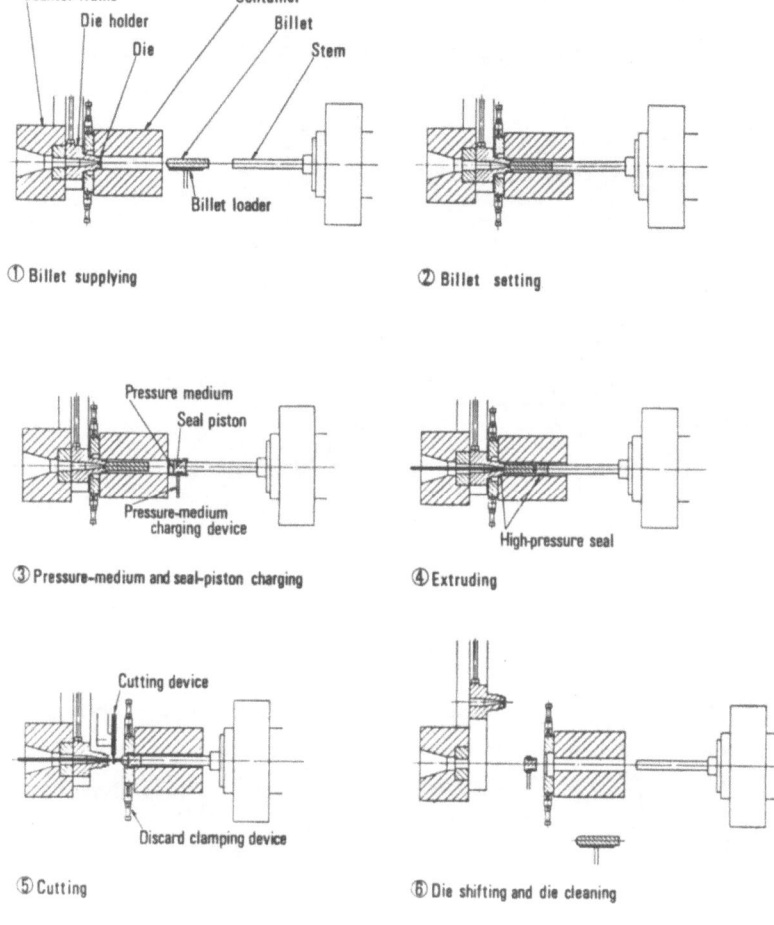

Fig. 9. Movement sequence of the press using
a visco-plastic-type pressure medium.

4. Structure of Hydrostatic Extrusion Press

As the preceding illustrations of typical sequences of movements suggest, it is necessary to provide various devices, as shown in Fig. 10, other than the press proper to perform the extrusion work. The major devices are explained below.

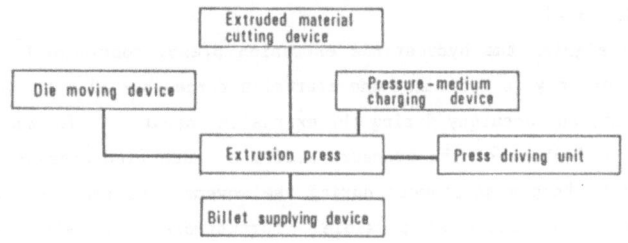

Fig. 10. Composition of extrusion press.

4.1 Press Proper

Horizontal presses of the straight unit type, shown in Fig. 5, are mostly used as the basic type of the press proper. The general structure of the press proper in practical use is shown in Fig. 11.

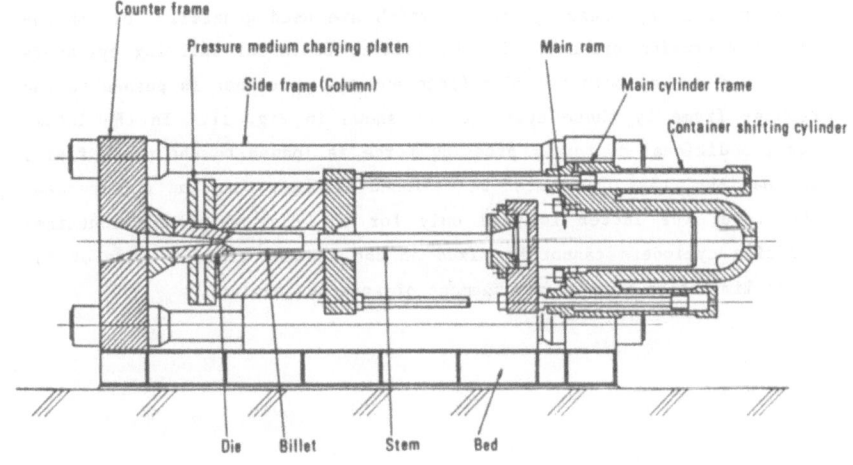

Fig. 11. Column-type hydrostatic extrusion press.

The extrusion force applied to the main ram by pressure in the main cylinder is finally applied to the die. The press frame, which holds the shape of the press proper while receiving this force, is generally composed of a main cylinder frame, a counter frame and side frames. The press frame of the column-type, as shown in Fig. 11, is generally used, but types other than this, such as are shown in Fig. 12, Fig. 13 and Fig. 14 are also used.

On designing the hydrostatic extrusion press, consideration should be given not only to resisting the extrusion force, but also to maintaining rigidity and accuracy during the extrusion operation. To ensure that there is no leakage of the pressure medium at such high pressure, it is necessary to keep misalignment during the movement of the container and the stem to as little as possible. Furthermore, even when complete extrusion occurs at the end of the extrusion stroke, the press must be sufficiently rigid to ensure that its accuracy is not adversely affected by shock.

To ensure sealing between a die and a container, it is necessary to hold a die between a counter frame and a container by large force. There are two methods for such sealing, i.e, one is to pull the container to the counter frame and the other is to push the container to the counter frame. In the former, container shifting cylinders are fixed on the counter frame as shown in Fig. 6 and the container is pulled to the counter frame by these cylinders, which are used generally for conventional extrusion presses. In the latter, container shifting cylinders are fixed on the main cylinder frame and the container is pushed to the counter frame by these cylinders as shown in Fig. 11. In the latter case, additional container pressing force is loaded to the press frame, so that the press frame must be designed stronger than the former case. Therefore, the latter is used only for the case which the container shifting cylinders cannot be fixed on the counter frame because of the space limitation from an arrangement of attached devices.

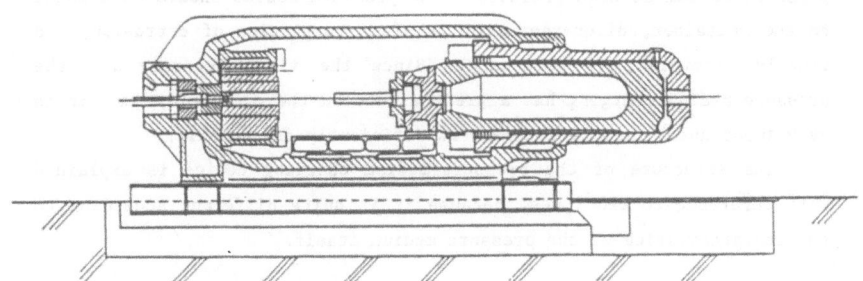

Fig. 12. Single-piece frame type hydrostatic extrusion press.

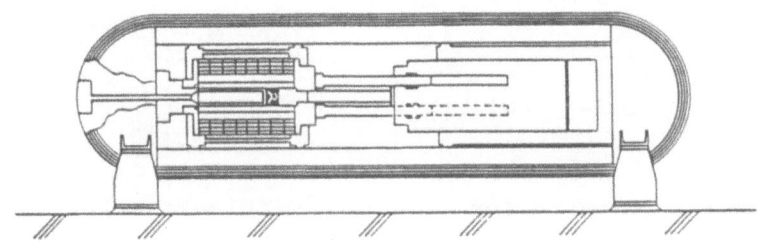

Fig. 13. Wire-winding frame type hydrostatic extrusion press [3].

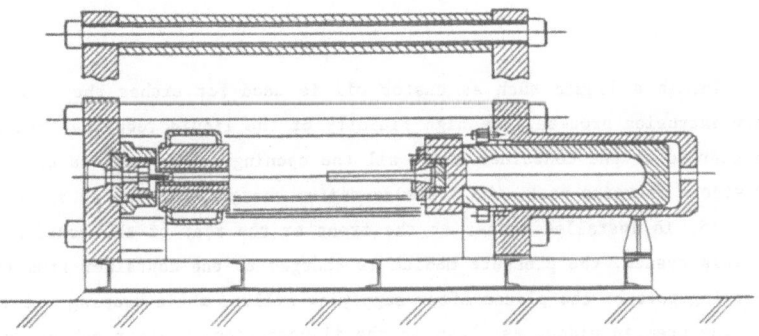

Fig. 14. Tie-rod connecting frame type hydrostatic extrusion press.

4.2 Pressure-Medium Charging Device

The hydrostatic extrusion press extrudes billets by using the pressure medium at high pressure. The pressure medium should be charged to the container, discharged from it after completion of extrusion, and finally recovered, if necessary. Since the time necessary for the pressure-medium charging has a great effect on the one-cycle time, it is very important to complete this work efficiently and quickly.

The structure of the pressure-medium charging device is explained with reference to two types, because it is quite different according to the characteristics of the pressure medium itself.

(a) Charging Device for Liquid-Type Pressure Medium

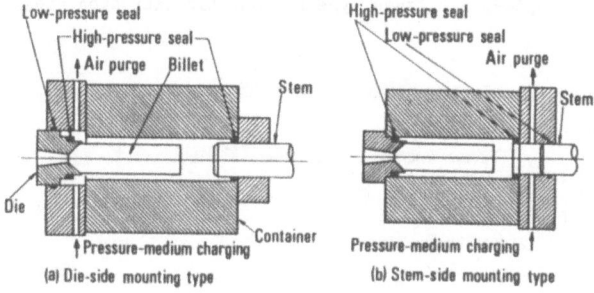

Fig. 15. Pressure-medium charging platen for liquid-type pressure medium.

Though a liquid such as castor oil is used for either the cold or warm extrusion process, the high fluidity of the liquid requires that it is charged to the container after all the opening points are sealed. To provide a charging port, the pressure-medium charging platen, as shown in Fig. 15, is installed either at the front or the rear of the container. In this system, the pressure medium is charged to the container from the charging port of the platen after supplying billet, while keeping the die and the stem in place, as shown in the illustrations. After making sure that the container is filled with the pressure medium, the die or the stem is sealed.

Of the two types shown in Fig. 15, the stem-side mounting type requires the stem to be extracted from the high-pressure sliding seal when charging the pressure medium, though repeated extraction and insertion of the stem seriously affects the high-pressure seal's life.

In the die-side mounting type, the high-pressure seal is of a fixed type, simple in structure, cheap in cost and easy to replace compared with the sliding type.

(b) Charging Device for Visco-Plastic-Type Pressure Medium

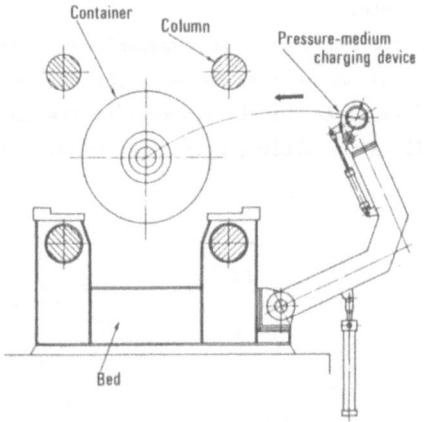

Fig. 16. Pressure-medium charging device for
a visco-plastic-type pressure medium.

Since the visco-plastic pressure medium used for the hot extrusion process does not flow at room temperature, a simple device similar to the billet loader is available. After charging the pressure medium into the container, it should be sealed by the seal piston before the pressure medium comes into contact with a hot billet in order to prevent it from burning.

As shown in Fig. 9, the pressure medium is pushed into the container by the stem together with the seal piston after supplying a billet, and then extrusion will start immediately.

An example of the swing-type pressure-medium charging device is shown in Fig. 16. This unit is constructed so as to allow the required quantity of pressure medium to be charged into the box at the outside position of the press.

4.3 Extruded Product Cutting Device

As shown in Fig. 8 and Fig. 9, after completion of extrusion, the extruded product is cut to separate an extrusion discard, and then the product should be drawn out of the die. For this reason, a cutting method which does not deform the cut portion excessively should be used. Generally, either a saw or a shear is used as a cutting device taking into consideration the kind and shape of the extruded product, the cutting position, etc.

A saw-type cutting device is available for almost all materials, when the cutting is to be done between the die and the container, as shown in Fig. 17 (a). As for the movement of the saw, the swing type is also used as well as the sliding type shown in the illustration.

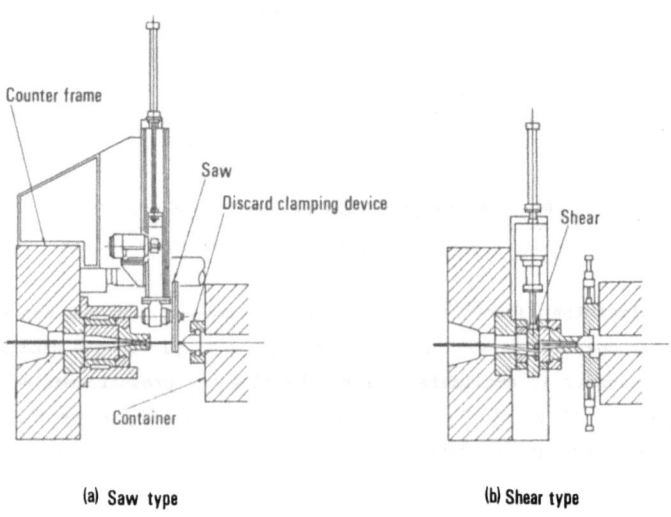

(a) Saw type   (b) Shear type

Fig. 17. Extruded product cutting device.

The cutting tool of the shear type should be flat, since cutting work of this type is done at the back of the die as shown in Fig. 17 (b), and it is generally used for cutting aluminium and copper products. However, the saw type should be employed for extruded products whose shape at the shearing face is deformed by shearing, because the product must be drawn off through the die after shearing.

4.4 Die-Moving Device

When extrusion has been completed, the extruded product and the discard are separated by cutting. Subsequently, the die is moved to the outside position of the press for cleaning and changing, if necessary.

The device used for the above purpose is the die-moving device, typical examples of which are shown in Fig. 18. The traverse type is superior to the rotary type since it is rigid structure and has good accuracy in extrusion.

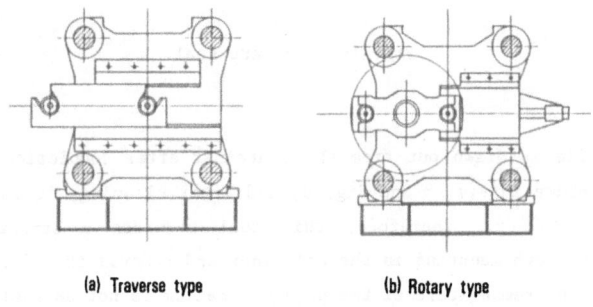

(a) Traverse type   (b) Rotary type

Fig. 18. Die-moving device.

5. Main Parts and Tools for Hydrostatic Extrusion Press

As described before, the structure of each part of the hydrostatic extrusion press is, in many cases, similar to that of the conventional ram-type extrusion presses. However, since this press uses a pressure medium at high pressure, high-pressure seals, extrusion tools, etc. have to be specially designed.

5.1 High-Pressure Seal

To seal the pressure medium at high pressure in the container, seals
are provided at two points of the die and stem parts, and fixed and
sliding seals are used, respectively. Moreover, mandrel-part seal is
necessary for a double-acting press which has moving mandrel.

(a) Die-Part Seal

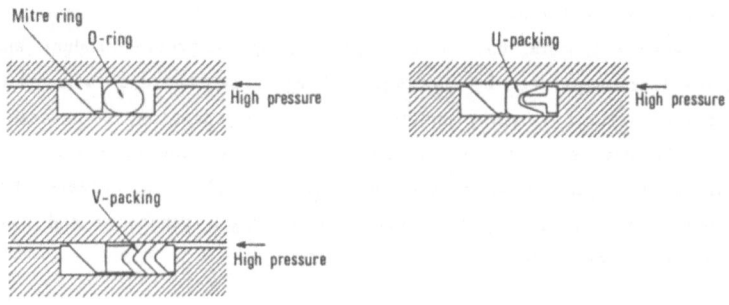

Fig. 19. Die-part seal.

The die is drawn out from the container after completion of extrusion, as shown in Fig. 8 and Fig. 9, and after cleaning, it is remounted
in the container. Therefore, this seal requires a structure which
facilitates both mounting in the container and removal from it.

When the temperature of the pressure medium is not excessively high,
an O-ring, U-packing or V-packing can be used, as shown in Fig. 19.
However, to prevent the packing from being extruded due to the high
pressure causing an enlargement of the container's inside diameter the
packing should be backed up with mitre rings, as shown in the illustrations. When using mitre rings, force applied in an axial direction can
keep the clearance between the ring and the container small by enlargement of the high-pressure side ring. Synthetic rubber is used for each
packing, and for the mitre ring, copper alloy is used.

In hot hydrostatic extrusion, in which visco-plastic material is
required as a pressure medium, mitre-ring type packing of copper alloy is
employed, because sealing materials such as rubber cannot be used.

181

(b) Stem-Part Seal

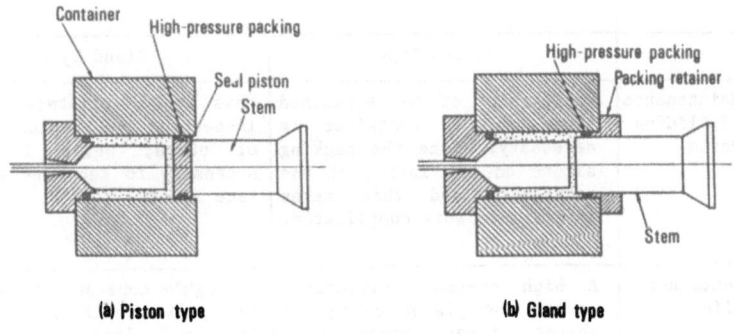

(a) Piston type   (b) Gland type

Fig. 20. Stem-part seal.

There are two methods of sealing the sliding parts of the container
and the stem, using a piston type and a gland type, as shown in Fig. 20.

Both types have advantages and disadvantages and a comparison of
their characteristics is shown in Table 1. An examination of which type
is more advantageous is given below on the basis of this comparison
table.

The low pressure, at which a low leaking speed may not damage the
container's interior by erosion, and may cause only slight local stress
on the container, allows the piston type to be used. The piston type, in
which the press center accuracy is not so necessary as the gland type,
makes its maintenance work very easy.

The piston type is used for hot hydrostatic extrusion, since the
high viscosity of a visco-plastic pressure medium causes less leakage,
and the high billet temperature allows a sufficient extrusion ratio of a
billet to be given at a relatively low pressure of 1 GPa or less.

In cold hydrostatic extrusion, however, the pressure medium is
generally a low viscous liquid such as castor oil, and usually a practicable extrusion ratio cannot be achieved without the use of very high
pressure. Accordingly, the gland type is generally employed with the
cold hydrostatic extrusion press.

Table 1. A Comparison between Stem-Part Seal Types.

	Piston Type	Gland Type
Maintenance of Sliding Parts	Replacement of the scratched liner of the container is necessary, since the packing slides on the inside of the container, and this makes maintenance work complicated.	This type facilitates the inspection and replacement of parts, while it is necessary to keep the stem face smooth.
Container Life	A high pressure difference across the piston causes an abrupt stress change in a local area of the container which travels in an axial direction. This adversely affects the life of the container.	A roughly constant pressure is applied uniformly over the full length of the container during extrusion, thus having a good effect on its life.
Required Centering Accuracy	A floating type separating the piston and the stem facilitates centering of the press.	Full care should be taken when centering the stem and the container.
Packing Gland	A piston mounting the packing does not need the packing gland, thus simplifying the structure of the press.	To facilitate the packing replacement by providing the packing gland, large force for retaining packing is required, which results in a complicated structure of the press.
Strength of Stem	As the extrusion process advances, the stem is affected by buckling which causes weakening.	Because the stem inside the container is backed up by the pressure of the presure medium, the strength of the stem improves as the extrusion process advances.
Materials of Sliding Parts	Abrasion resistant materials suitable for the inside of the container are limited by dimensions, etc.	It is easier to give abrasion resistance to the stem surface than the container's interior.

Both the piston and gland types use either unsupported area type or mitre-ring type packing. Each example of the piston type is shown in Fig. 21.

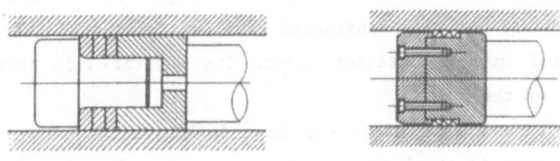

Fig. 21. Piston type seal.

(c) Mandrel-Part Seal

There are also two methods of sealing the sliding parts of the stem and mandrel, applying to a piston type and a gland type stem-part seal. Concerning the piston type stem-part seal, an additional packing is installed in the seal piston for the mandrel-part seal, as shown in Fig. 22.

On the other hand, with regard to the gland type stem-part seal, the packing is generally held in the seal groove on the mandrel, as shown in Fig. 25 (d), in order to prevent the decrease in the strength of the stem by grooving inside surface of stem bore. The type of packing is the same as the die-part seal shown in Fig. 19.

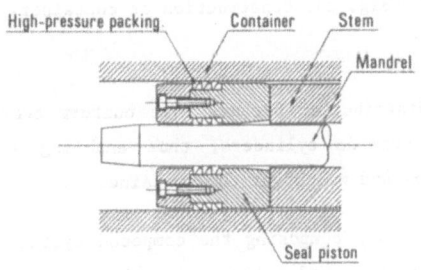

Fig. 22. Mandrel-Part Seal.

5.2 Container

The container is one of the most important components in the hydrostatic extrusion press. The container, stressed by repeated internal
pressure reaching a maximum 1.5 GPa in practical use, should be constructed with a view to extending its fatigue life. The operating
conditions should be fully considered when determining the structure and
materials used in the container, since its shorter life increases the
running cost of the press.

Various types of container are shown in Fig. 23.

The pressure in the hydrostatic extrusion process exceeds 1 GPa and
this requires that the structure of the container is that of a compound
cylinder, as shown in the illustrations. This structure has the following features.

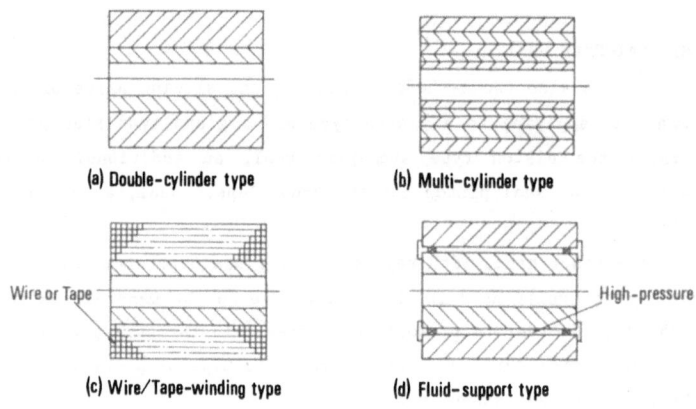

(a) Double-cylinder type   (b) Multi-cylinder type

(c) Wire/Tape-winding type   (d) Fluid-support type

Wire or Tape

High-pressure

Fig. 23. Construction of container.

(1) The stress distribution becomes more uniform over all the walls of
the inside and outside cylinders, thus enabling a reduction in the
outside dimensions and weight of the container.

(2) Each unit cylinder composing the compound cylinder is thin compared
with the monoblock one, and this permits the mass and size effect to be
reduced.

(3) Cracks are delayed from spreading over each cylinder by the boundary surfaces. This is good for safety.

(4) Materials of various characteristics are available in an effective combination.

(5) The fatigue life is improved by applying residual compressive stress to the inside cylinder of the container.

When using very high pressure of 1.5 GPa for cold hydrostatic extrusion, the container should have a special compound structure, such as the fluid-support type (Fig. 23 (d)), and the outside diameter and length of the container should be very large compared with the conventional ram-type extrusion presses.

On the other hand, the use of a pressure level of about 1 GPa for the hot hydrostatic extrusion process allows the container to withstand the pressure with the same shrinkage fit, which is composed of double or triple cylinders, as is used for the conventional ram-type extrusion presses.

The weight of a press for hot hydrostatic extrusion is largely reduced, since the press weight depends on the outside dimensions of the container, thus reducing the equipment cost.

For the innermost cylinder exposed to the severest stress conditions, maraging or tool steel is used. Hydrostatic extrusion, having no contact between the billet and the inside wall of the container, spares the container from being worn by the billet. Therefore, the problems of container wear will not occur if the sliding-part seal is of the gland type. The piston type causes friction between the packing and the container, but the extent is too small to affect the life of the container.

5.3 Stem

The absence of friction between the billet and the container in the hydrostatic extrusion process enables the use of infinitely long billets, but in practice, the length of the billet is limited according to the strength of the stem because of buckling.

The stem generally requires a length as long as the billet, but because compressive stress of a maximum of 1.5 GPa in the axial direction is applied to it, it should have sufficient strength to withstand buckling. The length-diameter ratio of the stem can be taken as around ten when using high-strength steel, and consequently, the length-diameter ratio of the billet can be taken as around 10. On the other hand, in the conventional ram-type extrusion process the length-diameter ratio of the billet is only 2 to 3.

As compared with this ratio, a very long billet is available, as a feature of the hydrostatic extrusion process.

Since the stem is subjected to very large compressive stress, as stated above, the structural materials should be selected for especially high strength and toughness to prevent scattering of debris in the case of breakage.

In the gland type seal, in which the high-pressure seal packing slides on the surface of the stem, the surface requires to be particularly smooth and wear resistant, obtained by grinding after hardening.

When using the piston type seal, it is necessary to keep the front end face correctly at right angles and to chamfer the corner to prevent crack generation, since the front end face pushes the piston. The neck of the stem, provided that the diameter changes considerably in the axial direction, should be designed to have a smooth curve to prevent stress concentration.

When extruding hollow products, it is necessary to provide a screwed hole on the top of the stem to mount the mandrel or make the stem hollow to let the mandrel pass.

Examples of stems for solid and hollow products are shown in Fig. 24.

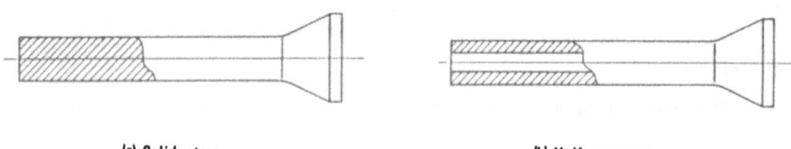

(a) Solid stem   (b) Hollow stem

Fig. 24. Stem.

5.4 Mandrel

To manufacture a hollow product, it is necessary to use a hollow billet, place the mandrel in the die hole, and then provide a clearance between the mandrel and the die. The required product is then extruded from this clearance.

Various types of mandrels are devised as shown in Fig. 25, and types (c) and (d) are the ones used practically.

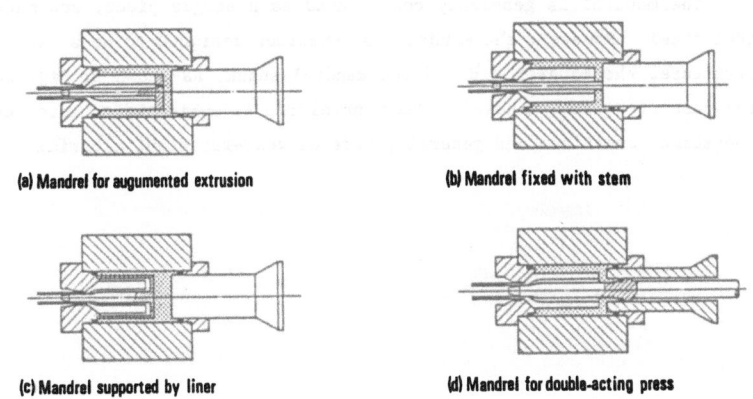

(a) Mandrel for augumented extrusion

(b) Mandrel fixed with stem

(c) Mandrel supported by liner

(d) Mandrel for double-acting press

Fig. 25. Arrangement of mandrel.

A mandrel of the fixed type keeps the position of the die and the mandrel unchanged during extrusion, and the forming part of the mandrel does not require more than the part near the die to extrude the hollow product into the specified shape. Therefore, the mandrel stated above can be used as a mandrel with a tip, and it is suitable for products of complex shape and for slender tubes. Further, the friction between the extruded product and the mandrel is limited to the tip part only, thus favorably reducing the friction force.

However, the need to fix the mandrel with a support liner imposes the limitation that no billet other than one which is shorter than the container length can be used.

The double-acting mandrel type is used only when the press is double-acting. The double-acting press, allowing the mandrel to move independently of the stem, provides an alternative method of moving the mandrel simultaneously with the stem, i.e. the travelling-mandrel type, or a method of fixing it to the die, i.e. the fixed mandrel type. In either case, non-use of the mandrel support liner which is mounted in the container, as shown in Fig. 25 (c), enables the application of long and large billets fully utilizing the container inside space. Therefore, from the viewpoint of hollow material production, this is the most efficient type.

The mandrel is generally constructed as a single piece, and made of tool steel. However, the mandrel is sometimes designed to be a two-piece structure, the mandrel tip and the mandrel shank, as shown in Fig. 26 in the case of a fixed mandrel. The mandrel tip is subjected to more severe operating conditions and generally made of wear-resistant material.

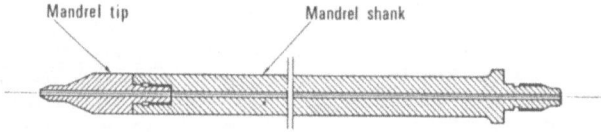

Fig. 26. Mandrel with tip.

When the billet temperature is over 800°C in the hot extrusion process, inside and outside cooling are required from the viewpoint of the strength of the mandrel. To cool the mandrel's interior, cooling water is fed through a hole machined at its center, as shown in Fig. 26 and to cool it further, water is sprayed on its exterior after completion of extruding.

5.5 Die

The die mounted on one end of the container is subjected to a large force while forming a billet into a product, and in the hydrostatic extrusion process, it should also act as a high-pressure seal.

The method of setting a die in a container is shown in Fig. 27.

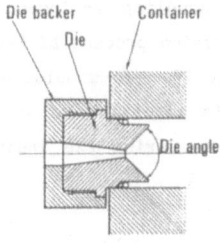

Fig. 27. Setting of die.

The approach of the die inlet is conical to facilitate the metal to flow during extrusion. In hydrostatic extrusion the friction between the billet and the die, being small at the conical part compared with conventional ram-type extrusion owing to the lubricating action of the pressure medium at high pressure, enables the conical angle to be adjusted to ensure that the billet is deformed most efficiently. The die angle depends on the kind and shape of the extruded material, but it is generally within the range of 90° to 120°.

The die, exposed to the high processing pressure from inside, is shaped so as to balance with the high pressure of the pressure medium applied to the outside wall of the die, as shown in the illustration.

The die is generally of tool steel, but it is necessary in the hot extrusion process, in particular, to extend its life by setting an insert of abrasion resistant material in the bearing part of the die, as shown in Fig. 28.

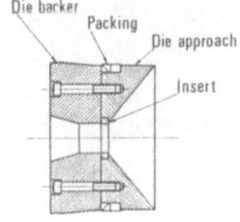

Fig. 28. Die for hot hydrostatic extrusion.

6. Auxiliary Equipment

The hydrostatic extrusion plant includes the billet-making equipment and the extruded product-handling equipment apart from the extrusion press, and in the hot extrusion process billet-heating equipment is also required, as shown in Fig. 29. To manufacture products efficiently in the extrusion plant, it is important to arrange the related equipment efficiently as well as considering the individual performance of each piece of equipment.

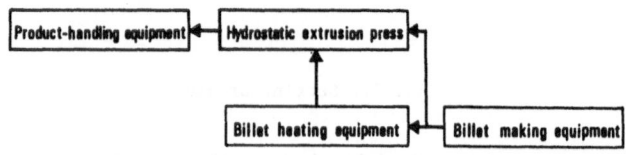

Fig. 29. Auxiliary equipment for a hydrostatic extrusion plant.

An example of arranging the equipment is shown in Fig. 30. Photo 1 shows a part of a modern hydrostatic extrusion plant which has been operating since 1983.

Photo 1. Hydrostatic Extrusion Plant for Copper Alloys.

handle products which are usually very long and easy to coil, like wire or slender tube, the coiler system is suitable, but to handle products which are relatively short or deformed by bending, e.g. shaped or large-section products, the runout table system should be adopted.

The runout table and its associated equipment are explained below in accordance with Fig. 30.

The products are first extruded forwards onto the runout table and cooled on the cooling table. Then, the products are subjected to crop-end cut and regular-size cut, collected in the stock yard and finally shipped, after deoiling, if required.

This process is almost the same as that for the conventional ram-type extrusion press.

7. Conclusions

Hydrostatic extrusion presses were considered to have a very pecul-iar structure at the initial stage of development, but subsequent adapta-tion has led to a similar structure to the conventional ram-type extru-sion presses as stated before. Furthermore, the cycle time in the hot hydrostatic extrusion process has been gradually shortened to a dry-cycle time of approximately 60 seconds excluding the extrusion time, and this has enabled the press to be used industrially. It is expected that an increase in the materials most suitable for the hydrostatic extrusion process will increase the use of this press.

In this paper, devices having almost the same structures as those of the conventional ram-type extrusion presses, such as a press driving unit, a billet supplying device, etc. are omitted because the length of this paper is limited. For a detailed description of the tools, the strength and the life of the container, etc., refer to Section 2.

REFERENCES

1 S. Yamamoto, T. Fujita and Y. Yamaguchi, U.S. Patent 3722245 (1973) "Hydrostatically operated extruding machine".

2 D. Green, "An experimental high speed machine for the practical exploitation of hydrostatic extrusion "J. Inst. Metals, 93-3 (1964/65) 65

3 H. Lundström, "Development in hydrostatic extrusion at ASEA" SME Technical Paper MF69-167 (1969)

SECTION 2
TOOLING FOR HYDROSTATIC EXTRUSION

Y. YAMAGUCHI

KOBE STEEL, LTD.

1. Introduction

The essential tools for hydrostatic extrusion are : (1) high pressure container, (2) stem, (3) die and mandrel. The container is subjected to very high cyclic pressures up to about 2.0 GPa, and a large tangential stress is imposed at the bore. In most cases, it is designed to be kept within elastic limits under service conditions, but usually suffers from fatigue failure. The stem is under a uniaxial compressive load, and its life is much longer than that of the container, but if it becomes buckled, catastrophic failure will occur. The die and the mandrel have considerably shorter lives than that of the container, since they are affected by large stresses and high temperatures as well as the wearing action of the materials extruded. The variable cost and safety in operation of hydrostatic extrusion presses are mostly dependent on the design considerations, including material selection and processing technique for these tools. The main concern of this section is, therefore, to describe the strategy of tool design which is necessary to extend tool life and to ensure the safe operation of the presses.

2. Materials for tools
1) Materials and their static mechanical properties

The primary consideration in tool design is that all components of a tool are practically elastic under service conditions. Large plastic deformation of a container is obviously undesirable from the standpoint of a bore sealing, and excessive deformation of a die or a mandrel is detrimental to the dimensional accuracy of extruded products. Consequently as the demand increases for higher working pressures, tool designers are resorting to higher strength materials.

Table 1 summarizes some of materials commonly used or having potential uses in hydrostatic extrusion tooling. A nickel-chromium-molybdenum steel, a typical high strength alloy steel, is usually specified having a yield

Table 1. Typical materials for hydrostatic extrusion tooling.

Material	Typical chemical composition (%)	Remarks
Ni-Cr-Mo steel	C 0.40, Si 0.20, Mn 0.80 Ni 1.80, Cr 0.80, Mo 0.20,	Forged, quenched & tempered. High strength & high toughness.
Medium C-5%Cr tool steel	C 0.40, Si 0.90, Mn 0.40 Cr 5.0 , Mo 1.2 , V 1.0	Forged, quenched & tempered. Very high strength & reasonable toughness.
High C-Cr tool steel	C 1.50, Si 0.30, Mn 0.30 Cr 12.0, Mo 1.00, V 0.30	Forged, quenched & tempered. High hardness & low toughness.
High speed steel	C 0.85, Si 0.30, Mn 0.30 Cr 4.0 , Mo 5.0 , V 2.0 W 6.0	Forged, quenched & tempered. High hardness at elevated temperature & low toughness.
18%Ni maraging steel	C<0.01, Ni 1.80, Mo 5.0 Co 7.50, Ti 0.80, Al 0.08	Forged, solution annealed & aged. Very high strength & reasonable toughness.
Cemented Carbide	Co 5~15%, WC bal.	Sintered. Very high hardness at high temperatures. High Young's Modulus.
Heat resistant alloy	Ni base: Cr 20, Mo 4, Co 14, Ti 3, Ni bal. Co base:Fe 2.5, C 0.9, Cr 30, W 4, Co bal.	Ni base; forged, Co base; cast. High strength at high temperatures.

strength of about 1.2 GPa, so as to give a sufficient ductility and toughness. This kind of steel is widely used for high pressure equipment, for instance, the outer ring of a container. A 5% chromium tool steel has a yield strength of over 1.4 GPa and a reasonable ductility. This steel can be used for the inner liner of a container and the stem under relatively low working pressures. 18% nickel maraging steels have yield strengths from 1.5 GPa to 2.5 GPa. As these steels are reasonably ductile and tough, they are favored for very high pressure uses. Tool steels with a high carbon content are high in hardness and relatively low in ductility and toughness. The high hardness is coupled to high compressive yield strength. Cemented carbide, which conventionally contains tungsten carbide as the hard constituent and cobalt as the binder phase, has a very high hardness up to 1800 VHN but a poor ductility. These hard materials may be used for the inserts of dies and for the tips of mandrels. For high temperature uses, some heat-resisting alloys are used in combination with other kinds of materials such as tool steels and cemented carbide.

Fracture toughness of a material is the important factor in the safe design of the tools, especially the high pressure container. Figure 1 shows the fracture toughness K_{IC} for several materials [1,2,3] . The values of K_{IC} for steels rapidly fall with increase in yield strength. The typical values of K_{IC} for harder materials are as follows;

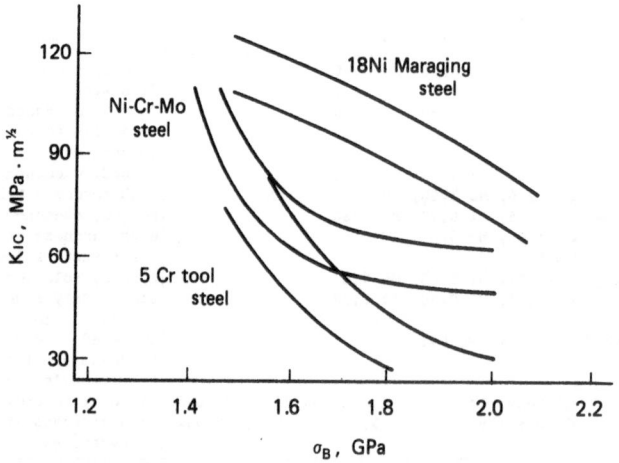

Fig. 1. Fracture toughness K_{IC} vs. ultimate tensile strength σ_B.

High speed steel	$15 \sim 25$ $MPa.m^{\frac{1}{2}}$
Cemented carbide	$9 \sim 20$ $MPa.m^{\frac{1}{2}}$
Ceramics	3 $MPa.m^{\frac{1}{2}}$

2) Fatigue Strength

Under cyclic loading conditions, extension of fatigue life is the main concern in tool design, where the first step is the optimum selection of materials. For the purpose of obtaining some basic information about fatigue properties of tool materials, the present author carried out fatigue and static tests on six different high-strength steels [3]. The chemical composition and the static mechanical properties of these steels are summarized in Tables 2 and 3.

Table 2. Chemical composition of steels tested.

Steel	Chemical composition												
	C	Si	Mn	P	S	Ni	Cr	Mo	V	W	Co	Ti	Al
Ni-Cr-Mo steel[1]	.40	.20	.75	.015	.015	1.80	.80	.20	–	–	–	–	.030
18%Ni-maraging steel[2]	.01	.06	.04	.007	.005	17.80	–	4.86	–	–	7.50	.78	.075
18%Ni-maraging steel[3]	.01	.10	.10	.006	.006	17.50	–	3.75	–	–	12.50	1.70	.150
5%Cr tool steel[4]	.39	.92	.42	.020	.023	.10	4.91	1.12	.84	–	–	–	–
High C-Cr tool steel[5]	1.55	.36	.38	.023	.008	.11	11.54	.83	.26	–	–	–	–
High speed steel[6]	.85	.30	.30	–	–	–	4.00	5.00	2.00	6.00	–	–	–

1) JIS SNCM439, 2) $\sigma_{0.2}$ 1.7GPa class, 3) $\sigma_{0.2}$ 2.4GPa class, 4) JIS SKD61
5) JIS SKD11, 6) JIS SKH51

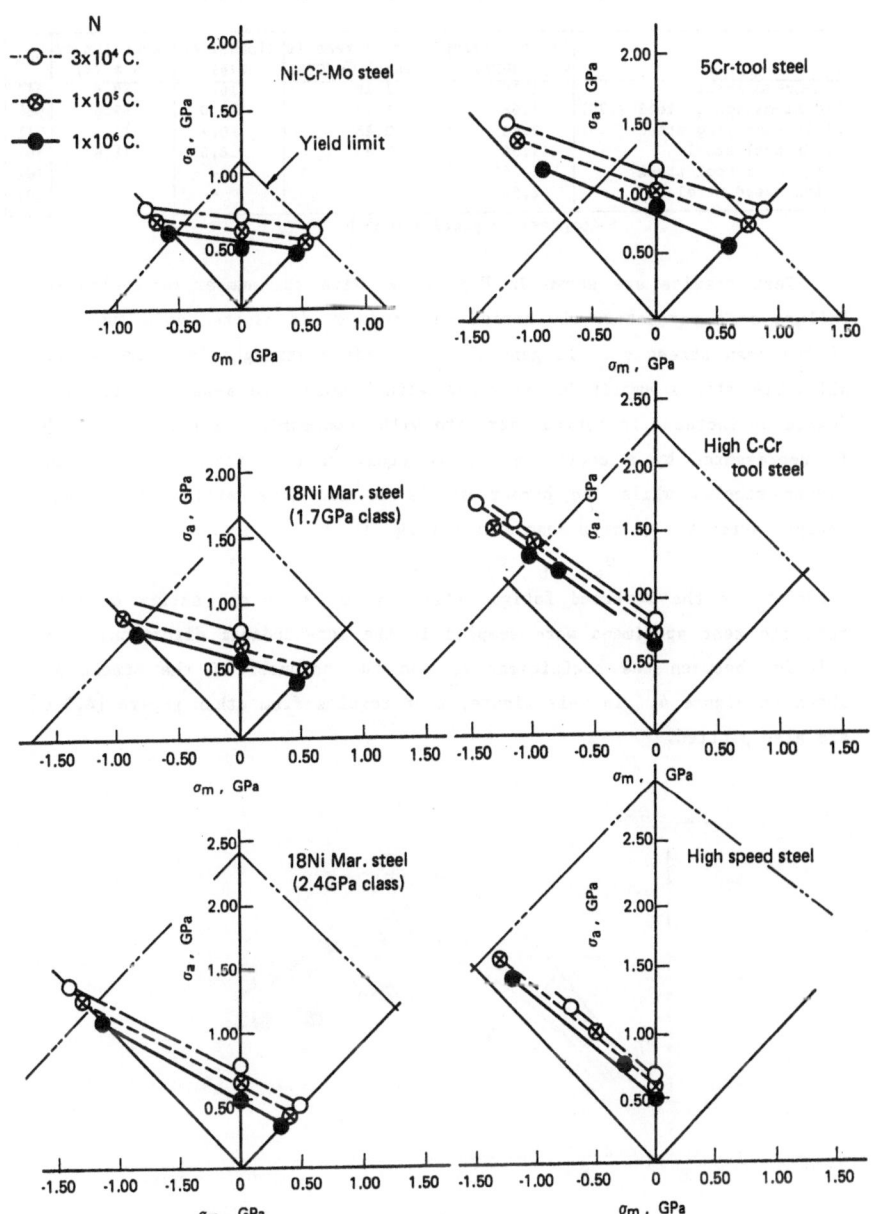

Fig. 2. Effects of mean stress σ_m on the uniaxial fatigue strength of steels in terms of stress amplitude σ_a.

Table 3. Mechanical properties of steels tested.

	Yield stress $\sigma_{0.2}$ (GPa)	Ultimate tensile strength σ_B (GPa)	Elongation ε (%)	Reduction of area ϕ (%)	VHN
Ni-Cr-Mo steel	1.10	1.19	16.5	59.4	360
18%Ni-maraging steel (170)	1.68	1.73	14.0	59.8	555
18%Ni-maraging steel (240)	2.40	2.45	10.9	41.3	711
5%Cr tool steel	1.50	1.80	14.0	41.6	561
High C-Cr tool steel	2.35*	-	-		762
High speed steel	2.94*	-	-		804

* Compressive yield strength

Test results are shown in Figure 2. With the number of cycles to failure as a parameter, the stress amplitude σ_a is plotted as a function of the mean stress σ_m. In general, the fatigue strength in terms of the allowable stress amplitude increases with compressive mean stress. The degree of increase in fatigue strength with compressive mean stress, which is represented by a coefficient λ in equation (1), becomes larger for harder steels, while the harder steels do not necessarily have higher fatigue strengths at zero mean stress (Fig. 3).

$$\sigma_a + \lambda_L \cdot \sigma_m = \sigma_{w \cdot L} \tag{1}$$

where σ_w is the reversed fatigue strength ($\sigma_m = 0$) and suffix L shows that the test specimens were sampled in the longitudinal direction. The relation between the coefficient λ_L and the hardness of the steels is shown in Figure 4. In this figure, test results from other papers [4,5,6] are also plotted.

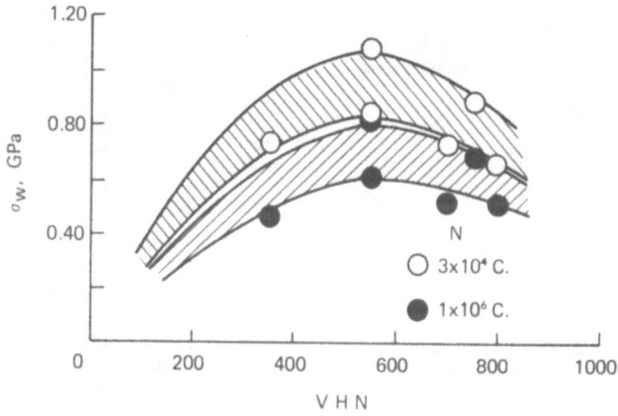

Fig. 3. Reversed fatigue strength σ_w vs. Vickers hardness number VHN.

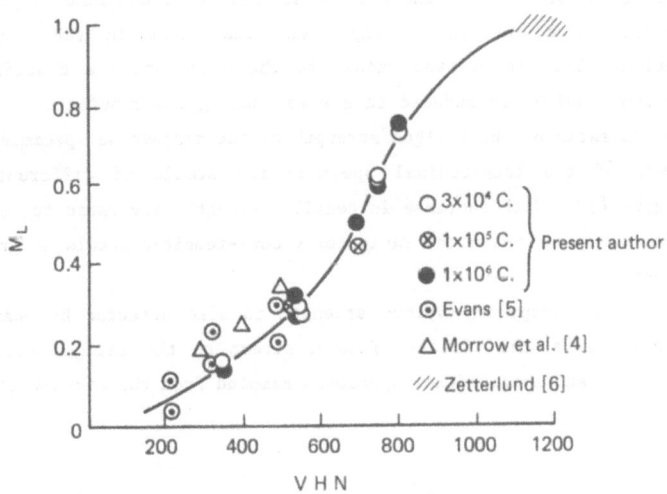

Fig. 4. Dependence of fatigue strength on mean stress for materials with different degrees of hardness.

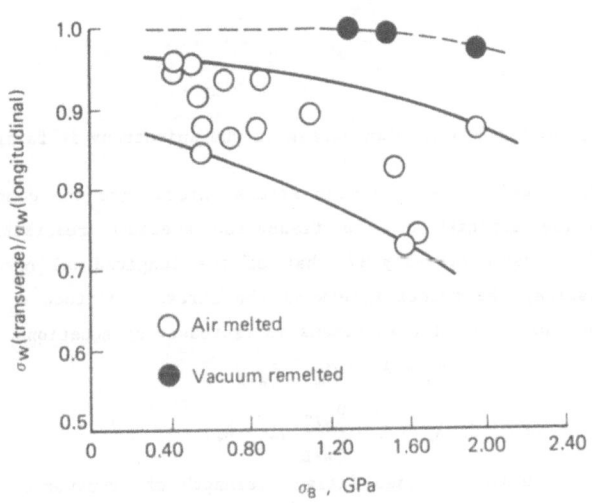

Fig. 5. Anisotropy in fatigue strength.

Another important feature of the fatigue strength of high-strength steels is the anisotropy; the fatigue strength of a specimen sampled in the transverse direction is usually lower than that in the longitudinal direction. This is related mainly to the shape and the distribution of inclusions, which are induced in a steel during its manufacture. Figure 5 shows the ratio of the fatigue strength of the transverse specimen compared to that of the longitudinal specimen for steels of different tensile strengths [3]. With increase in tensile strength, the ratio for air-melted steels decreases, while the ratio for vacuum-remelted steels is practically constant.

The anisotropy in fatigue strength is also affected by mean stress. Figure 6 shows the effects of mean stress on the fatigue strengths of transverse and longitudinal specimens sampled from the air-melted 5% chro-

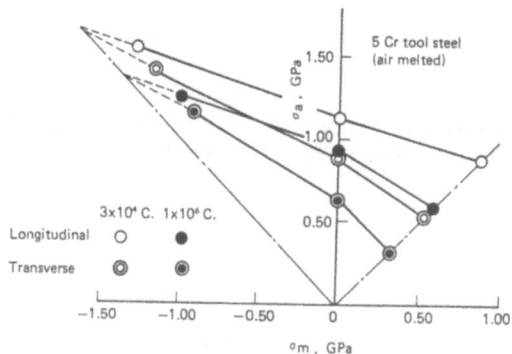

Fig. 6. Effects of mean stress on the anisotropy in fatigue strength.

mium tool steel. When the mean stress enters into the compressive side, the fatigue strength of the transverse specimen remarkably increases, approaching asymptotically to that of the longitudinal one. From these test results, the relation between the stress amplitude σ_a and the mean stress σ_m for transverse specimens is expressed by equations

$$\sigma_a + \lambda_T \cdot \sigma_m = \sigma_{w \cdot T} \tag{2}$$

$$\lambda_T = 1 - \frac{\sigma_{w \cdot T}}{\sigma_{w \cdot L}} (1 - \lambda_L)$$

where $\sigma_{w \cdot T}$ is the reversed fatigue strength of transverse specimens. In equation (2), the ratio $\sigma_{w \cdot T}/\sigma_{w \cdot L}$ can be roughly estimated from Figure 5, and the coefficient λ_L from Figure 4.

These test results indicate the importance of minimizing or avoiding tensile stresses when using harder materials such as tool steels.

The fatigue behaviour of high-strength steels mentioned above is as follows:

Fatigue is a three-stage process involving crack initiation, crack propagation, and final fracture. When there are no significant defects, crack initiation is mainly governed by the shear strain range. Therefore, a harder steel will have a longer crack initiation stage. During this stage, however, most of cracks initiate at inclusions. Nonmetallic inclusions in a longitudinal specimen are elongated and arranged in parallel with applied stress, while those in a transverse specimen are perpendicular to the applied stress. Therefore, the inclusions in the transverse specimen act as stress concentrators, shortening the crack initiation stage in the fatigue life. If these inclusions are minimized during the manufacturing process of steels, for instance by vacuum remelting, the anisotropy in the fatigue strength will be eliminated. At the crack propagation stage, a harder steel will be unsuitable, because the lower fracture toughness associated with the higher hardness reduces the critical crack depth to the final fracture. The inclusions in the transverse specimen will further aggravate the problem. Although the transverse specimen is practically similar to the longitudinal one in respect of the crack propagation characteristics (Fig. 7), the range of

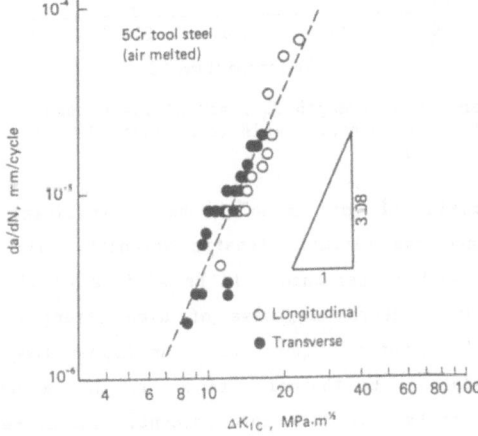

Fig. 7. Fatigue crack propagation rate da/dN as a function of the range of stress intensity factor in longitudinal and transverse specimens.

stress intensity factor for the crack initiated along a elongated inclusion in the transverse specimen will be much larger than that in the longitudinal specimen. On the other hand, as the crack propagation rate is effectively suppressed at the compressive side, the fatigue strength of a harder steel, especially a transverse specimen, is remarkably increased at compressive mean stresses.

3) Mechanical properties at elevated temperatures

At moderate temperatures, the yield strength is the property that primarily limits the use of a tool material. For high strength steels, these temperatures are usually below about 350°C. Figure 8 shows the change in tensile properties of a nickel-chromium-molybdenum steel with the test temperature up to 350°C [7]. In those temperature ranges, the strain

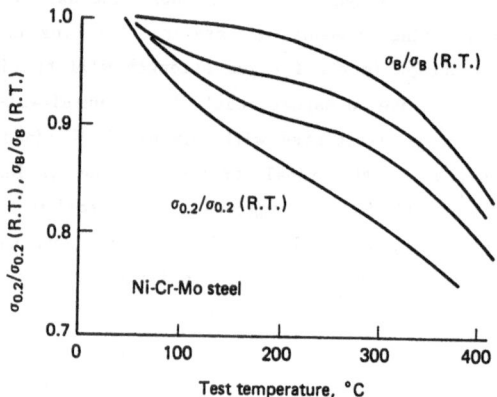

Fig. 8. Changes in yield strength $\sigma_{0.2}$ and ultimate tensile strength σ_B with test temperature (each is normalized by the value at room temperature).

rate and the duration of loading do not have significant effects on the yield strength and the ultimate tensile strength. The change in the fatigue strength with temperature will be similar to that in the yield strength, and the fracture toughness of high strength steels usually increases with the service temperatures. At higher temperatures, creep strength and resistance to thermal fatigue are the important properties that tool designers have to take into account. Therefore, even for hot hydrostatic extrusion, it is recommended that the temperature at the bore of a container is kept lower than 350°C.

Die inserts and mandrel tips may be subjected to temperatures above 500°C for hot hydrostatic extrusion. Unfortunately, stress and thermal histories of these tools in actual extrusion cycles are so complicated that it is difficult to evaluate their durability on the basis of laboratory data. Currently, therefore, material selection is a matter of trial and error.

3. High pressure container

High pressure containers are the most essential tools in hydrostatic extrusion processes. From the standpoint of practical operations, the design considerations are as follows:

(1) All parts of the container must be elastic under service conditions.

(2) The liner of the container has to have a fatigue life in the order of 10^4 cycles.

(3) The container must be proof against catastrophic failure.

(4) The liner can be easily replaced.

1) Static design and prestressing

Containers are usually designed to be elastic to ensure bore sealing. Experiences have also shown that cyclic plastic deformation at the container's bore shortens its life.

For an open-ended simple monobloc cylinder, the stress state at the inner bore is

$$\sigma_\theta = \frac{K^2 + 1}{K^2 - 1} P_i \tag{3}$$

$$\sigma_r = - P_i, \quad \sigma_z = 0$$

where P_i is the internal pressure, σ_θ the tangential stress, σ_r the radial stress, σ_z the axial stress, and K the diameter ratio. According to the Tresca criterion, the pressure P_y at which the first yield occurs at the bore is

$$P_y = \frac{\sigma_y}{2} \left(1 - \frac{1}{K^2} \right) \tag{4}$$

where σ_y is the yield strength. For a large diameter ratio, the pressure P_y cannot exceed the pressure of half the yield strength for a monobloc cylinder.

When the pressure exceeds P_y, plastic deformation will spread through the wall of the cylinder. If the pressure is removed, a residual compressive stress (tangential) is induced at the bore. Thus, it will behave elastically under the subsequent pressure cycles up to the original excess pressure. This overstraining, "autofrettage", produces beneficial residual stresses and increases the pressure P_y. The autofrettage pressure P_a to cause yield at the radius C_p and the residual stresses can be calculated by equations (5), (6), and (7), provided that the material is elastic ideally plastic.

$$P_a = \frac{\sigma_y}{2} \left\{ 1 + 2 \ln \frac{C_p}{r_0} - \left(\frac{C_p}{r_1}\right)^2 \right\} \tag{5}$$

$$r_0 \leq r \leq C_p$$

$$\sigma_\theta' = \frac{\sigma_y}{2} \left\{ \left(\frac{C_p}{r_1}\right)^2 - 2 \ln \frac{C_p}{r} + 1 \right\} - \frac{\left(\frac{r_1}{r}\right)^2 + 1}{K^2 - 1} P_a$$

$$\sigma_r' = \frac{\sigma_y}{2} \left\{ \left(\frac{C_p}{r_1}\right)^2 - 2 \ln \frac{C_p}{r} - 1 \right\} + \frac{\left(\frac{r_1}{r}\right)^2 - 1}{K^2 - 1} P_a \tag{6}$$

$$C_p \leq r \leq r_i$$

$$\sigma_\theta' = \frac{\sigma_y}{2} \left(\frac{C_p}{r_1}\right)^2 \left\{ 1 + \left(\frac{r_1}{r}\right)^2 \right\} - \frac{\left(\frac{r_1}{r}\right)^2 + 1}{K^2 - 1} P_a$$

$$\sigma_r' = \frac{\sigma_y}{2} \left(\frac{C_p}{r_1}\right)^2 \left\{ 1 - \left(\frac{r_1}{r}\right)^2 \right\} + \frac{\left(\frac{r_1}{r}\right)^2 - 1}{K^2 - 1} P_a \tag{7}$$

where r_0 is the inner radius, r_1 the outer radius, σ_θ' the residual tangential stress, and σ_r' the residual radial stress. From equation (6), the condition that the cylinder material behaves completely elastically during unloading is

$$\frac{K^2 + 1}{K^2 - 1} \left\{ 1 - \left(\frac{C_p}{r_1}\right)^2 + 2 \ln \frac{C_p}{r_0} \right\} - \left\{ 1 + \left(\frac{C_p}{r_1}\right)^2 - 2 \ln \frac{C_p}{r_0} \right\} \leq 2 . \tag{8}$$

If a cylinder is overstrained through the whole wall, equation (8) gives

$$\frac{K^2}{K^2 - 1} \ln K \leq 1 \text{ and } K \leq 2.218 .$$

For a monobloc cylinder with the diameter ratio of 2.218, the bore will start to yield at the pressure 0.46 σ_y, while the bore of a cylinder autofrettaged through the whole wall will not yield until pressure exceeds 0.80 σ_y. Autofrettage is a simple method to produce beneficial residual stresses, but the residual stresses are sometimes unstable under cyclic loading.

Another method of achieving a more advantageous residual stress is the compound shrinkage by thermal means or by force fitting. In this method, several individual layers are assembled with appropriate degrees of interference. When a liner and an external supporting cylinder are shrink-fitted with the radial interference, the tangential residual stress at the bore of the liner $\sigma_\theta{}^*$ is calculated by equation (9), provided that each has the same Young's modulus.

$$\sigma_\theta{}^* = -2E \cdot \frac{\delta}{r_1} \cdot \frac{k_1^2}{k_1^2 - 1} \cdot \frac{1}{\dfrac{k_1^2 + 1}{k_1^2 - 1} + \dfrac{k_2^2 + 1}{k_2^2 - 1}} \tag{9}$$

where E is Young's modulus, δ the amount of radial interference, r_1 the outside radius of the liner, k_1 the diameter ratio of the liner, and k_2 the diameter ratio of the external cylinder.

When the overall diameter ratio of a container is fixed and the amount of interference is constant, the residual compressive stress induced at the inner bore increases with a decrease in the diameter ratio of the liner (Fig. 9).

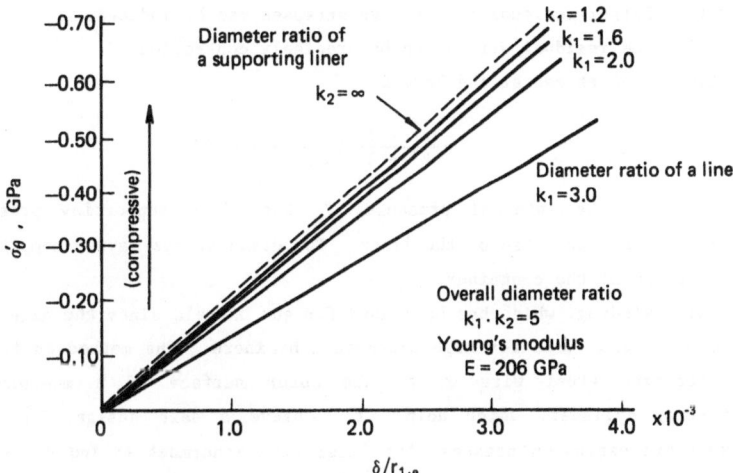

Fig. 9. Residual tangential stress $\sigma_0{}'$ at the bore of a liner with different interferences.

The amount of the interference given by the thermal means is

$$\frac{\delta}{r_1} = \alpha \, (T_s - T_R) - \frac{\delta_0}{r_1} \tag{10}$$

where α is the coefficient of linear thermal expansion, T_s and T_R the heating temperature of the external cylinder and room temperature, r_1 the outside radius of a liner, and δ_0 the clearance for fitting operations. The temperature difference, $T_s - T_R$, is usually limited to within $350°C$, and the value of δ/r_1 is $3/1000$ or so.

In force fitting, the force requirement is estimated by equation (11)

$$F \simeq \frac{\pi E \cdot \dfrac{\delta}{r_1}}{\dfrac{k_1^2 + 1}{k_1^2 - 1} + \dfrac{k_2^2 + 1}{k_2^2 - 1}} \ (\ 2r_1 + L \tan\theta \ ) \ L \ (\ \mu + \tan\theta \ ) \tag{11}$$

where μ is the coefficient of friction, 2θ the taper angle of mating surfaces, and L the length of the liner. The amount of interference is limited by the press force or compressive yielding of the liner.

An alternative design for reducing the tangential tensile stress at the bore is the static fluid supporting. The advantages of this design are as follows:

(1) The liner can be easily replaced after failure.

(2) Larger residual compressive stresses can be induced.

(3) The residual stress can be precisely controlled.

The tangential stress at the bore is

$$\sigma_\theta = \frac{k^2 + 1}{k^2 - 1} \ P_i - \frac{2k^2}{k^2 - 1} \ P_e \tag{12}$$

where P_i is the internal pressure, P_e the fluid supporting pressure, and k the diameter ratio of the liner. The disadvantage is the complicated construction of the container.

Wire winding, which has been used for gun barrels since the nineteenth century, is also used for high pressure containers. The method is to wind high-strength steel wire on to the outer surface of a monobloc or multi-ring cylinder, which helps to achieve a safe design [8]. For hydrostatic extrusion presses, the liner, the innermost cylinder, is most likely to fail owing to fatigue, troubles in sealing, or large scratches incurred during operation, and, having a limited life, its replacement has to be simplified. A liner directly wire wound on its outer surface could not be replaced, therefore this method can be used for strengthening the external supporting cylinder with which the liner is shrink-fitted.

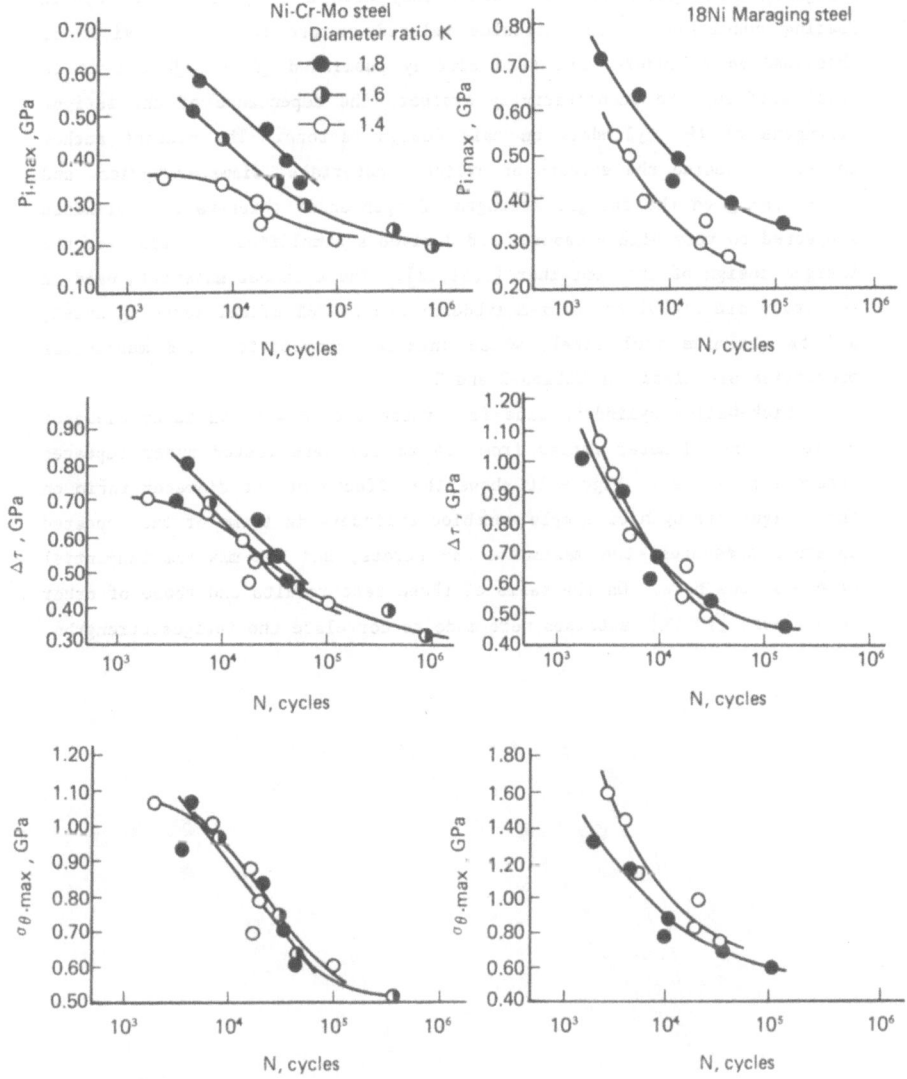

Fig. 10. Effects of diameter ratio on the fatigue strengths of simple
monobloc cylinders in terms of max. internal pressure $P_{i.max}$,
shear stress range $\Delta\tau$, and max. tangential stress $\sigma_{\theta.max}$.

2) Fatigue design

In general, there are no significant problems in designing containers for static requirements. On the other hand, consideration must be given in designing these containers to ensure they have a long life under cyclic loading conditions. Many articles related to the fatigue behaviour of thick-walled cylinders have been already published [9 ∿ 20], but it is still difficult to quantitatively evaluate the dependence of the fatigue strengths of the cylinders on main design factors. The present author therefore studied the effects of cylinder materials, diameter ratios, and prestressing on the fatigue strength of open-ended thick-walled cylinders subjected to very high pressure, and derived a simplified criterion for the fatigue design of the container [3,21,22]. The cylinder materials used in the study are nickel-chromium-molybdenum steel, 18% nickel maraging steel, and 5% chromium tool steel, whose chemical compositions and mechanical properties are listed in Tables 2 and 3.

Thick-walled cylinders made from these steels with an inner diameter of 20 mm and diameter ratios from 1.4 to 3.2 were tested under repeated internal pressures. Figure 10 shows the effects of the diameter ratio on the fatigue strength of simple monobloc cylinders in terms of the repeated internal pressures, the maximum shear stress, and the maximum tangential stress at the bore. On the basis of these test results and those of other researchers [10,13], attempts were made to correlate the fatigue strengths

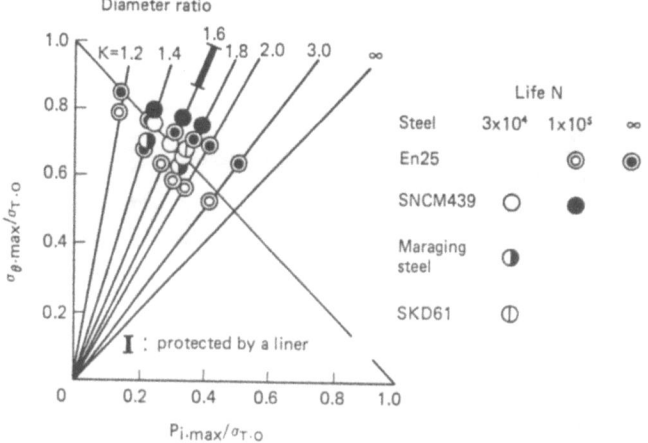

Fig. 11. Correlation between the fatigue strengths of simple monobloc cylinders and those of uniaxial specimens.

of the cylinders with those of uniaxial test specimens (Fig. 11). In this figure, the repeated internal pressures $P_{i.max}$ and the maximum tangential stresses at the bore $\sigma_{\theta.max}$ are normalized by the repeated tension fatigue strengths of the transverse specimens $\sigma_{T.o}$, which are estimated by using the relation (2) when it is not determined by experiments. From Figure 11 an empirical relation is derived as

$$\sigma_{\theta.max} + P_{i.max} = \frac{2K^2}{K^2 - 1} P_{i.max} = \sigma_{T.o}. \tag{13}$$

Theoretical consideration based on fracture mechanics also leads to the similar relation. The rate of fatigue crack propagation da/dN is described by the well-known equation

$$\frac{da}{dN} = A \cdot \Delta K_I^{\,m} \tag{14}$$

where ΔK_I is the range of the stress intensity factor, and A and m are material constants. The stress intensity factor K_I for a shallow semi-elliptical crack at the bore of a thick-walled cylinder subjected to internal pressure is given by Underwood [23] as approximately

$$K_I = \frac{1.1}{\phi} (\sigma_{\theta.max} + P_{i.max}) \cdot (\pi a)^{1/2} \tag{15}$$

where ϕ is the elliptical integral function, $P_{i.max}$ the maximum internal pressure, $\sigma_{\theta.max}$ the maximum tangential stress at the bore, and a the crack depth. If the crack depth is small in comparison with the wall thickness and the value of ϕ remains constant during crack propagation, $\sigma_{\theta.max} + P_{i.max}$, the range of the maximum difference of principal stresses, at the crack propagation lift N is given as

$$\sigma_{\theta.max} + P_{i.max} = \frac{\phi}{1.1\pi^{1/2}} \left\{ \frac{1}{AN} \int_{a_i}^{a_f} a^{-\frac{m}{2}} da \right\}^{\frac{1}{m}} = \text{const.} \tag{16}$$

where a_i is the initial crack depth and a_f is the final crack depth. The fatigue strength of a cylinder with a diameter ratio of 1.0 may be regarded as the repeated tension fatigue strength of a transverse specimen of the cylinder material, provided that the effects of the curvature, the stress gradient, and the chemical attack of a pressure medium are neglected. Consequently, equation (16) will be identical to the relation (13).

The physical meaning of the relation (13) was confirmed by a fatigue test of a low alloy steel cylinder with a thin copper alloy liner. In this

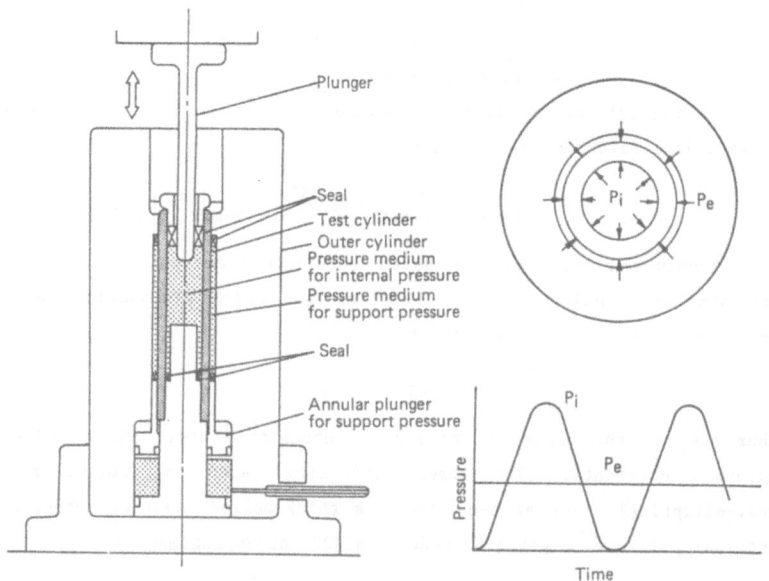

Fig. 12. Schematic illustration of fatigue test of static fluid supported cylinders.

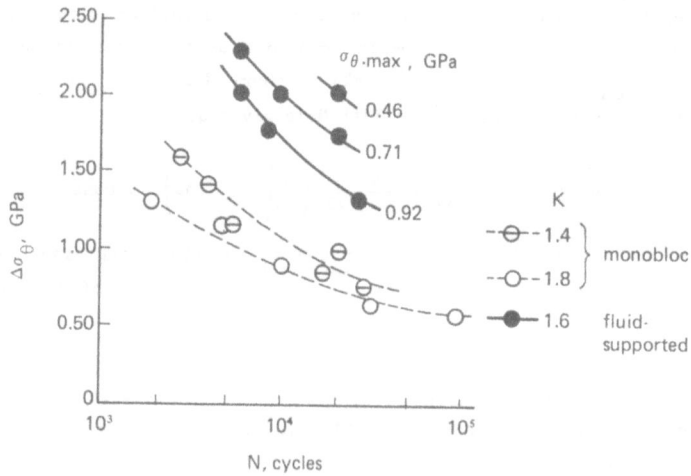

Fig. 13. Typical results of fatigue tests on fluid supported cylinders (material: 18% Ni maraging steel).

test, the bore of the cylinder was completely protected against the pressure medium by the liner. The result is also shown in Figure 11. In terms of maximum tangential stress at the bore of the steel cylinder, the fatigue strength is considerably increased by the liner, being apparently coincident with $\sigma_{T.0}$. These results strongly suggest that the detrimental effect of pressure media such as oils is mainly due to their penetration into cracks in the bore.

Subsequently the effects of prestressing were studied by using test cylinders supported by static fluid pressures (Fig. 12). Typical results are shown in Figure 13. The effects of mean stress on the fatigue strengths of cylinders and of uniaxial specimens are summarized in Figure 14. In this figure, the fatigue strengths of cylinders are expressed in terms of tangential stress at the bore, because most of the fatigue life of a cylinder will be spent for the crack propagation near the bore. It is noteworthy that the fatigue strength of a cylinder considerably increases and asymptotically approaches that of a uniaxial specimen when the mean tangential stress is decreased by prestressing. These results also suggest that the effects of a pressure medium can be almost eliminated with a large compressive mean tangential stress.

Under repeated internal pressures, fatigue cracks in simple monobloc cylinders usually propagate in a radial direction until they reach the

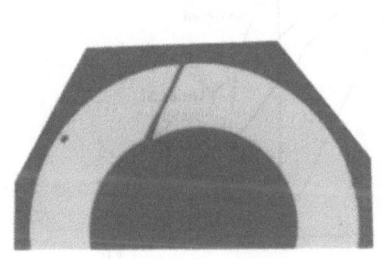
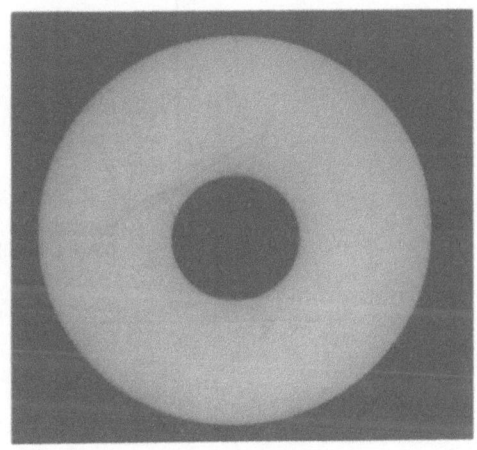

(a) Diameter ratio K=1.6   (b)   K=3.2

Photo. 1. Fatigue cracks of fluid-supported cylinders made from 5% Cr tool
steel.

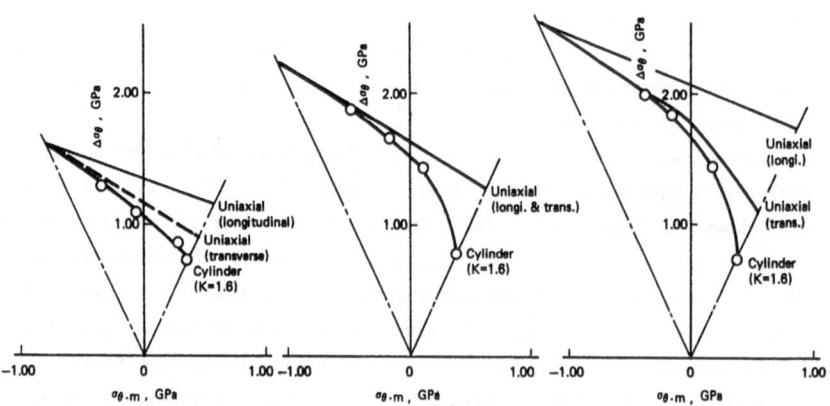

(a) Ni-Cr-Mo steel   (b) 18% Ni maraging steel (c) 5% Cr tool steel

Fig. 14. Effects of mean stress on the fatigue strengths of cylinders in
terms of tangential stress at their bores (cycles to failure: 3
x 10^4 cycles).

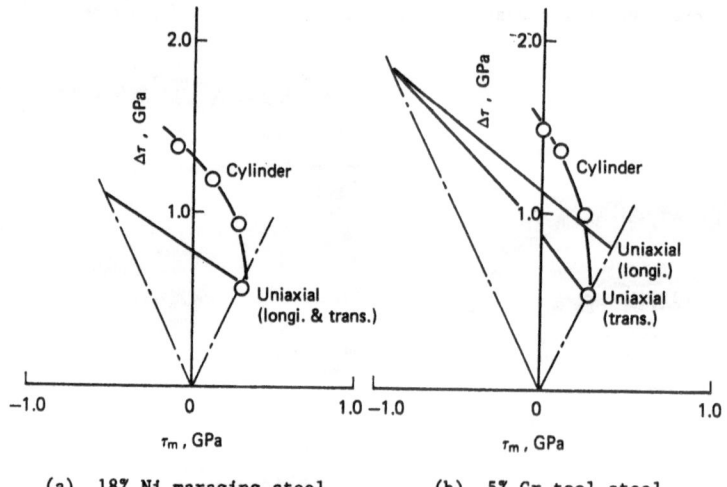

(a) 18% Ni maraging steel   (b) 5% Cr tool steel

Fig. 15. Effects of mean stress on the fatigue strengths of cylinders in
terms of shear stress range at their bores (cycles to failure:
3 x 10^4 cycles).

critical crack depths governed by the fracture toughness K_{IC} of the cylinder materials. On the other hand, it was observed that fatigue cracks deviated from radial direction and propagated through the whole-wall thickness at a large compressive mean stress (Photo.1). This phenomenon was discussed on the basis of the fracture mechanics [3,22].

It is generally said that the fatigue strength of a material under triaxial stress conditions will be determined by the maximum shear stress at the bore. In Figure 15 the fatigue strengths of the cylinders are compared with those of the uniaxial test specimens in terms of the maximum shear stress. The effects of prestressing on the fatigue strength of a cylinder are apparently underestimated by the shear stress criterion. Therefore, another simple criterion for fatigue design of very high pressure containers with finite lives was derived [21]. Figure 16 is a schematic diagram for the fatigue design of a cylinder in terms of tangential stress at the bore. The line AB indicates the fatigue strength of uniaxial specimens transversely sampled from a cylinder material. The stress conditions along the line OA are for repeated tensions (uniaxial)

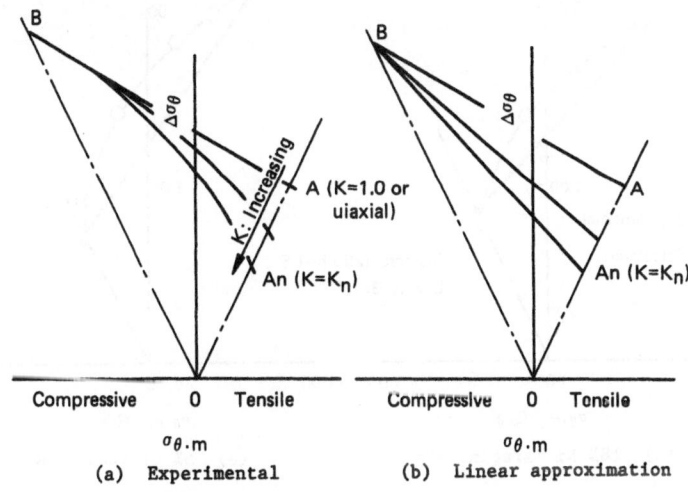

Fig. 16. Schematic diagram for fatigue design of containers.

or for repeated internal pressures (cylinder). In these conditions, the fatigue strength of a cylinder can be evaluated by the relation (13), while the fatigue of the cylinder on the condition $\sigma_{\theta.max} = 0$ is assumed to be equal to the repeated compression fatigue strength of the uniaxial specimen at the point B. In the region of $|\sigma_{\theta.m}| < 1/2 \, \Delta\sigma_\theta$ the straight line A_nB will give a conservative fatigue strength of a cylinder with a diameter ratio K_n at various mean tangential stresses. Under the condition of $|\sigma_{\theta.m}| \leq 1/2 \, \Delta\sigma_\theta$, the relationship is formulated as follows;

$$\Delta\sigma_\theta = \frac{2(\alpha - \beta)}{\alpha + \beta} \sigma_{\theta.m} + \frac{2\alpha\beta}{\alpha + \beta} \sigma_{T\cdot O} \tag{17}$$

where $\alpha = \dfrac{K^2 + 1}{2K^2}$, $\beta = \dfrac{1 + \lambda}{1 - \lambda}$, and λ is defined in equation (1) or (2). The stress condition for a cylinder to be elastic during pressure cycling is the prerequisite for this design criterion. The fatigue strengths predicted by (17) are well fitted with the test results (Fig. 17). Thus,

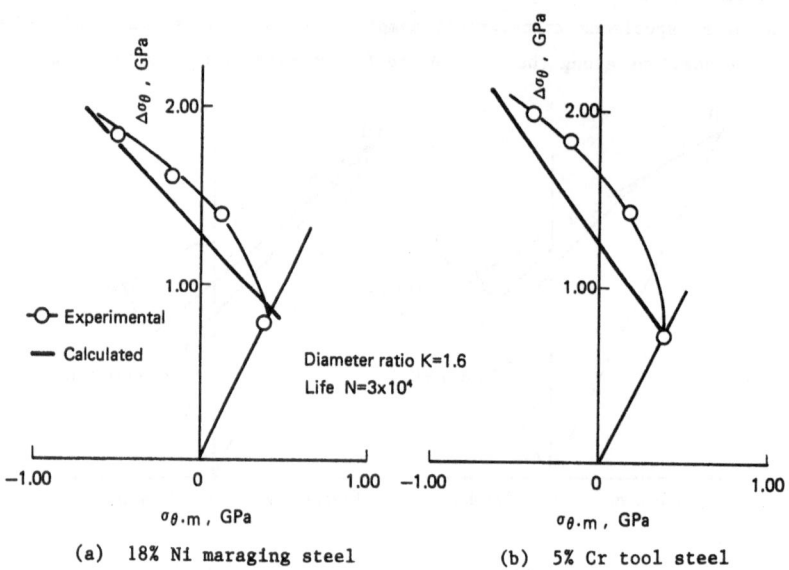

(a) 18% Ni maraging steel   (b) 5% Cr tool steel

Fig. 17. Comparison between calculated fatigue strengths and experimental values for cylinders subjected to internal pressures.

compressive mean tangential stresses by prestressing are essential for designing a very high pressure container with a long life, but several problems associated with prestressing have to be taken into account.

Autofrettage, by which a large residual compressive stress is easily induced, is a common method of prestressing. It is noticed, however, that the residual stress initially induced by this method cannot necessarily be maintained during service [21]. Figure 18 shows the changes in the residual tangential stress at the bore of autofrettaged cylinders during pressure cycling. The relaxation of the residual stress depends on the operating pressures, the number of pressure cycles, and the cylinder materials. The residual stress is rapidly relaxed at the early stage of pressure cycling and flattens out to a lower level after pressure cycles of about 10 percent of the total life. The relaxation rate increases at higher operating pressures. This tendency is more noticeable when the cylinder is made from the maraging steel, which is of cyclic strain softening. In evaluating the fatigue strength of an autofrettaged cylinder, it is recommended that the settled residual stress after the relaxation is taken into account. The residual stress induced by shrink fitting or fluid supporting will be firmly maintained in service.

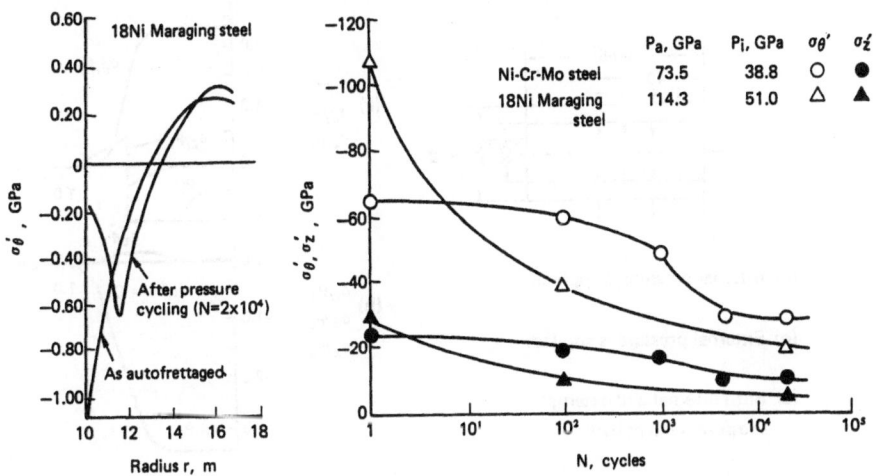

Fig. 18. Changes in residual stresses of autofrettaged cylinders during pressure cycling (inner diameter: 20 mm, diameter ratio: 1.8).

Another problem is the circumferential cracking at their sealing positions of cylinders made from hard and brittle materials such as high carbon tool steels. Photo. 2 shows this kind of cracking. Even when a cylinder is strengthened in the tangential direction by prestressing, axial

 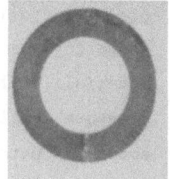

Photo. 2. Circumferential cracking of a cylinder at the sealing position.

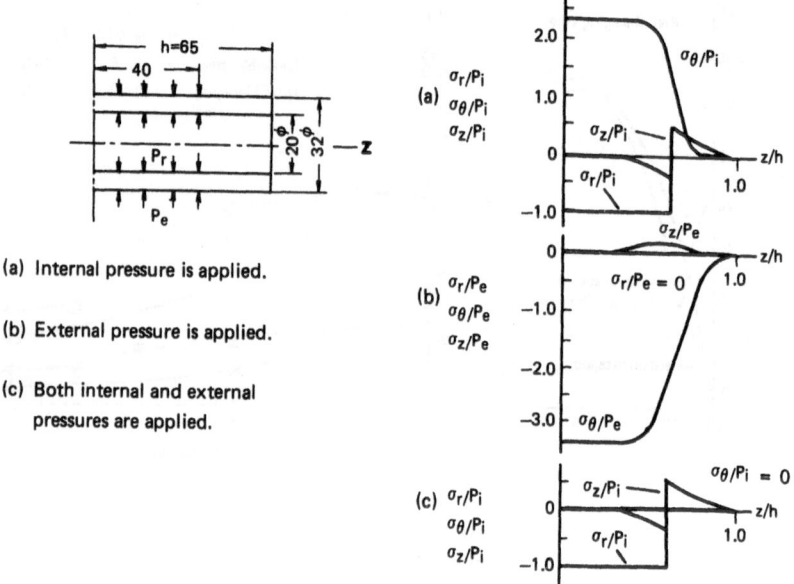

(a) Internal pressure is applied.

(b) External pressure is applied.

(c) Both internal and external
    pressures are applied.

Fig. 19. A stress analysis on a finite cylinder along its axis.

tensile stress may be induced under internal pressure at the sealing position. Figure 19 illustrates the distribution of stresses along the axis of an open-ended cylinder, which is subjected to both internal pressure and external supporting pressure. These serve to emphasize the importance of retaining hydrostatic compressive stress states for liners made from hard and brittle materials.

4. Stem

As a stem is essentially a member subjected to a uniaxial compressive load, the design consideration is primarily focused on its buckling strength. In hydrostatic extrusion presses, the length to diameter ratio of stems and the applied compressive stress will usually be much larger than in conventional extrusion presses. To clarify the design criterion for stems, the author conducted some buckling tests on a small scale test rig (Fig. 20). In these tests, axial strains were monitored on compressive loading by four strain gauges on the cylindrical surface of each test specimen. The difference of the axial strains ($\Delta\varepsilon_1$ or $\Delta\varepsilon_2$) is rapidly increased beyond a certain compressive stress, which is defined as the buckling stress (Fig. 21).

Figure 22 shows the dependence of the buckling stress on the length to diameter ratio and the compressive yield strength of the stem materials.

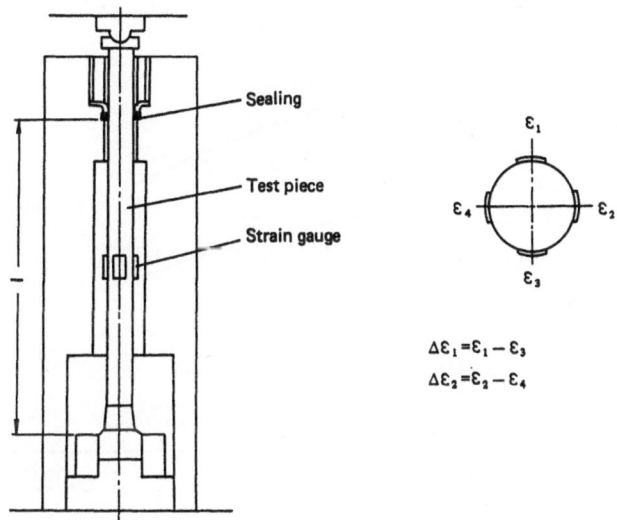

Fig. 20. Test rig for determining the buckling stress of stems.

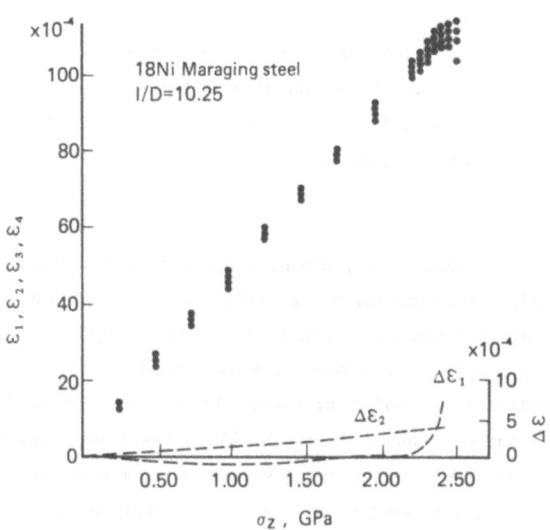

Fig. 21. A typical result on buckling.

With a decrease in the length to diameter ratio, the buckling stress increases approaching to the compressive yield strength. These test results are apparently well fitted with the curves calculated by equation

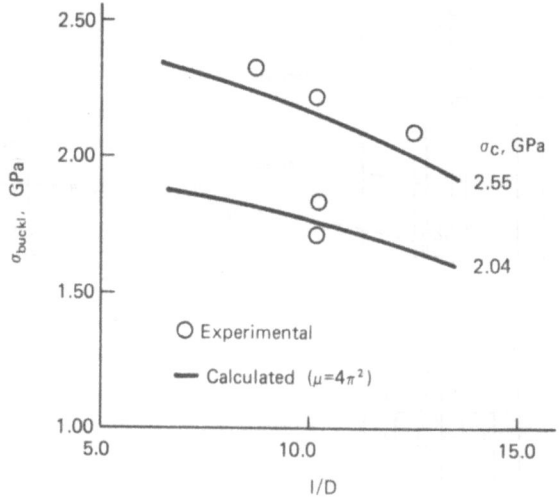

Fig. 22. Buckling stress vs. length to diameter ratio.

(18), which was proposed for the buckling stress of relatively short bars by J.B. Johnson,

$$\sigma_{buckle} = \sigma_c - \frac{\sigma_c^2}{4\mu E} \left(\frac{41}{D} \right) \tag{18}$$

where σ_{buckl} is the buckling stress, σ_c the compressive yield strength, E Young's modulus, 1 the effective length, D the diameter, and μ a constant determined by the end condition.

Another design consideration is fatigue under cyclic loading conditions. As the fatigue strengths of high-strength steels are extremely high under cyclic compressive stress, a stem made from these steels will not suffer from fatigue failure provided that all the stresses in the stem are within elastic limits. When, however, yielding occurs at stress concentrations such as at a fillet contour in a stem under compressive loads, tensile stress occurs on unloading. Stress cycles may accordingly not be entirely on the compressive side, and the fatigue strength of the stem will be decreased. The influence of the local yielding on the fatigue strength is schematically illustrated in Figure 23. As fatigue cracks in a stem may lead to catastrophic failure, considerable care must be taken to prevent local yielding in the stem.

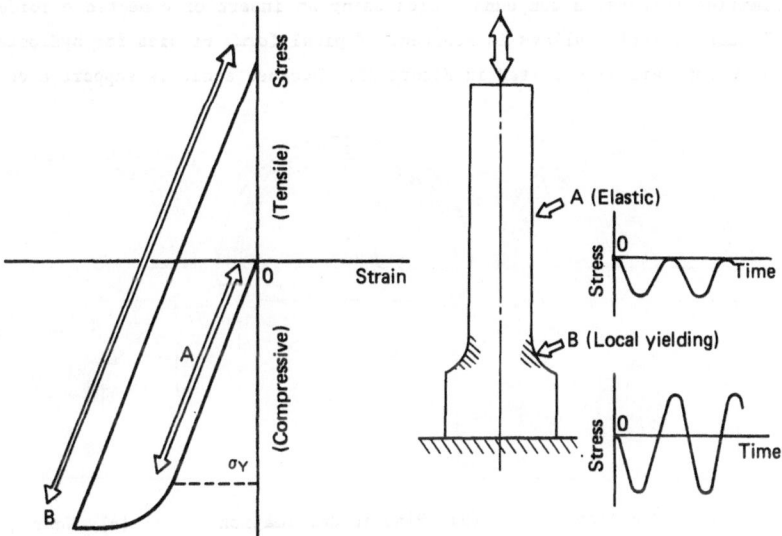

Fig. 23. Influence of local yielding on the fatigue strength of a stem (schematic illustration).

5. Die and mandrel

Dies and mandrels being in direct contact with work materials in extrusion cycles are subjected to extremely large stresses, high temperatures, and severe wearing actions. Consequently, these tools usually have a shorter life than containers. Tool failure will involve the following processes:

    (1) plastic deformation,

    (2) cracking by thermal and/or mechanical stress cycles,

    (3) brittle fracture,

    (4) excessive wear.

Although the qualitative significance of these four processes can often be evaluated in a given case, no definite design criteria for these tools have been established. In fact, the conditions which cause failure are so varied that it is necessary to describe individual cases.

In this section, some guide-lines for tool design will be briefly described from the viewpoint of tool life.

1) Die

The appearance of dies after failure are schematically illustrated in Figure 24. When wear and plastic deformation are the principal life-limiting factors, a compound design using an insert of cemented carbide or of heat-resisting alloys is adopted. Typical forms of dies for hydrostatic extrusion are illustrated in Figure 25. Because a die is supported on its

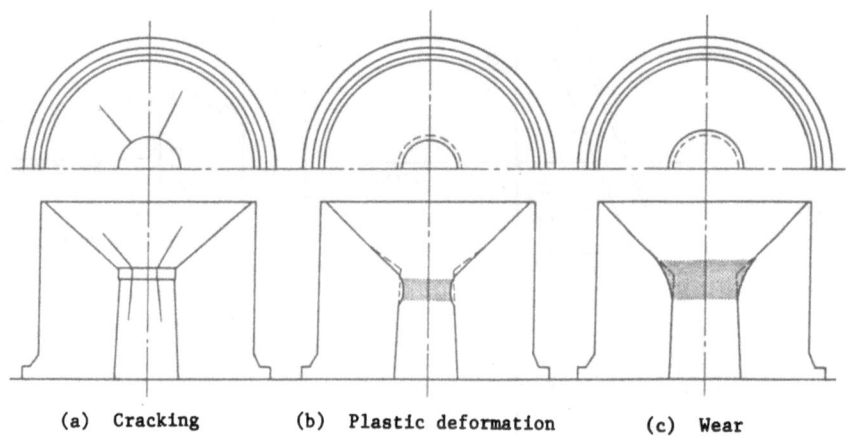

    (a) Cracking   (b) Plastic deformation   (c) Wear

Fig. 24. Appearances of dies at failure.

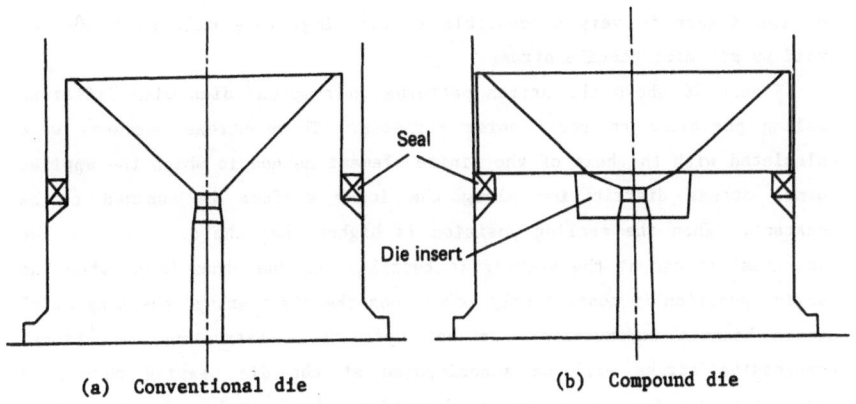

(a) Conventional die   (b) Compound die

Fig. 25. Dies for hydrostatic extrusion.

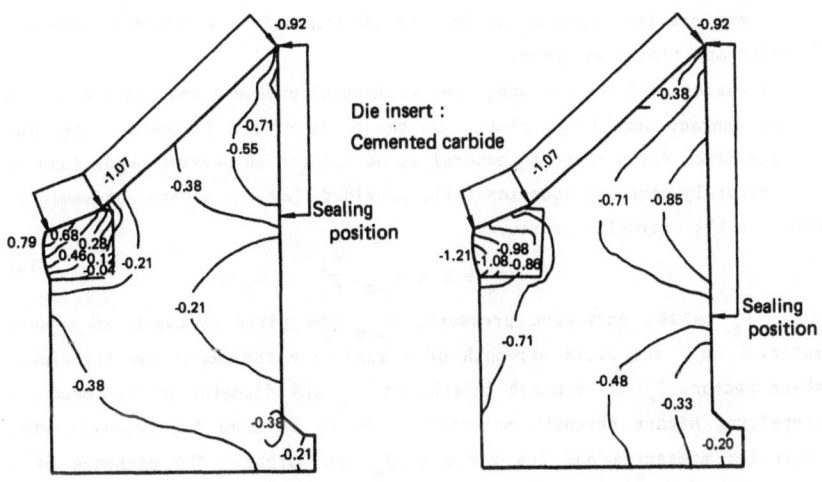

Fig. 26. Patterns of tangential stress in compound dies with different
         sealing positions.

outer cylindrical surface by hydrostatic pressure, tangential stress on its inner surface can generally be decreased. If, however, the material used for the insert is very susceptible to cracking, care will be needed to avoid or minimize tensile stress.

Figure 26 shows the stress patterns in compound dies with different sealing positions on their outer surfaces. These stress patterns were calculated with the help of the finite element method in which the applied normal stress distribution along the inner surface is assumed to be constant. When the sealing position is higher than the die bearing, the tangential stress at the bearing is tensile. On the other hand, when the sealing position is considerably lower than the die bearing, the tangential stress becomes compressive. In hot hydrostatic extrusions, additional compressive stress will be superimposed at the die bearing through a temperature gradient in the radial direction. Accordingly, it seems likely that cracks in inserts may be initiated through two causes : one is the tensile stress on loading, and the other is plastic deformation on the compressive side on loading, resulting in tensile stress on unloading.

2) Mandrel

There are two types of mandrel for hydrostatic extrusions of tubes : floating and tipped mandrels.

Floating mandrels are subjected to lateral pressure and tensile stress at the contact zone (Fig. 27-a). On the basis of the Tresca's criterion, the condition for a floating mandrel to be elastic in service conditions is approximately given by equation (19), provided that the lateral pressure is equal to the extrusion pressure.

$$P_e + 2 \ m \ \sigma_{y \cdot m} (\frac{l_c}{d_m}) \leq \sigma_y \tag{19}$$

where P_e is the extrusion pressure, $\sigma_{y \cdot m}$ the yield strength of a work material, σ_y the yield strength of a mandrel material, m the frictional shear factor, l_c the contact length, and d_m the diameter of the mandrel. Therefore, higher strength materials have to be used for mandrels when extrusion pressures and the ratio l_c/d_m are larger. The presence of a large tensile stress will also require materials to be sufficiently tough. Floating mandrels, because of their travelling contact zone, are less affected thermally in hot hydrostatic extrusions than tipped mandrels.

If wear is the primary factor in selecting material for the tool, a mandrel equipped with a tip made of a wear-resistant material such as cemented carbide, may be used (Fig. 27-b). In this case, it is necessary to design the tip so as to minimize or avoid tensile force at its connection with the mandrel bar. In this design, tensile force is approximately

$$F_t = \pi(r_p^2 - r_m^2)P_e + \frac{1}{2\tan\beta}\, \pi m\, \sigma_{y.m}(r_c^2 - r_i^2) + \pi m \sigma_{y.m} L_p$$

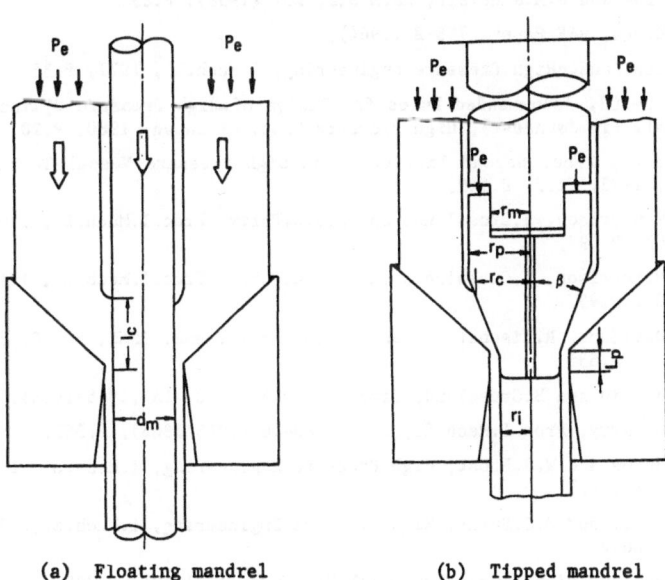

(a) Floating mandrel   (b) Tipped mandrel

Fig. 27. Mandrels for hydrostatic extrusion.

where P_e is the extrusion pressure, $\sigma_{y.m}$ the yield strength of work material, m the frictional shear factor, and ß, r_p, r_m, r_c and r_i are shown in Figure 27-b. When normal pressure on the surface of the tip is assumed to be equal to the extrusion pressure, compressive force is

$$F_c = \pi(r_p^2 - r_i^2)P_e.$$

Thus, the condition that the resultant force is not tensile gives

$$(r_m^2 - r_i^2)P_e - \frac{1}{2\tan\beta}(r_c^2 - r_i^2) + L_p\, m\, \sigma_{y.m} \geq 0. \tag{20}$$

The range of materials suitable for mandrels will be extensively widened under this condition.

REFERENCES

1   E.A.Steigerwald, ASTM STP, 463 (1970), P.102.

2   J.A.McMahon and G.Thomas, Proc. Third Int. Conf. on Strength of Metals and Alloys, Cambridge, Eng., Vol.1, 1973, P.180.

3   Y.Yamaguchi, Studies on the Fatigue of Very High Pressure Cylinder (in Japanese), Dr.Thesis, Osaka University, 1978.

4   J.Morrow and G.M.Sinclair, ASTM STP, 237 (1958). P.83.

5   W.P.Evans, SAE Paper, 793-B (1964).

6   B.Zetterlund, High Pressure Engineering, I.Mech.E., 1977, P.35.

7   HPIS-C-103, Recommended Rules for Design of High Pressure Cylindrical Vessels (in Japanese), High Pressure Inst. of Japan, 1980, P.20.

8   K.Zander, Proc. Second Int. Conf. on High Pressure Vessel Technology, ASME, 1973, Pt.2, P.549.

9   J.L.M.Morrison, B.Crossland, and J.S.C.Parry, Proc.I.Mech.E., 170 (1956), P.697.

10 J.L.M.Morrison, B.Crossland, and J.S.C.Parry, Proc.I.Mech.E., 174 (1960), P.95.

11 T.E.Davidson, R.Eisenstadt, and A.N.Reiner, Trans. ASME, Ser.D, 85 (1963), P.555.

12 B.A.Austin and B.Crossland, Proc.I.Mech.E., 180, 1A (1965/1966).

13 J.S.C.Parry, Proc.I.Mech.E., Pt.1, 180-16 (1965/1966), P.387.

14 D.J.Burns and W.J.Frost, High Pressure Engineering, I.Mech.E., 1967, Paper No.28.

15 W.J.Frost and D.J.Burns, High Pressure Engineering, I.Mech.E., 1976, Paper No.29.

16 G.H.Haslam, High Temperature-High Pressure, 1 (1969), P.705.

17 G.H.Haslam, J.Mech.Engng. Sci., 13-3 (1971), P.130.

18 G.H.Haslam, Trans. ASME, Ser.D, 94 (1972), P.284.

19 D.C.Harvey and B.Lengyel, High Temperature-High Pressure, 5 (1973), P.515.

20 K.Nishioka and K.Hirakawa, High Pressure Engineering, I.Mech.E., 1977, P.325.

21 M.Nishihara, Y.Yamaguchi, and S.Hattori, High Pressure Engineering, I.Mech.E., 1977, P.297.

22 M.Nishihara, Y.Yamaguchi, and S.Hattori, Proc. Third Int. Conf. on Pressure Vessel Technology, ASME, 1977, P.723.

23 J.H.Underwood, ASTM STP, 513 (1972), P.59.

SECTION 3

HYDROSTATIC EXTRUSION PLANTS IN INDUSTRY

A. KOBAYASHI

HITACHI CABLE, LTD

1. Introduction

As a practical example of the industrial application of hydrostatic
extrusion presses a typical press employed in hydrostatic extrusion plants
is described along with its auxiliary equipment. The manufacture of solid
products is primarily discussed but that of hollow products is also
touched upon briefly.

When compared with that of conventional presses, the plant of
hydrostatic extrusion presses does not differ much from the former in so
far as the billet preparation equipment for solid products is concerned.
The billet preparation equipment for solid products is used to prepare the
billet before it is loaded into the press. The only difference between
hydrostatic and conventional extrusion plants is that in the former special
care is needed in handling the billet because it is provided with a conical
nose.

In the extrusion of composite materials good metallurgical bonding
between the different materials is necessary. For this purpose a rather
involved process to treat the surface and adjust the materials according
to their properties is essential. With the conventional press it is
difficult to extrude products with homogeneous and good bonding
characteristics, and the manufacturing of composite materials can be said
to be a main feature of the hydrostatic extrusion press.

From the viewpoint of the type of extrusion, multiple extrusion is
difficult for composites and a single extrusion with high reduction ratios
is uniquely employed. The receiving equipment for the extrudate for
postextrusion treatments is required to handle an extremely long extrudate
efficiently within the cycle time. For this reason special consideration
is needed for this type of press.

2. General Description of 4,000 tonf Hydrostatic Extrusion Press

   2.1 Press Operation

The appearance of the 4,000 tonf hydrostatic extrusion press is shown
in Fig. 1, its basic dimensions in Fig. 2, and its primary specifications
are given in Table 1. In the table two containers of 16 kb and 14 kb
capacity are described, which are interchangeable according to need.

Fig. 1 4,000 tonf hydrostatic extrusion press

Table 1.   Main specifications of 4,000 tonf
hydrostatic extrusion press.

	Item	Specification	
Main body of press	Maximum extrusion force	4,000 tonf	
	Main ram pressure	436 kgf/cm^2	
	Maximum stroke of main ram	1,900 mm	
Container	Maximum extrusion pressure	16 kb	14 kb
	Outer diameter	1,200 mm	1,200 mm
	Inner diameter	180	190
	Length	2,200	2,200
	Billet diameter (Maximum)	160	170
	Billet length (Maximum)	1,200	1,200

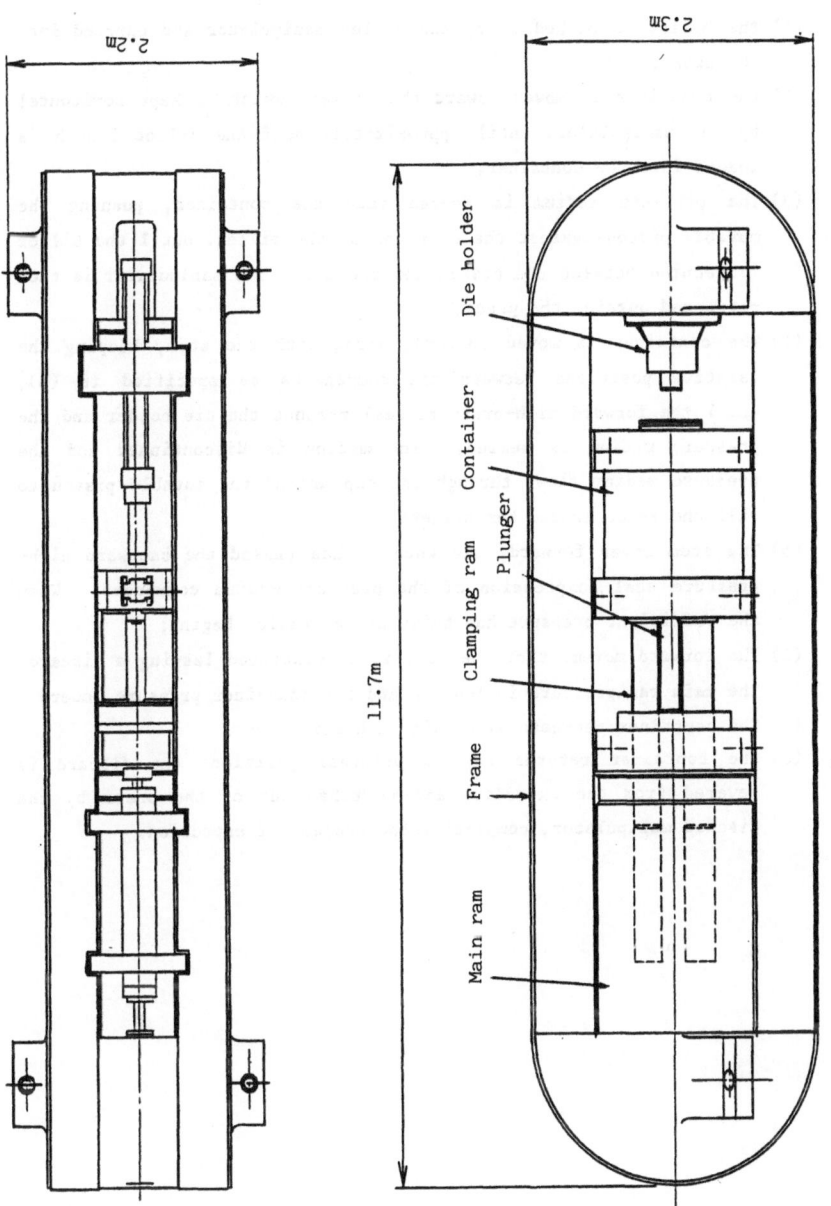

Fig. 2 Structure of 4,000 tonf extrusion press

The operation of the press is described in Fig. 3, which is explained in detail as follows:

(1) The billet is picked up by the billet manipulator and carried into the press;

(2) The container is moved toward the billet, which is kept horizontal by the manipulator, until approximately half the billet length is loaded into the container;

(3) The pressure medium is poured into the container, pushing the movable piston against the rear end of the billet, until the billet is secured between the piston and the die. The manipulator is then retracted outside the press;

(4) The container is moved forward, along with the stem, keeping the relative positions between the components as specified in (3), until the forward high-pressure seal reaches the die holder and the pressure medium is sealed. The motion is discontinued and the pressure medium flows through the gap around the movable piston to fill the space around the billet;

(5) The stem moves forward, and when it has passed the backward high-pressure seal compression of the pressure medium commences. When the sufficient pressure has built up, extrusion begins;

(6) The forward movement of the stem is discontinued leaving a discard. The main ram pressure is lowered and the container pressure lowers.

(7) The container retreats along with the main ram;

(8) The container returns to its original position; the discard is severed from the extrudate and is taken out of the press by the discard manipulator, completing the process of extrusion.

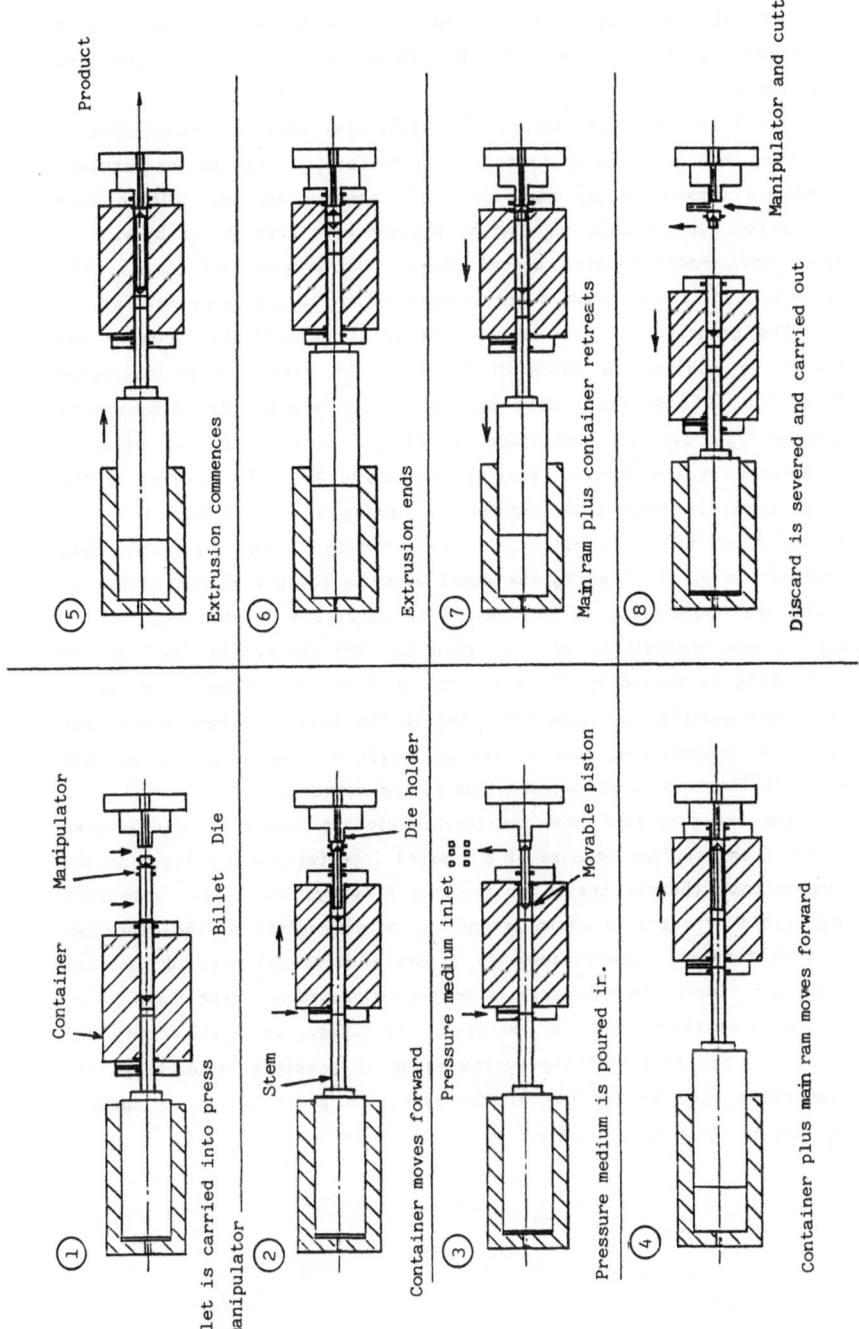

Fig. 3 Press operation

2.2 Main Features of Structure

The size of the press is kept to the minimum by adopting a prestretch winding system in the construction of its frame and container.

The frame is a combination of semicircular yokes and four columns, as shown in Fig. 4, in contrast to the horizontal 3-column or 4-column structure conventionally employed. The members are not jointed with each other but are held together with pre-compressive stress by winding steel wire under tension around them. The stress variation in the frame members is designed to remain within the range of compression.

The principle of pre-stretch winding is as follows. Suppose the member is subject to force of P_w when the press is in operation (Fig. 4). If the steel wire has been wound around the columns with tension P_p, elastic elongation of AB is present in the wire as determined by the force - displacement curve AD. The column, on the other hand, is subject to compressive stress and is shortened by the amount of BC as determined by the force - displacement curve CD. Upon application of the load P_w the steel wire is further elongated by A_1, while the compression of the column is recovered by the same amount. Due to the application of the force P_w, therefore, the load on the steel wire is varied by GE, while the load on the column is varied by EF. Consequently, the load variation in the column is kept within the range of compression. An additional merit of the structure is that there is little stress concentration in the members.

For the container, the pre-stretch winding system is also adopted on its outer surface because of a limited interference per layer of the conventional thermal insert or taper-fitting system. Steel wire with high yield strength is wound around the outer surface of the container at high tension. Simultaneously a tapered liner is pressed into the container, imparting extremely large pre-compression to the inner liner. At peak internal pressure in extrusion the liner is still free from tension. The taper-fitting system makes it possible to exchange the liner with ease by the use of the extrusion press itself in case of failure of the liner.

(1) Wire-wound press frame

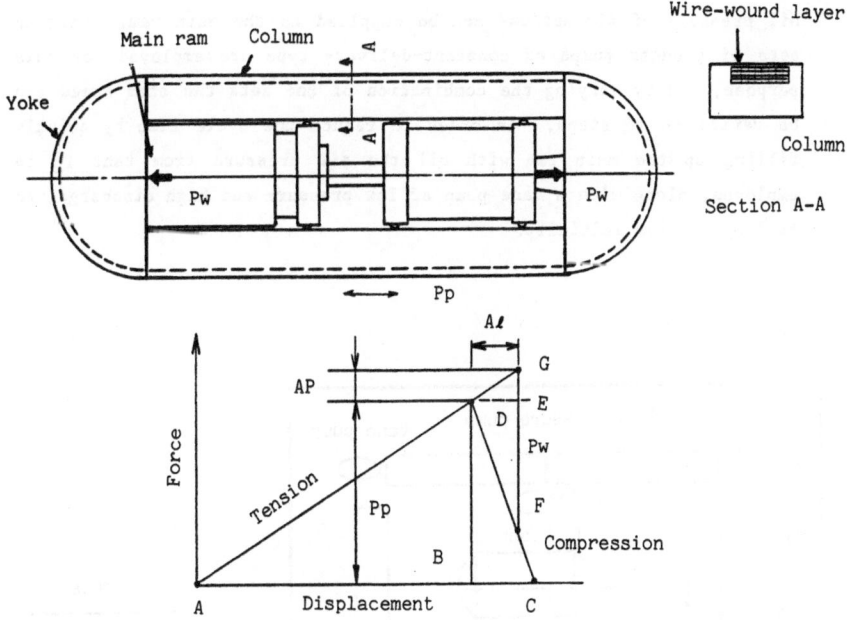

(2) Wire-wound container

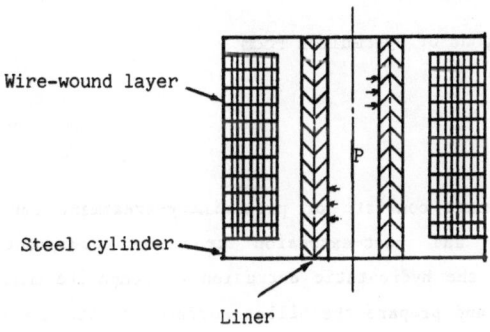

Fig. 4   Structure of press frame and container

2.3 Hydraulic System

The layout of the hydraulic system is shown in Fig. 5. Maximum oil pressure of 436 kgf/cm^2 can be supplied to the main ram. Sixteen sets of plunger pumps of constant-delivery type are employed for this purpose, and by varying the combination of the sets the stem speed can be varied in 16 steps. In order to reduce the cycle time by quickly filling up the main ram with oil the air pressure from tank T_2 is employed, along with a vane pump of low pressure and high discharge, to fill up the ram initially.

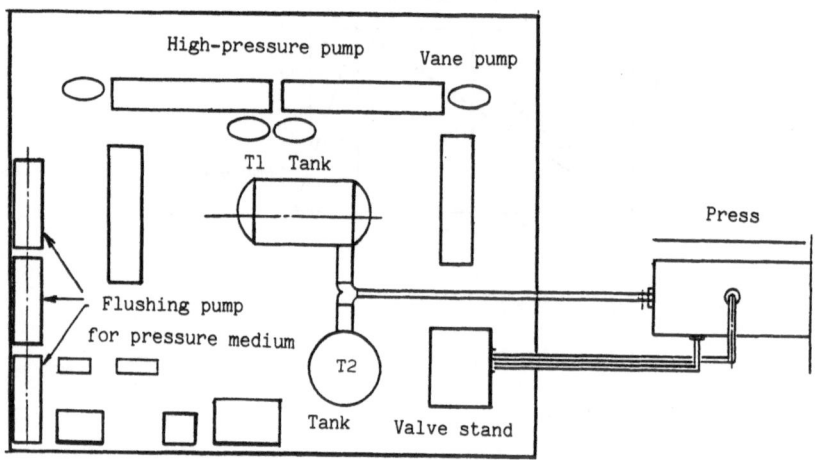

Fig. 5 Layout of hydraulics room

3. Auxiliary Equipment

The auxiliary equipment consists of preliminary-treatment equipment for billet preparation and post-extrusion treatment equipment for extrudate finishing. In the hydrostatic extrusion of composite materials it is necessary to clean and prepare the billet surface to obtain a strong bond between different metals and also to assemble the composite billet. The preliminary treatment equipment is required for this purpose,

233

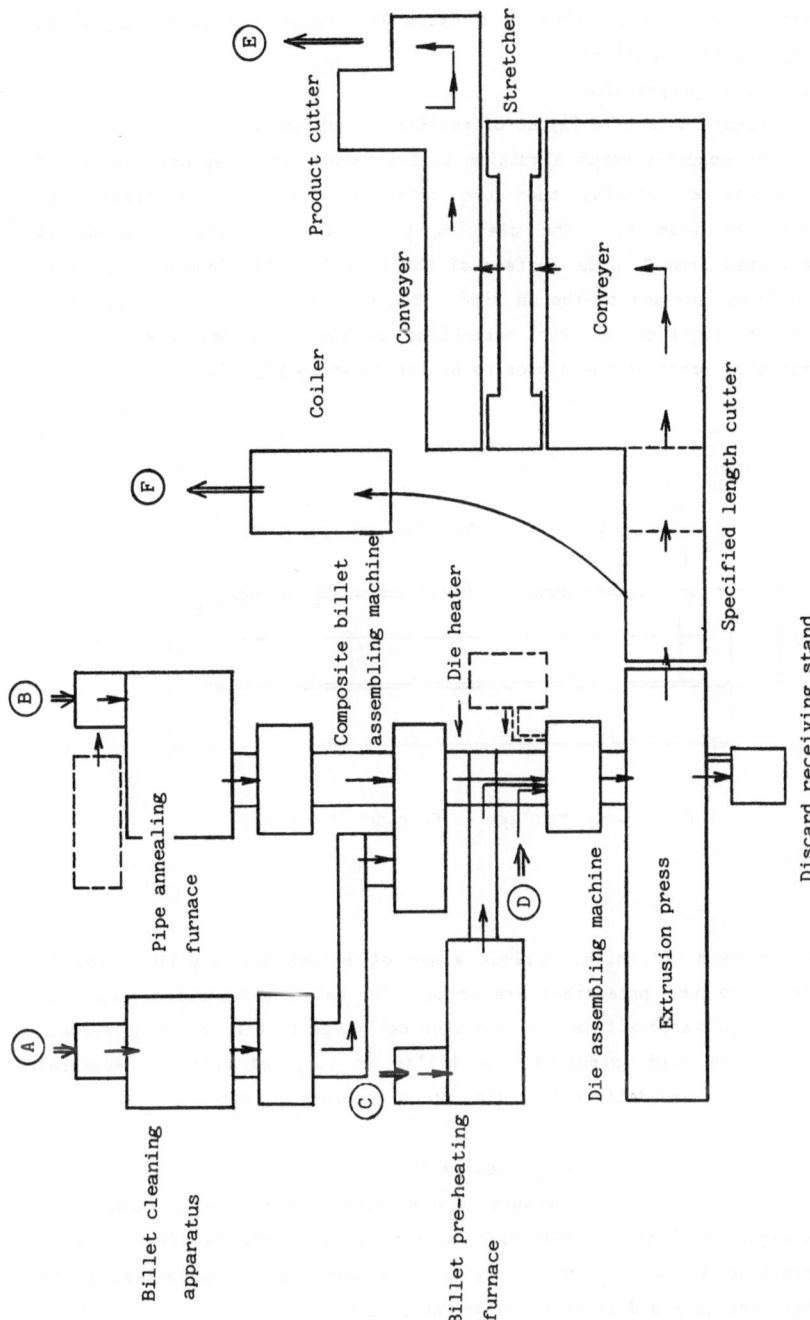

Fig. 6 Layout of auxiliary equipment

rendering the process rather complicated when compared with the extrusion of single-material billets.

3.1 Billet preparation

Figure 6 shows a layout of auxiliary equipment.

(1) For copper-covered aluminium billets shown as a typical example of a composite material the core material, i.e. the Al billet, is suppplied from A. The cladding material, i.e. the Cu pipe, is delivered from B. The surface of the Al billet is cleaned as also is the inner surface of the Cu pipe. The Cu pipe and Al billet are then brought together by the assembling machine and delivered to the extrusion press as the composite billet shown in Fig. 7.

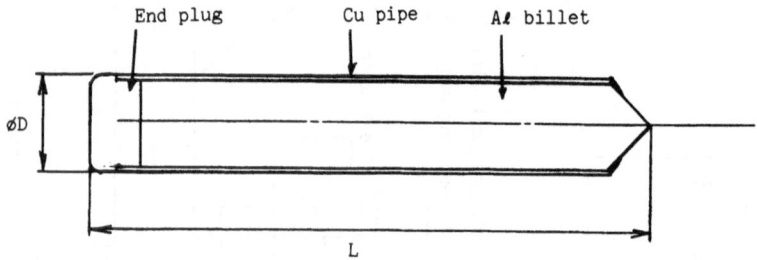

Fig. 7 Configuration of Cu clad Aℓ billet

(2) In warm extrusion, various kinds of billet are supplied from C, heated to the prescribed temperature by the low-frequency induction heater, provided with a die, and then delivered to the extrusion press.
(3) In the cold extrusion, the billet is supplied from D, assembled with a die, and delivered to the extrusion press.

3.2 Post-Extrusion Treatment Equipment

When the extruded product is a straight rod it passes through the straight-rod line, is delivered to E in Fig. 6, and is finished to a specified length. When the extruded product is to be coiled, it is delivered to F and is taken up by the coiler.

Figure 8 shows the layout of the straight-rod treatment equipment. In the line, the extrudate from the press is first cooled by water in the cooling zone to prevent a change in colour and also to remove subsequent handling problems due to high temperature. The cooling process is necessary because the extrudate has been heated by the deformation process, even when extrusion is performed at room temperature.

The extrudate subsequently enters into the runout table. In hydrostatic extrusion a single rod is normally extruded with a high reduction ratio, as stated above. The length of the extruded product may reach 80 to 100 m, depending on its cross-sectional area. Since the process line has a limited length the extrudate may extend beyond the controllable limit. It is, therefore, sheared at intervals of approximately 20 m maximum by the rotary flying shear while being extruded.

The sheared extrudate is accelerated by the runout table from 1.5 to 1.6 times faster than the rate of extrusion to keep it at a distance from the following extrudate. When the extrudate reaches the end of the runout table, it is transferred by the transfer mechanism (Fig. 9) to the transverse conveyer where it is pulled straight by the stretcher and subsequently severed by the saw to specified lengths of 3 - 6 m.

It is to be noted in the post-extrusion treatment equipment of the hydrostatic extrusion press that, unlike the conventional extrusion press, a long product of single extrusion should be sheared by the flying shear while being extruded and be fed to the stretcher at uniform intervals within the cycle time.

The extrusion liner or the press is shielded by wall to safeguard against the possible jetting-out of extrudates or pressure medium in case of complete extrusion or other anomalies during extrusion.

4. Solid-Rod Extrusion with 4,000 tonf Press

Extrusion characteristics of the 4,000 tonf extrusion press are illustrated in Figs. 10-15 for various materials [1]. Fig. 10 shows the stroke - pressure diagram of the cold extrusion of 99.5 % pure Al. At the outset of extrusion a rather high starting pressure P_{start} appears, but after the peak the extrusion pressure is extremely stable. In the warm extrusion of oxygen-free copper, shown in Fig. 11, the pressure decreases from P_{start} to $P_{stationary}$ but thereafter steadily increases to P_{end}, where extrusion is completed.

236

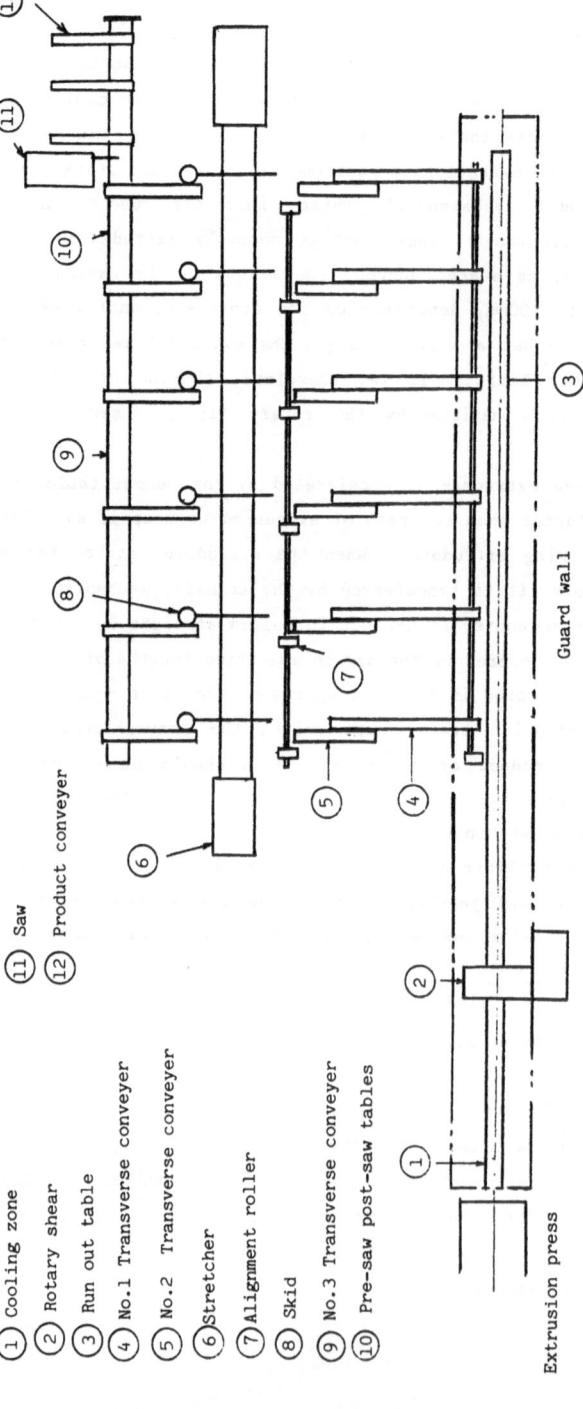

① Cooling zone
② Rotary shear
③ Run out table
④ No.1 Transverse conveyer
⑤ No.2 Transverse conveyer
⑥ Stretcher
⑦ Alignment roller
⑧ Skid
⑨ No.3 Transverse conveyer
⑩ Pre-saw post-saw tables
⑪ Saw
⑫ Product conveyer

Fig. 8 Straight-rod treatment

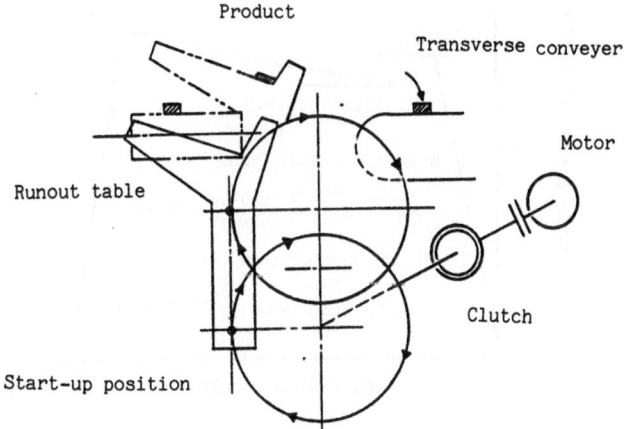

Fig. 9 Transfer apparatus from runout table
to transverse conveyer

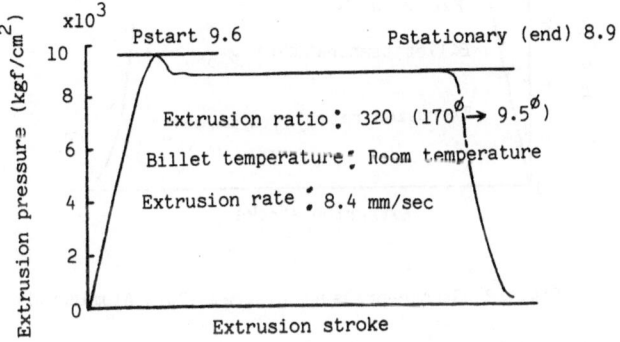

Fig. 10 Stroke-pressure diagram of cold extrusion
(Commercially pure aluminium)

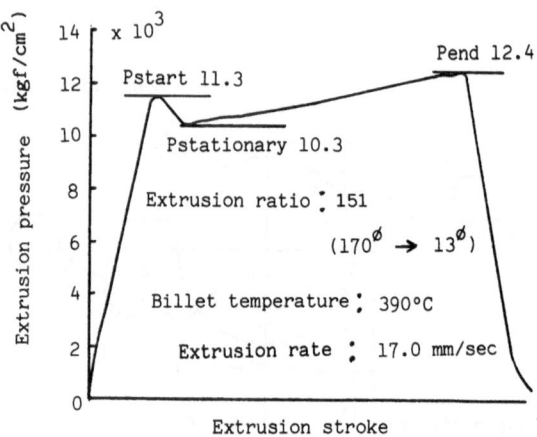

Fig. 11 Stroke-pressure diagram (Oxygen-free copper)

Similarly in the stroke - pressure diagram of the warm extrusion of 7075 Al alloy, Fig. 12, approximately the same tendency as in oxygen-free copper can be seen.

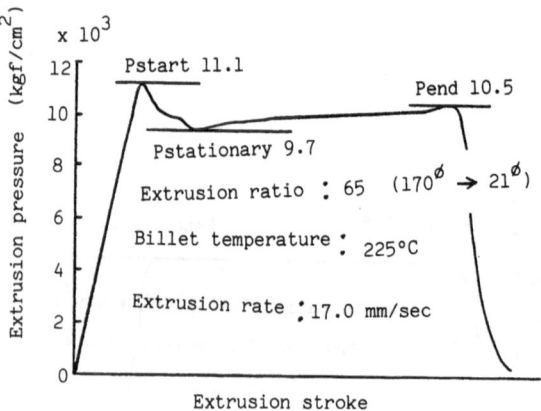

Fig. 12 Stroke-pressure diagram (7075 Aluminium)

From these figures the following can be concluded:

(1) In warm extrusion the pressure gradually increases during steady extrusion;

(2) A dynamic instability, known as the 'stick-slip phenomenon', which is peculiar to hydrostatic extrusion, never occurs.

The pressure rise of (1) can be attributed simply to the decrease of billet temperature during extrusion. Absence of the stick-slip phenomenon of (2) is a main feature of the production-type press and can be considered as mass-effect due to the use of a long billet, as explained by Nilsson and Laker [2]. In principle, the stick-slip phenomenon is caused by the pressure medium with rather low spring constant between billet and stem, and also by instability in the lubricating effect of high-pressure fluid. In the production-type press, which extrudes long billets, the large mass of the billet reduces acceleration caused by the instability phenomenon. In laboratory-type hydrostatic extrusion the stick-slip phenomenon occurs rather frequently and is one of the points to be seriously considered in the industrial application of this technique. When it occurs, the surface quality and dimensional accuracy of the product are degraded, and the press itself will be adversely affected. In the 4,000 tonf extrusion press, however, there is scarcely such a problem.

Figure 13 shows the stroke – pressure diagram of the cold extrusion of copper-clad aluminium product with 15% Cu coverage. The Al top of the billet is first extruded, resulting in a low P_{start}, which is followed by another P_{start}, higher than the first one, when the Cu covered Al billet begins to extrude. Thereafter the extrusion pressure stabilizes at $P_{stationary}$.

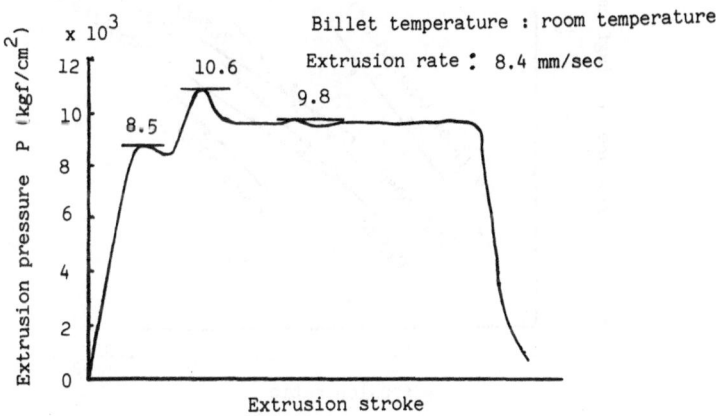

Fig. 13 Stroke-pressure diagram (15% Cu clad Al)

Figure 14 shows the extrusion ratio – extrusion pressure diagram of the cold extrusion of a Cu clad Al composite with 99.5% Al and 10 and 15% Cu. Figure 15 shows a similar diagram of the warm extrusion of tough-pitch Cu, oxygen-free Cu, phosphorous Cu, and 5056 and 7075 Al alloy. The cold-extrusion data of tough-pitch copper is also included in the figure for comparison. Phosphorous Cu and 5056 and 7075 Al alloy are known to be extremely difficult to extrude at room temperature, but in warm extrusion phosphorous Cu and tough-pitch Cu are rather easier to extrude than oxygen-free Cu. These results are to be interpreted to be due to the temperature dependence of the flow stress of the material.

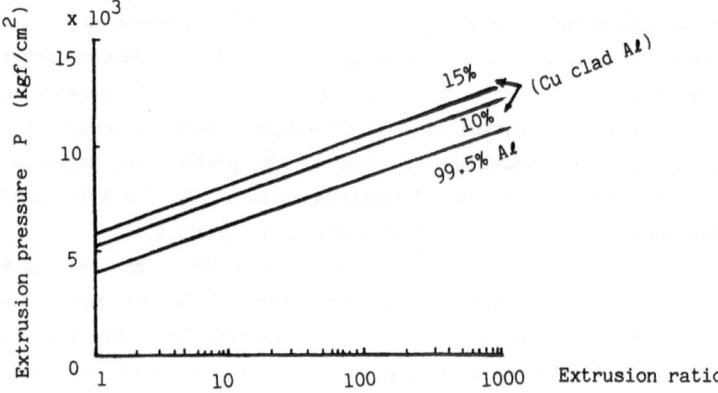

Fig. 14 Extrusion ratio vs. pressure in cold extrusion

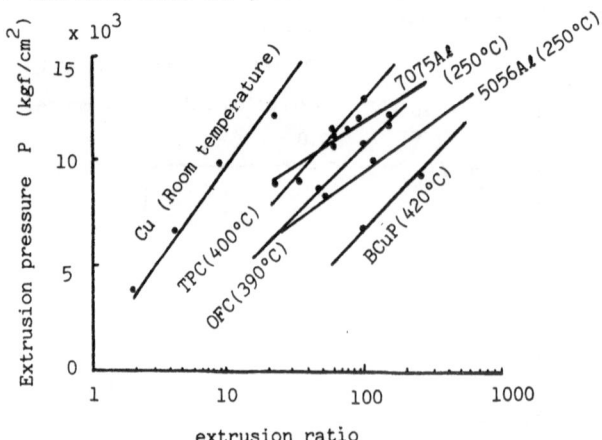

Fig. 15 Extrusion ratio vs. pressure in warm hydrostatic extrusion

5. Pipe Extrusion with 4,000 tonf Extrusion Press

5.1 Hydrostatic Extrusion of Pipe

Hydrostatic extrusion of pipes can be classified into 2 types according to the method of supporting the mandrel [3]: one is called fixed-mandrel extrusion, in which the mandrel is fixed relative to the die by some means; and the other is called floating-mandrel extrusion, in which the mandrel moves in the axial direction as the billet is extruded. From the viewpoint of extrusion apparatus the traveling mandrel extrusion is simpler, while the extrusion mechanism is simpler with the fixed-mandrel extrusion as shown in Fig. 16.

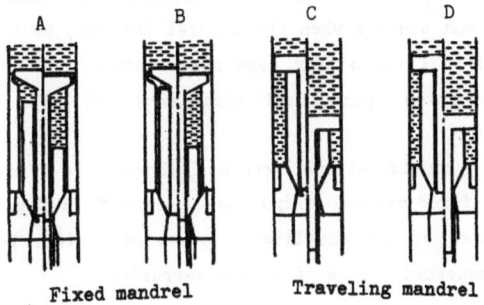

Fig. 16 Hydrostatic extrusion of pipe

The extrusion ratio is limited by the maximum allowable pressure of the extrusion apparatus and varies with billet materials. In principle, fixed-mandrel extrusion can be performed with any extrusion ratio so long as the pressure required is below the upper limit of the allowable pressure. There are three types of mandrel configuration, A, B, C shown in Fig. 17. Type A is employed to reduce the outer dimension only and, being simpler in configuration than types B and C, is mostly used during the initial period of pipe forming; type B is employed to reduce outer and inner dimensions simultaneously; and type C is employed to increase outer and inner radii.

The mandrel is loaded with the axial load, given by the product of the cross-sectional area of the mandrel by the difference between the pressure inside the pressure vessel and the outside pressure below the die, when approximately estimated. For the fixed-type mandrel the load is borne by the jig fixing the mandrel but is not applied to the billet. For the floating-type mandrel, on the other hand, the load is borne by the billet causing compression in the axial direction of the billet: the differential pressure works as augmentation. As a result, if the extrusion ratio is the same, traveling-mandrel extrusion is said to require lower extrusion pressure than fixed-mandrel extrusion.

In traveling-mandrel extrusion the mandrel is extruded out of the die, along with the billet, as extrusion proceeds (Fig. 17). Consequently, friction between the press and the mandrel tends to increase. A small converging taper on the mandrel lessens the friction increase to a greater or lesser degree [3], but causes inexactness of the inner diameter. Also with the floating-type mandrel it is not always easy to remove it from the extrudate after extrusion. This presents a serious problem when the mandrel deforms, even slightly, due to imbalance of friction or for some other reason. If this happens in industrial production operational efficiency of yield is seriously affected.

From the considerations stated above, in industrial pipe extrusion, the fixed-mandrel extrusion of type B in Fig. 17 is mostly employed in spite of the complicated press mechanism, because of its capability of repeated cycles of stable extrusion.

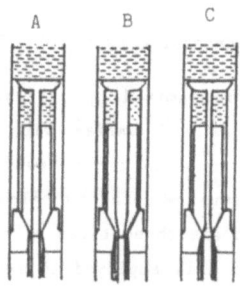

Fig. 17 Configuration of fixed-type mandrel

5.2 Pipe Extrusion system with 4,000 tonf Press

The general principle of pipe extrusion using the 4,000 tonf press is shown in Fig. 18 [4, 5].

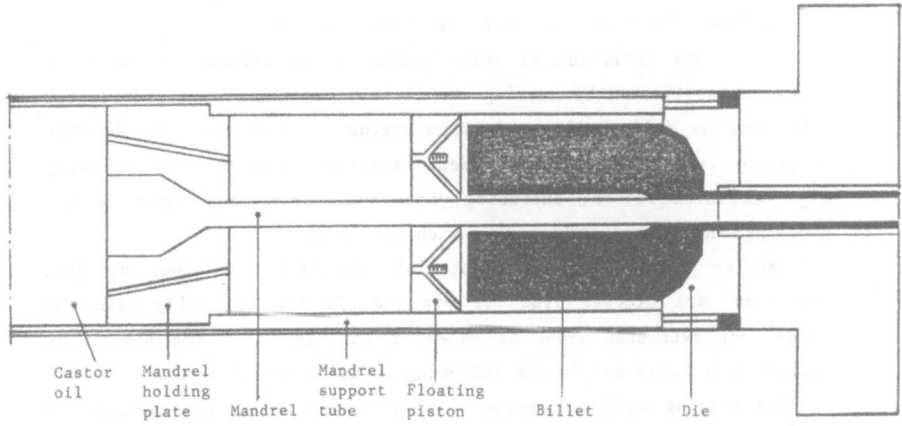

Fig. 18 Principle of pipe extrusion with 4,000 tonf extrusion press

The hollow billet with tapered end is placed in the pressure
vessel along with the die. The bore diameter of the hollow billet is
made slightly larger than that of the extrudate, as shown in the
figure. Around its forward end, however, the bore diameter of the
billet matches that of the extruding pipe. The mandrel is subject to
the driving force of the extrusion but the force is borne by the press
frame through the mandrel hold plate, mandrel support tube, die, and
die holder.

The pressure medium flows through the passage provided in the
floating piston and fills up the space outside the outer surface of the
billet and also the bore of the hollow billet, that is, outside the
mandrel. The floating piston supports the end of the billet so that
the billet may be kept as horizontal as possible.

When the container pressure rises due to forward movement of the
stem the billet begins to extrude through the orifice between the die
hole and outer surface of the mandrel. By accurate determination of
the stroke of the stem, extrusion can be precisely discontinued so that
a short length of discard is left unextruded. In warm extrusion, the
billet and the die are heated before entering into the container.

5.3 Equipment for Warm Hydrostatic Extrusion of Pipe

As in the extrusion of solid products, hydrostatic extrusion of pipes is facilitated by heating the billet. The merits of hydrostatic extrusion are fully utilized in warm extrusion which makes it possible to produce pipes with high efficiency and fewer processes by employing high extrusion ratios, extruding pipes with dimensions close to the final product, and finishing them with one draw.

As an example of the industrial production of pipes by warm extrusion, the process line for pipes of Cu and its alloy with the 4,000 tonf extrusion press is shown in Fig. 19 [4]. The processing equipment is explained in the following as the material flows:

(1) The milling machine provides a short taper around the forward end of the billet;

(2) The billet is provided with a hole by the deep-hole drilling machine;

(3) The billet is cleaned at the degreasing station;

(4) In order to provide a seal between the mandrel and the die at the beginning of extrusion the forward end of the billet is upset;

(5) Lubricant is smeared onto the billet;

(6) The billet is pre-heated by induction heating;

(7) The die is heated by the infra-red heater;

(8) The billet and the die are set in the press by the manipulator;

(9) The container pressure rises and extrusion starts;

(10) The extrudate passes through the guide tube and is water-cooled in the cooling zone;

(11) The extrudate is wound up by the rotary coiler or taken up in a straight length.

The extrudate is subsequently transferred to the draw-bench or bullblock to receive one-pass or two-pass drawing, and is finished to the size of the final product.

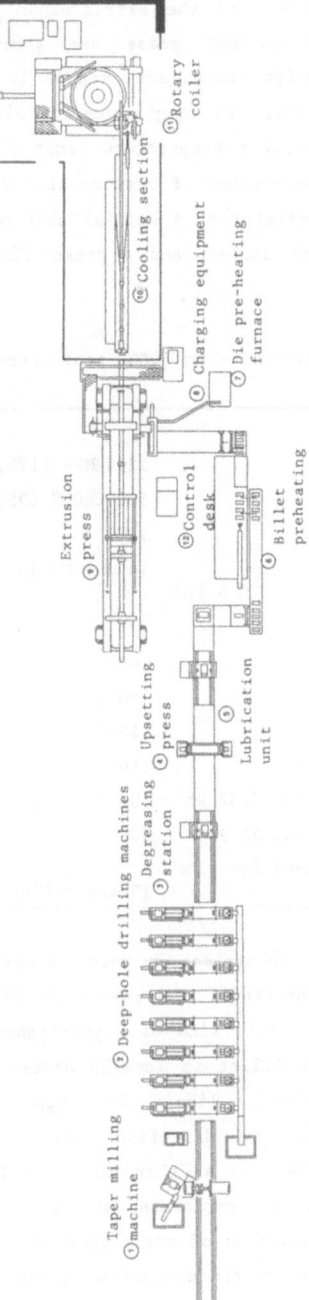

Fig. 19 Process line and equipment for pipe extrusion by 4,000 tonf press

5.4 Pipe Extrusion with 4,000 tonf Press

The process parameters in the extrusion of various non-ferrous materials with the 4,000 tonf press are shown in Table 2 [4]. Considerably high extrusion ratios are seen to be obtained by heating the billet to 500 - 600°C. As long as castor oil is employed as the pressure medium, however, the temperature limit is around 500 - 600°C due to burning and carbonization of castor oil. If extrusion is to be performed at higher temperatures, a special heat-resistant property is required of the pressure medium, and a grease-like substance will be needed.

Table 2.   Pipe extrusion by 4,000 tonf extrusion press

Process parameters	
Operating pressure:	1200 MPa (170,000 psi)
Preheat temperature:	500-600°C (950-1100°F)
Billet taper angle:	75°C
Tube dimensions: (0.55 x 0.040 to 2.4 x 0.4 in)	14 x 1 to 60 x 10 mm
Reduction ratios	
DHP copper	500:1
63-37 brass	250:1
Aluminium brass	140:1
Admiralty brass	160:1
Leaded brass (58% Cu. 2.5% Pb)	500:1
Leaded brass (60% Cu. 2% Pb)	250:1
Maximum extrusion speed for coiling:	15 m/s (3000 ft/min)

In Table 3 [6] a comparison is made between hot conventional extrusion and warm hydrostatic extrusion of 2014 Al alloy. In conventional extrusion of this kind of high-strength Al alloy it is a serious problem that the billet is locally heated by friction between it and container or die and a liquid phase appears, resulting in the occurrence of cracks in the direction normal to the longitudinal direction of the billet - a phenomenon called hot shortness. Consequently, extrusion is performed at an extremely slow rate, resulting in a drastic lowering of efficiency, or water cooling of the die is required. In hydrostatic extrusion on the other hand, there is no direct contact between container and billet, and the die is being lubricated during extrusion. Consequently the problem of hot shortness is resolved and extrusion can be performed at a considerably higher rate, when compared with conventional extrusion.

With the 4,000 tonf press, if the billet is heated to about 250°C, an extrusion ratio of 400 : 1 can be employed without difficulty. In Table 3, however, the comparison is made for pipe extrusion when each type of extrusion is performed with an equal extrusion ratio of 100 : 1. Hydrostatic extrusion demonstrates that it is possible to attain a considerably higher rate of extrusion in a remarkably reduced cycle time. Moreover, a longer billet can be used and the unit weight of the coil of the extruded product can be increased.

Table 3. Comparison between conventional hot extrusion and hydrostatic extrusion of Al alloy pipe

		Hot Extrusion	Hydrostatic Extrusion (200,000 psi)
Alloy		AA 2014	AA 2014
Press force	Short tons	1,760	4,400
	tonnes	1,600	4,000
Billet size	inches	6 x 20	6 x 46
	mm	155 x 500	155 x 1100
Product	inches	1 x 1	1 x 1
(tube OD x wall thickness)	mm	25 x 2.5	25 x 2.5
Reduction ratio		100:1	100:1
Extrusion speed	ft./min.	5	800
	m/min.	1.5	240
Length of product	ft.	130	300
per extrusion cycle (Approx.)	m	40	90
Time of extrusion	min.	26	375
Time for compression of fluid	sec.	-	15
Time for billet handling, loading and unloading of press, etc.	sec.	30	62.5
Total time for one cycle	min.	26.5	1.7
Weight of product	lbs.	45	100
per cycle	kg	20	45
Maximum amount of product	lbs.	100	3600
per hour	kg.	45	1620

6. Conclusion

The 4,000 tonf extrusion press and its auxiliary equipment are described along with their applications. It is over 10 years since hydrostatic extrusion presses were first employed for industrial production in several countries in the world. Techniques for utilizing the process have been accumulated, new fields of application are being developed, and demands for the extrusion of special composites are increasing. Further improvements in practical technology, wider applications and future development of the hydrostatic extrusion process are therefore to be desired.

REFERENCES

1 S. Tsunokawa, Reports of High Pressure Committee No. 138, Japan Society for Promotion of Science, Feb. 7(1975)29-32.
2 J. Nilsson and H. Laker, J. Japan Society for the Technology of Plasticity, 15(1974)789-793.
3 Yamaguchi et al., J. High Pressure Institute, Japan, 7(1969)45.
4 ASEA Technical Pamphlet AQ12-103E Edition 2.
5 R. Hogland, S. Friborg, and D. Ermel, Metallwissenschaft und Technik, 5(1977)516.
6 R. Hogland, S. Friborg, und D. Ermel, ALUMINIUM 55(1979)224.

CHAPTER 6

HYDROSTATIC EXTRUSION OF VARIOUS MATERIALS

SECTION 1

CLAD COMPOSITE METALS

M. SEIDO AND S. MITSUGI

HITACHI CABLE, LTD.

1. Introduction

Clad composite metals are typical examples of products manufactured by the hydrostatic extrusion process [1]. Uniform flow can be obtained during the deformation of clad metals, because there is no friction between the metal and the container or the metal and the die. Relatively low temperature with high reduction ratio is the most appropriate environment for the manufacture of clad metal products and this is provided by the hydrostatic extrusion process. Clad metals made by this process have a very strong bonding between the metals because of the controlled diffusion that takes place during the process.

Moreover, a very high degree of dimensional uniformity can be achieved because of the 'uniform flow' that can be maintained during hydrostatic extrusion.

This section discusses billet making, extrusion method, extrusion pressure, workability limits, and the interface profiles of clad metals. Also discussed, as examples of products manufactured on commercial scale, are copper-clad aluminium products, lead-clad aluminium wires and superconducting wires, the production of which is expected to increase in the near future.

2. Method of Billet Making and Extrusion

It is extremely important to maintain a high standard in billet making as this has a direct bearing on the quality of the extruded rods.

Fig. 1 shows sectional views of some composite billets.

In billet making by insertion method (Fig. 1a), the core block is inserted into a sheath tube, and the top and bottom ends are sealed to prevent leakage of the pressure medium along the interface of end-joints during the extrusion process. The outside of the inserted block and the inside of the tube are properly cleaned to ensure good bonding between the surfaces. To seal both ends during cold extrusion, upsetting or suitable sealing tapes can be used, while in warm or hot extrusion can be done by soldering or welding.

In billet making by casting method (Fig. 1b) the billet is made by casting the low melting-point material around the core or into the sheath tube. The application of this method is limited only to billets whose components do not produce brittle compounds at the interface during the casting process. Lead-clad aluminium wires are, for example, manufactured by this method.

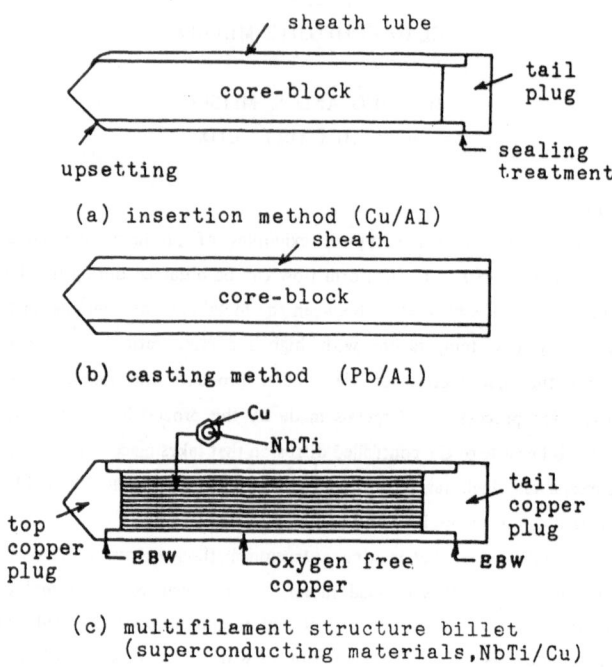

Fig. 1 Billet structures of various clad metals.

The multifilament structure billet (Fig. 1c), which is used for the production of superconducting wires, is also made by the insertion method (Fig. 1a). A bundle of single wires is inserted into a tube, both ends of which are sealed with plugs by electron beam welding, which seals the billet inside a vacuum.

The following extrusion process of clad composite billets is similar to that of solid billets. General industrial presses are equipped with induction heaters for the solid billets. These billets are heated homogeneously to the desired temperature in 2 or 3 minutes. As for clad composite billets, controlled slow heating should be used to heat the billets homogeneously. If rapid heating is applied to clad billets, the outer sheath would become over-heated while the core material would remain unheated.

3. Extrusion Pressure

In hydrostatic extrusion, friction between billet and container is almost zero, and slight on the die approach, so that it is rather easy to calculate the extrusion pressure of clad composite billets, compared with other extrusion methods. The extrusion pressure of composite metals with good lubricating conditions nearly equals the value calculated by the rule of mixtures.

3.1. Calculation of Extrusion Pressure

Three methods to calculate the pressure are used, as shown in the following.

3.1.1 Experimental equation using the hardness values of the metals [2].

The following equation is used to estimate the extrusion pressure for the hydrostatic extrusion of the non-composite billets,

$$P = (AHv + B)\ln R \qquad (MPa) \qquad (1)$$

where A, B = material constant, R = extrusion ratio, Hv = vickers hardness.

From Eq. 1, the extrusion pressure of composite materials P is introduced by the rule of mixtures in the case of uniform deformation of the outer sheath and the core materials.

$$P = \{(A_1 H_1 + B_1)f_1 + (A_2 H_2 + B_2)f_2\}\ln R \qquad (2)$$

where f is the volume fraction and subscripts 1, 2 mean the core and the sheath, respectively.

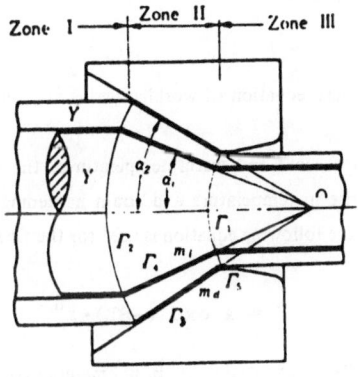

Fig. 2 A kinematically admissible velocity field for clad materials.

3.1.2 Equation by energy method.

Eq. 2 is not sufficient to estimate the effects of the friction and the die angle on the pressure. The equation by the energy method, which takes them into consideration, is effective and based on the following assumption.

(a) The billet consists of rigid-plastic materials obeying von Mises' yield criterion.

(b) The velocity fields are shown in Fig. 2. The velocity discontinuity surfaces Γ_1, Γ_2 are smooth and spherical through the core and sheath area.

(c) Along the die surface Γ_3, a constant shear friction coefficient is employed to calculate the friction losses.

(d) The friction losses at the die land is neglected.

After all, the following extrusion pressure equation of clad composite billets is derived according to Avitzur's solution [3] for solid billets.

$$P = [\{f(\alpha_2) - f_1 f(\alpha_1)\}\ Y_2 + f_1 f(\alpha_1)Y_1] \ln R$$
$$+ \frac{2}{\sqrt{3}}[\{h(\alpha_2) - f_1 h(\alpha_1)\}\ Y_2 + f_1 h(\alpha_1) \cdot Y_1]$$
$$+ \frac{1}{\sqrt{3}} m_d Y_2 \cot \alpha_2 \ln R \tag{3}$$

where

$$f(\alpha) = \frac{1}{\sin^2 \alpha}\{1 - \cos\alpha\sqrt{1 - \frac{11}{12}\sin^2\alpha}$$
$$+ \frac{1}{\sqrt{11 \cdot 12}} \ln \frac{1 + \sqrt{\frac{11}{12}}}{\sqrt{\frac{11}{12}}\cos\alpha + 1 - \sqrt{\frac{11}{12}}\sin^2\alpha}\}$$

$$h(\alpha) = \frac{\alpha}{\sin^2\alpha} - \cot\alpha$$

3.1.3 Extrusion pressure equation of workhardening and temperature softening materials [4], [5].

During extrusion at cold and warm temperatures, the flow stresses of billet metals vary depending on the rise in temperature and strain hardening, but :re almost independent of the strain rate. Then, the following equation is used for the flow stresses.

$$Y = A \exp(-BT) \cdot \varepsilon^n \tag{4}$$

where T: temperatue (°C), ε: strain, A, B, n: material constants. The temperature rise during extrusion is estimated by the following equation 5, where K is the fraction of the plastic deformation energy which causes the temperature rise of the materials.

$$\frac{dT}{d\varepsilon} = K \frac{Y}{J\rho C} \tag{5}$$

where J: mechanical equivalent of heat

ρ: density

C: specific heat of billet material

Deformation analysis of the material in the die using the above equations 4 and 5 introduces an equation of extrusion pressure of workhardening and temperature softening materials, which are suitable for conditions of high reduction at cold and warm temperatures during industrial hydrostatic extrusion.

3.2 Experimental Data of Extrusion Pressure

The relations between extrusion ratio and extrusion pressure are shown for several clad composite materials in Figs. 3 and 4. The extrusions were of copper-clad aluminium billets — copper volume ratio 15% and 170mm in diameter (Fig. 3), and lead-clad aluminium billets — various clad ratios and 20mm in diameter (Fig. 4). In these figures, the solid lines

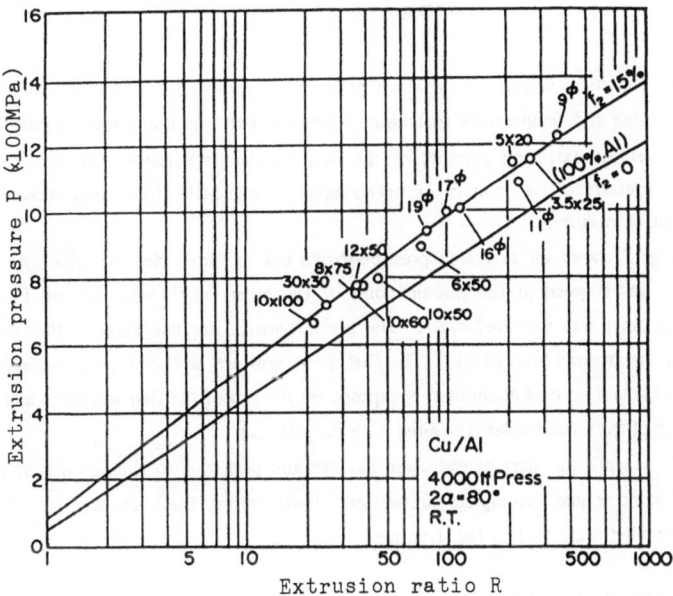

Fig. 3 Relation between extrusion pressure
and extrusion ratio of Cu/Al clad
metals.

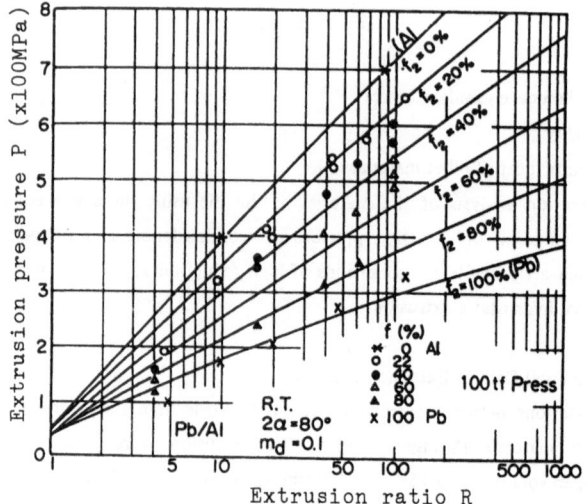

Fig. 4 Relation between extrusion pressure and extrusion ratio of Pb/Al clad metals.

show the anlaytical calculated extrusion pressure using the rule of mixture considering the workhardening and temperature softening of the material. As the extrusion ratio increases, so also does the extrusion pressure of the clad metals. It is found that the experimental extrusion pressure of clad metals are nearly equal to, or a little higher than, those calculated by the rule of mixture.

Fig. 5 shows an extrusion pressure-stroke line of Pb/Al clad material. At first, the extrusion was stopped at the middle point of the billet length and after on interval, the remaining billet was extruded again. The pressure-stroke line has peaks at the start of the first and the second extrusions. The first peak pressure is higher than the second one, because the hard core of aluminium is exposed on the top of the first full billet and the core on the top of the second billet is covered with the soft lead.

In general, the start of extrusion has pressure peak due to the absence of generated heat of working and incomplete lubrication. Furthermore, the peak pressure of the clad composite billet is affected by the structure.

4. Conditions for Sound Extrusion

4.1 Defects and Control Factors during Extrusion of Clad Composite Metals

The extrusion of clad composite metals is more difficult than that of solid metals,

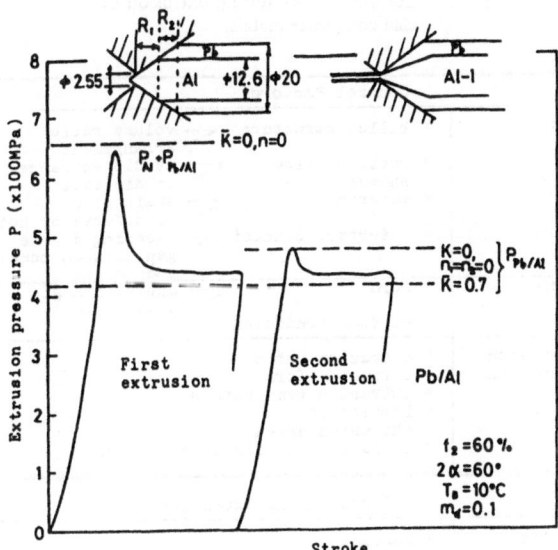

Fig. 5 Extrusion pressure — stroke lines of
Pb/Al clad metal (100 tf press, R=62).

and many defects are generated depending on the extrusion condition. Table 1 shows various
defects generated during the extrusion of clad composite materials and the control factors
used to prevent them. The control factors are; reduction of the flow stress ratio of the
component metals of the billet; improvement in lubrication to decrease friction between the
work and the die; and cleaning of the bonding surfaces of the components of the billet to
increase the interface friction.

Fig. 6 shows typical examples of extrusion defects due to the composite structure.
These are; Bulge (Fig. 6a) as a defect induced at the die entrance; and core or sheath fracture
(Fig. 6b, c) as defects of the extruded rod. These defects are induced by the flow imbalance
of the inner and outer layers.

4.2 Analysis of Extrusion Limit

4.2.1 Analysis of core or sheath fracture of extruded rods.

Fig. 7 shows the deformation models of extrusion defects of clad billets. These defects
are induced by the slipping at the interface of the inner and outer layers due to the flow
difference in the case of large flow stress ratio of the billet components. The deformation
patterns of these phenomena vary depending on where the billet structure is soft core-hard
sheath or hard core-soft sheath.

Table 1 Control factors during extrusion of
clad composite metals.

Control Factors		
Billet	* billet structure	volume ratio arrangement
	* ratio of flow stress	at die entrance at die exit
	* material	grain size no defects or not
	* interface condition	cleaning degree gap between bonding components
Die	* form	approach angle exit angle
	* surface condition	
Extrusion condition	* extrusion ratio * extrusion form * extrusion temperature * lubrication * extrusion speed * cooling condition of extrusion	
Defects during Extrusion		
Billet	* shaving * bulge deformation * dead metal	
Extrusion	Defects due to clad structure	* waving * burst of inner layer * burst of outer layer * swelling * eccentricity
	General defects	* stripe * wrinkle

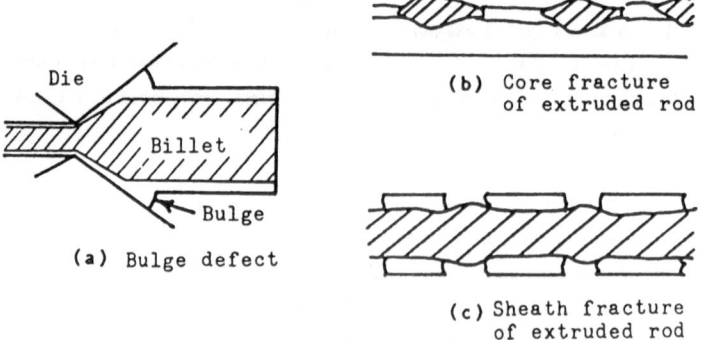

(a) Bulge defect

(b) Core fracture
of extruded rod

(c) Sheath fracture
of extruded rod

Fig. 6 Examples of extrusion defects.

In the case of a soft core-hard sheath billet, sheath fracture may occur as follows (Fig. 7a): The soft core is only extruded when slipping at the interface and compression at the die exit corresponds to the outer layer fracture stress. The extrusion pressure in this case is introduced by the energy method like equation 3 as follows.

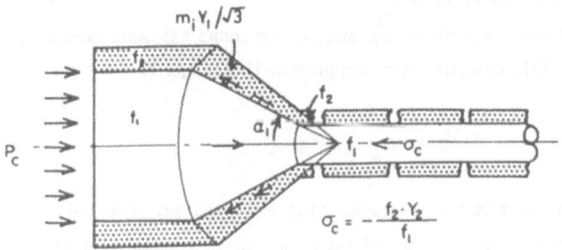

(a) Core extrusion (soft core - hard sheath)

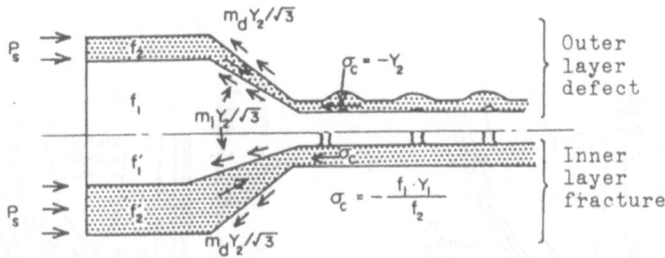

(b) Sheath extrusion (hard core - soft sheath)

Fig. 7 Deformation models inducing defects.

$$Pc = Y_1 \ln Y + \frac{2}{\sqrt{3}} Y_1 h(\alpha_1) + \frac{Y_2}{\sqrt{3}} m_i \cot\alpha_1 \ln R + \frac{f_2 Y_2}{f_1} \tag{6}$$

In the case of hard core-soft sheath billet, however, the soft sheath only is extruded and the hard core may act as a mandrel on the pipe extrusion. Compression, corresponding to the inner layer fracture stress, is applied at the die exit (Fig. 7b). The extrusion pressure is introduced as follows.

$$Ps = Y_2 \ln R + \frac{2}{\sqrt{3}} \frac{1}{f_2} Y_2 \{ h(\alpha_2) - f_1 h(\alpha_1) \}$$

$$+ \frac{1}{\sqrt{3}} \frac{1}{f_2} Y_2 m_d \cot \alpha_2 \ln R + \frac{1}{\sqrt{3}} \frac{f_1}{f_2} Y_2 m_i \cot \alpha_1 \ln R + A_1 \qquad (7)$$

where A_1 is the smaller value of Y_2 or $\frac{f_1}{f_2} Y_1$.

To make the extrusion sound, the value of the sound extrusion pressure (Eq. 3) must be smaller than that of the defective extrusion pressure (Eqs. 6 and 7).

$$P < P_c \text{ and } P < P_s \qquad (8)$$

Fig. 8 shows the maximum extrusion ratio for sound extrusion [6]. The solid lines are calculated values from Eq. 8. Sound extrusion can be achieved at a smaller ratio than that of the lines, and it varies significantly depending on the volume ratio of the outer layer, flow stress ratio Y_1/Y_2, die angle and friction conditions. As the flow stress ratio increases, the maximum extrusion ratio for sound extrusion gets smaller and the workable region becomes narrow.

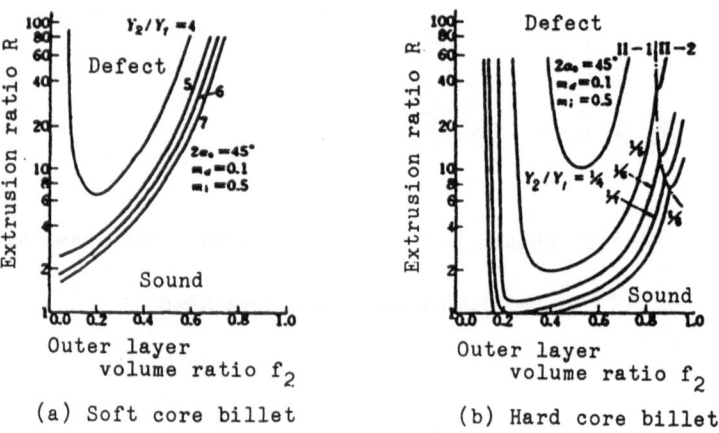

(a) Soft core billet   (b) Hard core billet

Fig. 8   Maximum extrusion ratio for sound extrusion of clad metal billets [6].

In Eqs. 3, 6, and 7, it is assumed that the material flow at the die is radial towards the focus. It is correct for the sound extrusion, but not exactly so for the defective extrusion. Avitzur obtained a stricter analysis by assuming that the flow is variable [7]. As the flow stress is not exactly constant during the extrusion process, a more accurate analysis was achieved by considering temperature elevation and strain hardening during extrusion [5].

The workable limits are also affected by the friction conditions between the metals and the die. Fig. 9 shows the analytical maximum extrusion ratios for sound extrusion which are calculated for the various billet-die shear friction coefficient m_d and inner-outer metals shear friction coefficient m_i. As a result, large m_i and small m_d provide a wide workable region. Therefore, it is important to clean the bonding areas of the billet and to maintain m_i high value. Hydrostatic extrusion, which has excellent lubricating conditions on the die, is therefore effective for the working of clad metals.

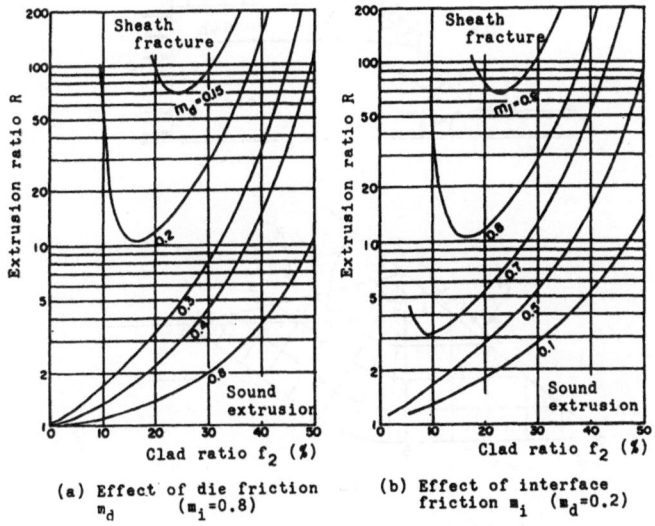

(a) Effect of die friction m_d (m_i=0.8)

(b) Effect of interface friction m_i (m_d=0.2)

Fig. 9 Effects of die friction (m_d) and interface friction (m_i) on the workability limits of soft core-hard sheath clad metals (Cu/Al, 2α=105°, R.T.).

4.2.2 Analysis of Bulge defect at the die entrance.

Bulge is a phenomenon which occurs during drawing with small reduction. Hill [8] analysed it by the slip-line field method. It has been reported by Kubo [9] and Avitzur [3] that Bulge also occurs at a small extrusion ratio and large die angle during the hydrostatic extrusion of a solid billet. In the case of hydrostatic extrusion of hard core-soft sheath clad metals, even at a high extrusion ratio, Bulge also occurs as an irregular deformation only on the soft material due to the high flow stress ratio of the metals.

Generation of the Bulge defect is discussed here using the slip-line field method, and it is assumed that it occurs only on the soft sheath metal at the die entrance.

Fig. 10 shows such a slip-line field. The pressure q_B over the die surface on Bulge area is derived from the slip-line field analysis by considering the surrounding pressure p for the extrusion, as follows.

$$q_B = \frac{1}{\sqrt{3}} Y_2 (1 + 2\phi + \sin2\lambda) + p \tag{9}$$

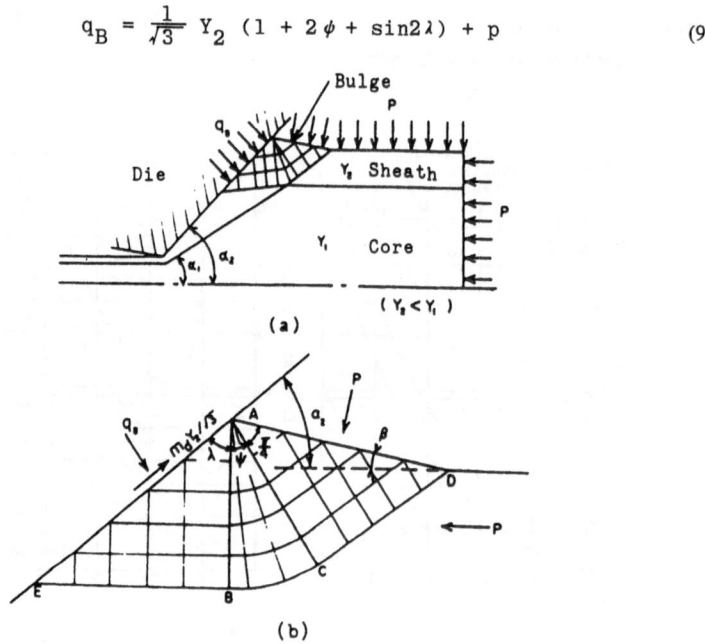

(a)

(b)

Fig. 10 Slip-line field at Bulge defect.

From the relations of the angles and the friction on the die,

$$2\phi = \frac{3}{2}\pi - 2\lambda - 2\alpha_2 - \beta \tag{10}$$

$$\cos2\lambda = -m_d \tag{11}$$

On the other hand, the pressure at the die entrance is approximately derived from the flow stresses of the clad billet, as follows.

$$q = f_1Y_1 + f_2Y_2 + p \tag{12}$$

It is considered that Bulge defects occur when the pressure q derived from the flow stresses of the billet exceeds the pressure q_B of Bulge deformation. Therefore, the condition of Bulge occurrence is

$$q > q_B \qquad (13)$$

By substituting Eqs. 9 and 12 into Eq. 13

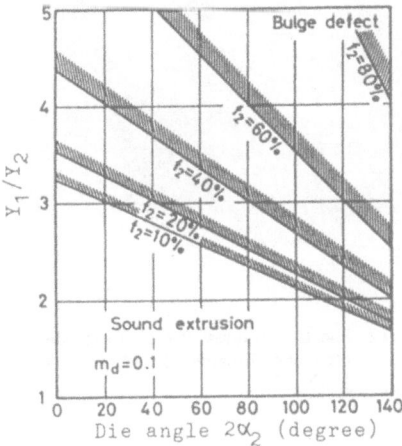

Fig. 11 Bulge occurrence condition of hard core-soft sheath clad metals.

$$\frac{Y_1}{Y_2} > \frac{1}{f_1} \left\{ \frac{1}{\sqrt{3}} \left(1 + \frac{3}{2}\pi - 2\lambda - 2\alpha_2 + \sin 2\lambda \right) - f_2 \right\} \qquad (14)$$

Thus Bulge defect occurs when the flow stress ratio of the metals exceeds the critical limit determined by the clad ratio, die angle and friction on the surface.

The calculated results are shown in Fig. 11. The defects occur in the region of large die angles and high flow stress ratio of the billet.

4.3. Experimental Data of Extrusion Limits

Cu/Al billets have been extruded experimentally, as examples of soft core-hard sheath clad metals.

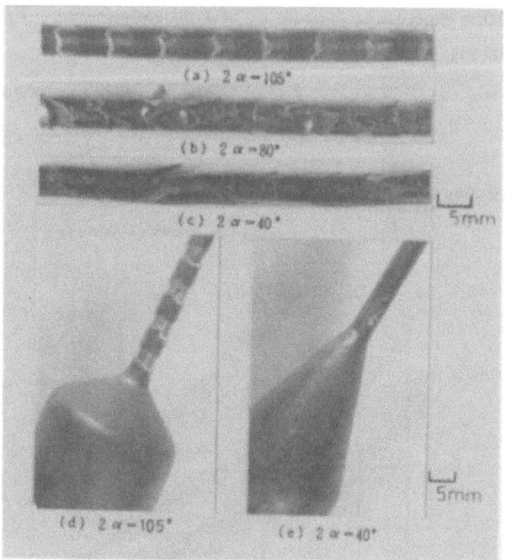

Fig. 12 Sheath fractures of Cu/Al extruded
rods and discards. (R=39, f_2=12%,
Y_2/Y_1=3.5)

Fig. 12 illustrates outer layer fractures of rods, caused by extrusion at a high reduction ratio. The fracture patterns vary depending on the die angle – at a large die angle of 105°, tensile fracture occurs vertically to the direction of extrusion at the die exit; at a small die angle of 40°, shear fracture has occurred already at the die approach, and fracture of the extruded rods inclines to the axis; at a medium die angle of 80°, the mixed type fracture occurs.

Fig. 13 shows the experimental results for the extrusion limits of annealed and hard Cu/Al billets at a die angle of 80°. In this figure, the solid line shows the calculated results of the anlaysis taking into consideration the temperature increase and strain hardening during extrusion and by choosing the suitable shear friction coefficients m_d, m_i, the results well coincide with the experimental data. The workable region is variable depending on the clad ratio and is narrow at ratios between 10 and 20%. Also, the annealed billet has small flow stress ratio Y_2/Y_1 and widens the workable region. The excellent agreement between the calculated and experimental data tends to be confined to the region of large die angles.

To obtain sound extrusion the lubricated condition of the die surface is very important, for good lubrication provides low extrusion pressure and sound extruded rods, as shown in Fig. 14.

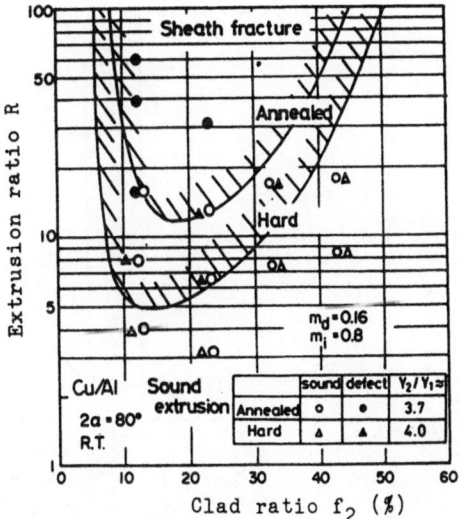

Fig. 13 Workability limits of Cu/Al clad metals.

For examples of hard core-soft sheath clad metals, the defects of Pb/Al extruded rods are shown in Fig. 15 and the comparison between the experimental data and analytical values of extrusion limits at various die angles and clad ratios is shown in Fig. 16.

When the clad ratio is in the region of 22 to 60%, the diameters of defective extruded rods are larger than that of the die hole, and the excess soft sheath is extruded as a pipe-like bamboo defect. When the clad ratio is 80%, the diameter of the extruded rod is equal to that of the die hole, and the hard core fractures. These experimental results support the deformation model shown in Fig. 7.

In Fig. 16, the outer layer defects occur at the higher extrusion ratio and larger die angle, independent of the extrusion ratio.

Hydrostatic extrusion with back pressure was also applied experimentally to work with clad metals, and it is obvious that back pressure prevents the occurrence of defects of hard core-soft sheath clad metals, as shown in Fig. 17 [10].

5. Thickness Distribution of Clad Layers

Clad composite metals are put to various uses, and the thickness distribution of the clad layer is one of the main parameters in producing articles of excellent quality. It is therefore

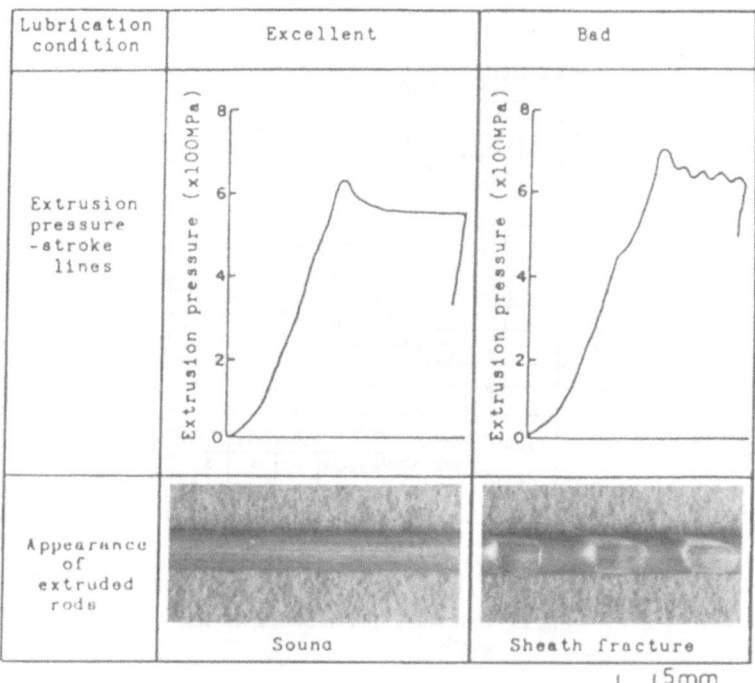

Fig. 14 Effect of lubrication condition on
extrusion pressure and appearance of
extruded rods (Cu/Al, f_2=12%,
R=15.7, $2\alpha_2$=80°).

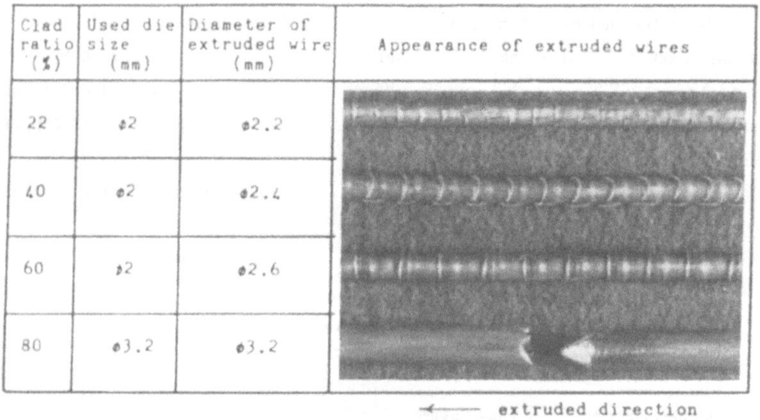

Clad ratio (%)	Used die size (mm)	Diameter of extruded wire (mm)	Appearance of extruded wires
22	φ2	φ2.2	
40	φ2	φ2.4	
60	φ2	φ2.6	
80	φ3.2	φ3.2	

◄──── extruded direction

Fig. 15 Appearance of Pb/Al extruded wires
with sheath defect and core fracture.
($2\alpha_2$=105°, R.T., Y_1/Y_2=5)

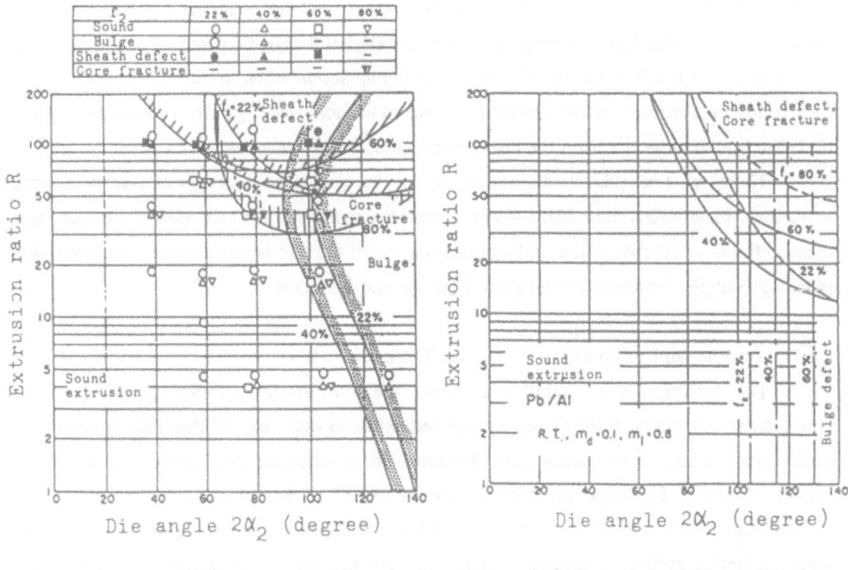

f_2	22%	40%	60%	80%
Sound	○	△	□	▽
Bulge	◐	△	–	–
Sheath defect	●	▲	■	–
Core fracture	–	–	–	▼

(a) Experimental data   (b) Calculated data

Fig. 16 Workability limits of Pb/Al clad
metals. (R.T., $Y_1/Y_2 = 2\sim5$)

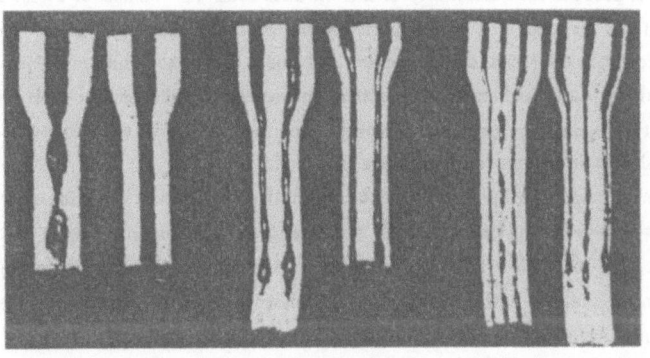

(a) $P_B = 0$ at.,   (b) $P_B = 0$ at.,   (c) $P_B = 0$ at.,
$P_3 = 1\,000$ at. $P_B = 1\,500$ at. $P_B = 2\,000$ at.
$r = 64\%$, $A_{f1} = 9\%$, $\alpha = 20°$

Fig. 17 Effect of back pressure on the
uniform flow during hydrostatic
extrusion of Al/Cu clad metals [10].

very important to maintain strict control of the thickness distribution when manufacturing these products. In the case of copper-clad aluminium used for electrical bus bars, constant thickness of the clad layer is necessary to prevent electrical corrosion, and of adequate depth especially if the bus bar will be subjected to a sliding action when in use. For multifilament superconducting wires, it is also important to maintain homogeneity in the cross section of the clad structure in order to obtain the high super-conducting characteristics.

There are two subjects to discuss about the clad layer thickness of clad composite extrusions: longitudinal uniformity – which provides high yield and high quality products; and cross-sectional uniformity – the uniform distribution of the clad layer on non-circular extruded rods, for example, hexagonal, rectangular, I-section and T-section.

5.1 Longitudinal Uniformity of Clad-Layer Thickness

The 'uniform flow' maintained during the hydrostatic extrusion process provides longitudinal uniformity in the dimensions of clad extruded rods. Fig. 18 shows the longitudinal distribution of clad layer thickness of hexagonal Pb/Al wire extruded hydrostatically from a cylindrical billet. The thickness is very constant except at the top end. Fig. 19 shows the longitudinal distribution of the outer copper layer thickness and copper ratio (copper area/ superconducting material area) of a superconducting rod which was extruded hydrostatically from the billet shown in Fig. 1 (c). Variations in the thickness and copper ratio are extremely small longitudinally, except at both ends.

Compared with hydrostatic extrusion, the conventional lubricated hot extrusion process would not produce such uniformity of the extruded rods. An example of a Cu/Fe clad metal extruded rod manufactured by the lubricated hot process is shown in Fig. 20. The clad ratio varies greatly in the longitudinal direction, and this is due to non-uniform deformation at the start of the extrusion and insufficient lubrication between the billet and the tools. Non-uniform deformation is caused by only axial compression on the billet before extrusion starts. Therefore the metal flow field is not constant during extrusion.

With respect to hydrostatic extrusion, lubrication between the billet and the tools is excellent and the extrusion starts with hydrostatic compression on the billet by the pressure medium. The metal flow field is constant and uniform from the beginning to the end of the extrusion, so that the extruded rods have similar sectional profiles to those of the billets and longitudinal uniformity.

5.2 Outer Layer Thickness Distribution of Square Profile Clad Products [11]

Outer layer thickness distribution was measured and analyzed for square profile clad products (width and thickness ratio: 4–12) extruded from a cylindrical billet in steady state. Fig. 21 shows the sectional view of a square profile Cu/Al clad product (size: 100mm × 10mm,

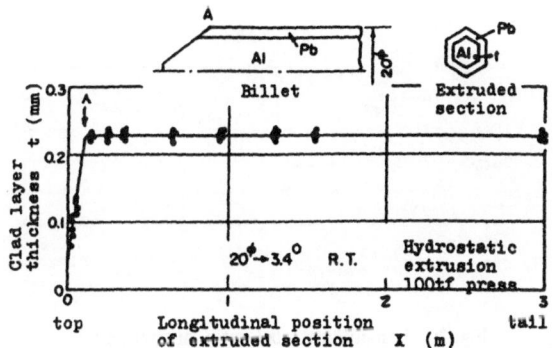

Fig. 18 Longitudinal clad-layer thickness distribution of a Pb/Al hexagonal extruded section.

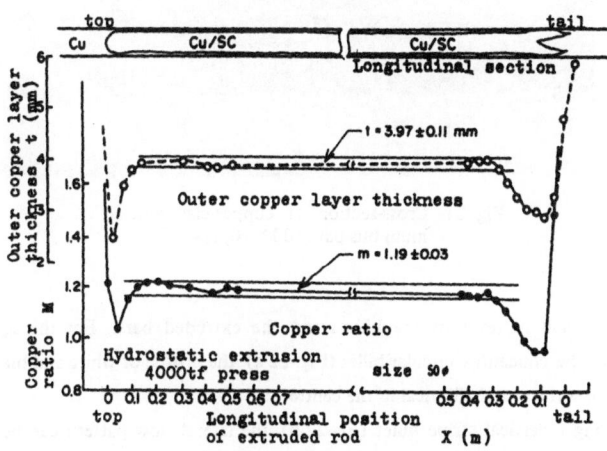

Fig. 19 Longitudinal distributions of outer copper layer thickness and copper ratio of the superconducting rod.

copper-clad ratio: 21%) extruded from a 170mm diameter cylindrical billet. The outer layer thickness is not constant, being thin along the length and thick at the ends of the square cross-section.

In order to observe the deformation state from a circular section to a square one, Plasticine billets were extruded with a small 100tf power hydrostatic extrusion press. Fig. 22

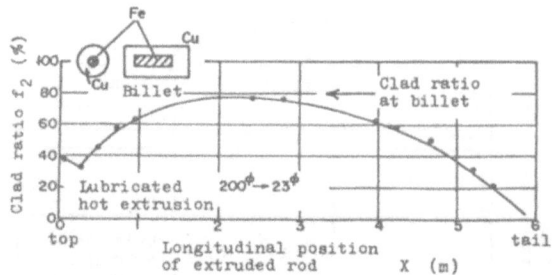

Fig. 20 Longitudinal clad ratio distribution of the Cu/Fe rod extruded by the lubricated hot extrusion process.

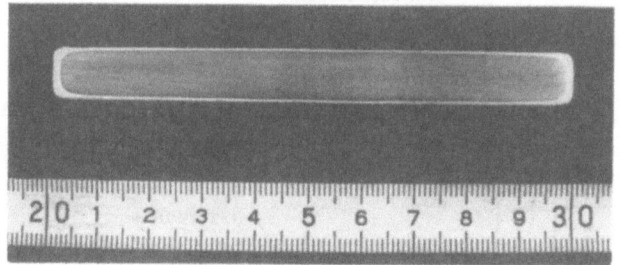

Fig. 21 Cross-section of copper-clad aluminium bus bar (100×10, $f_2 = 21\%$).

shows the sectional patterns of the billets and the extruded bars. For the square section extruded from the concentric circular billet (Fig. 22 a), the profile of white and black pattern is square near the surface, and elliptical in the centre.

By using a vertical stripe billet (Fig 22 b), the lateral flow pattern can be observed in detail. On the billet, the stripes are parallel to each other, while on the square section they are curved and not parallel. The stripes at the centre are swollen in the middle and look like a barrel, but those at both sides look like bows and are comparatively uniform in thickness.

With regard to deformation of the units, the area reduction ratio is constant in the section and the following equations are derived throughout the section according to the constant volume condition during extrusion.

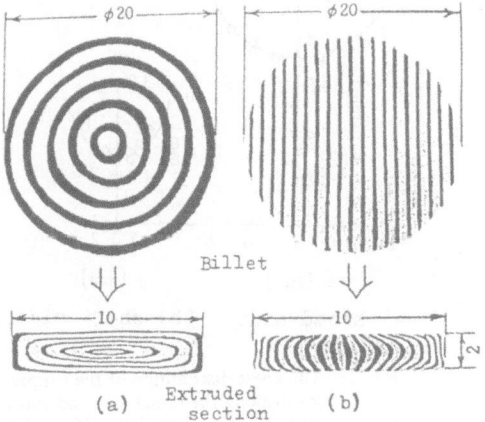

Fig. 22 Flow pattern of square profile products extruded from cylindrical
Plasticine billets by a small hydrostatic extrusion press.
(a) concentric circular billet,
(b) vertical stripe billet

$$\epsilon_z = \ln R \qquad (15)$$

$$\epsilon_x + \epsilon_y = -\ln R \qquad (16)$$

where ϵ_z is axial strain, ϵ_x and ϵ_y are width and thickness direction strains, respectively, and R is extrusion ratio. These equations hold true for hydrostatic extrusion in steady state, and are independent of the extruded product profiles.

On the basis of flow pattern observation and these equations, a theoretical analysis has been carried out for the deformation of square profile bars from cylindrical clad billets produced by hydrostatic extrusion. It can calculate the thickness distribution of the clad-layer on the square cross-section in detail.

The thickness distribution of the copper-layer along the broad and short sides of the bus bar (100×10, $f_2 = 15\%$) is shown in Fig. 23. The circles indicate measured values and the solid lines calculated ones. The copper-layer is thin at the neutral position and thick at the corners on both sides. Also, the layer along the short side is thicker than that along the broad side.

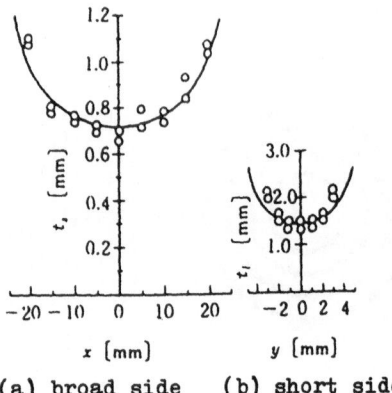

(a) broad side   (b) short side

Fig. 23 Thickness distribution of the copper-
layer along the broad side and short
side of the bus bar (100 × 10,
$f_2 = 15\%$): (a) along the broad side,
(b) along the short side.

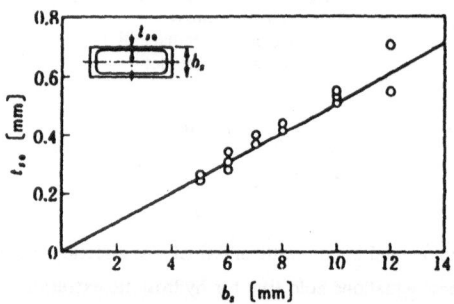

Fig. 24 Relation between the copper-layer
thickness at the neutral position of
the broad side and the thickness of
bus bars ($f_2 = 15\%$).

The relation between the copper-layer thickness at the neutral position of the broad
side and the thickness of bus bars ($f_2 = 15\%$) are shown in Fig. 24. Fig. 25 also shows the
relation between the copper-layer thickness at the neutral position of the short side and the
width of various bus bars ($f_2 = 15\%$). The relations are linear and the copper-layer thickness
increases proportionally.

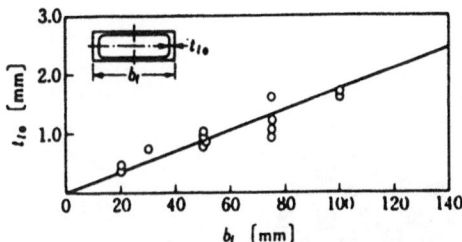

Fig. 25 Relation between the copper-layer
thickness at the neutral position of
the short side and the width of
various bus bars ($f_2 = 15\%$).

5.3. Material Flow and Clad-Layer Thickness Distribution of Complex Profile Products

5.3.1 Material flow of complex profile extruded sections.

Concentric and checkered Plasticine billets are extruded to more complex sections, such as I or T, in order to demonstrate the flow patterns during the deformation of these sections. The patterns are rather simple in spite of the complex profiles, as shown in Fig. 26. Using the concentric billet profiles the ouline of information is obtained about the deformation, while the checker pattern correlation between the billets and the extruded sections provides more detailed information about it. The extruded section area can be divided into several rectangular units. It is easy to understand the deformation states of complex profile extruded sections, by assuming them to be integrations of each square unit's deformation.

Fig. 27 shows the positional correspondence between the extruded sections and the billets derived from Fig. 26. This correspondence determines what shapes the divided rectangular units in the extruded sections are in the billet. As a result, the whole deformation in the section is understood as an integration of each unit's deformation. Even during these extrusions, equations 14 and 15 hold true, and the area reduction ratios from the billets to the extruded sections are constant in each rectangular unit. That is to say, the occupancy area ratios of P-1, P-2 and P-3 in the I-section, and of P-1, P-2, E-1, E-2 and E-3 in the T-section do not change between the extruded sections and the billets.

With respect to any other profile extrusion, it is possible to divide the extruded sections into several rectangular units, like an I or a T-section, as shown in Fig. 27, and to approximately predict the corresponding positions in the billets. The rule in the prediction is to keep the occupancy area ratios of each rectangular unit constant.

272

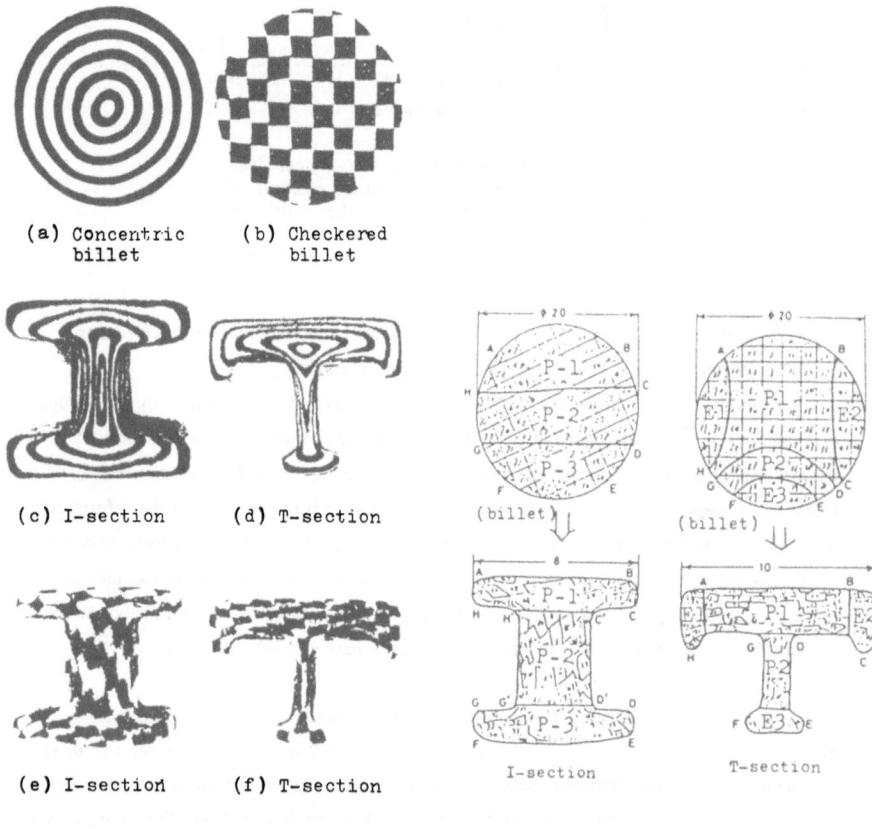

(a) Concentric billet

(b) Checkered billet

(c) I-section

(d) T-section

(e) I-section

(f) T-section

(billet)

(billet)

I-section

T-section

Fig. 26 Flow pattern of complex profile products extruded from cylindrical Plasticine billets.

Fig. 27 Correspondence between the position in the cross-section of cylindrical billets and complex profile sections.

5.3.2 Interface profiles of clad products.

Interface profiles of clad products can be predicted from the flow pattern during extrusion. Since the area reduction ratio is equal to $1/R$ everywhere in the section, the following equation about the clad-layer thickness t_e in the extruded section is derived, as shown in Fig. 28.

$$t_e = \frac{t_b}{R} \frac{a_b}{a_e} \tag{17}$$

where t_b is the clad-layer thickness at the billet, a_b and a_e are the distance $\overline{AB}$ at the billet and

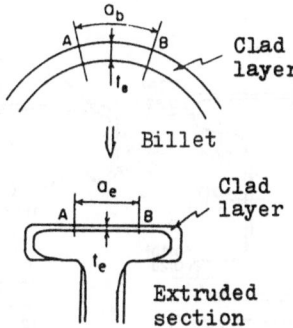

Fig. 28 Comparison between the clad-layer
thickness of a cylindrical billet and
a complex profile extruded section.

extruded section, respectively.

Since concentric billets are used commercially, as shown in Fig. 1, the clad-layer thickness t_b is constant. By predicting the value of a_b/a_e from the flow pattern, the thickness distribution of the clad-layer in the extruded sections can be calculated. Considering the deformation of the corresponding points $A - H$ on the outer-profiles of the billet and extruded section I, the change of the length CD during the extrusion is smaller than that of the length AB, and the clad-layer thickness at the position C'D' on the extruded section is thinner. Also, the position C on the billet corresponds to the line C'C on the extruded section and the clad-layer thickness along the line is especially thin.

For the T-section, the distance $\overline{CDE}$ spreads especially at the extruded section, and the clad-layer thickness becomes thin along the distance. Generally, the clad-layer thickness is thin on the concave sides of complex sections.

The thickness distribution of the clad-layer of the I and T-sections extruded from Cu/Al clad billets (clad ratio $f_2 = 16\%$) is shown in Fig. 29. In this figure, the values within parentheses indicate the calculated thickness from the deformations of each rectangular unit during extrusion, as shown in Fig. 28, using the Eq. 17, while the value without parentheses indicate the measured thicknesses on the sections. Both data are in good agreement, which means that the thickness distribution of the clad-layer on the complex section is approximately estimated by this method.

5.3.3. Factors which influence interface profiles.

There are many factors which affect the interface profiles of the complex sections: for example, the clad billet conditions (clad ratio, the flow stress ratio of metals and preparation of

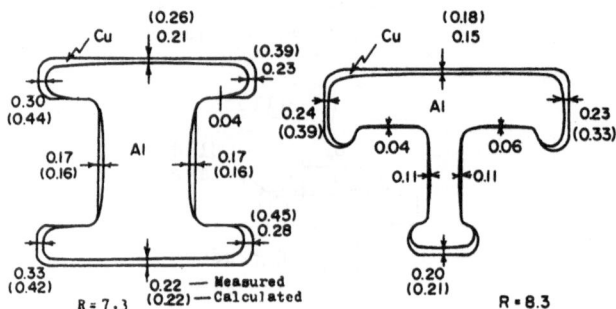

Fig. 29 Clad-layer thickness distribution of hydrostatically extruded I-section and T-section.

the interface); and the extrusion conditions (the die shape used, extrusion ratio, lubrication condition between billet and tools, temperature). Of these factors, the die shape used, is a main independent one from the billet specification.

The positional correspondence between the billets and the extruded sections, estimated from the flow pattern of Plasticine billets, as shown in Fig. 26, is affected especially by the die shape. The clad-layer thickness distribution on the complex sections is varied by the shape of the die. In general, conical approach and spherical or flat bottom dies are used in the extrusion of complex sections. Using such a die shape, the whole thickness distribution of the clad-layer can almost be determined, but the clad-layer thickness on the local concave sides of complex sections is varied by the change of the local die shape. For example, even the small radius at the corner of the bottom to the land improves the thickness distribution there. It is important to choose a die shape which makes the clad-layer thickness uniform.

If other working methods are applied to clad metals, the thickness distribution also varies. Fig. 30 shows the clad-layer thickness distribution of hexagonal Pb/Al wire manufactured by the drawing method compared with the hydrostatic extrusion method. In the hydrostatic extrusion method, the clad-layer thickness is more uniform on the section. While, in the drawing method, the interface profile is round, not hexagonal, and the clad-layer thickness is varied on the section. The extrusion ratio of the clad material is about 60, while the reduction ratio of drawing is about 20% (area ratio is about 1.25). It is considered that these differences of working ratio affect the clad-layer thickness distribution on the hexagonal section.

In order to produce excellent clad complex sections, it is important to make the clad-layer uniform by controlling the various process parameters referred to above.

6. Clad Composite Products

6.1. Copper-clad Aluminium Products

Copper-clad aluminium rods and shapes were among the first products to be manufactured industrially by hydrostatic extrusion [12]. If these clad products were extruded with the conventional hot press, the metals would react with each other to produce brittle compounds due to the high-temperature environment. On the other hand, by using the cold hydrostatic extrusion press, sound extrusion can be obtained without producing brittle compounds.

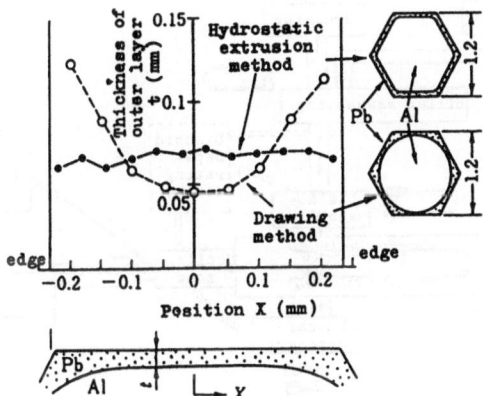

Fig. 30   Thickness distribution of outer-layer of lead-clad aluminium hexagonal wire.

The production line of copper-clad aluminium wires and bars is shown in Fig. 31. The clad billets are manufactured by the insertion method, and large aluminium rods and large copper pipes are used. The large aluminium rod is machined, cleaned, brushed on the surface and assembled into the large copper pipe the inner-surface of which is cleaned and brushed on another line. The top end of the assembled billet is upset, the tail end is sealed and the billet making process is completed. The billet is then extruded with hydrostatic extrusion press, for example 4000tf power press, to produce straight bars or a coil.

The appearance of various copper-clad aluminium products manufactured by the hydrostatic extrusion method is shown in Fig. 32. Copper-clad volume ratio is 10 or 15% usually, and $9 - 50$ round rods and $20 \times 5 - 100 \times 12$ bars are extruded at room temperature.

These copper-clad aluminium sections and wires are used for various conductor products.

6.2. Lead-clad Aluminium Wires

Lead-clad aluminium wires, coated with low melting-point metal, is also a product suitable for manufacture by hydrostatic extrusion. However, if the extrusion ratio is very high,

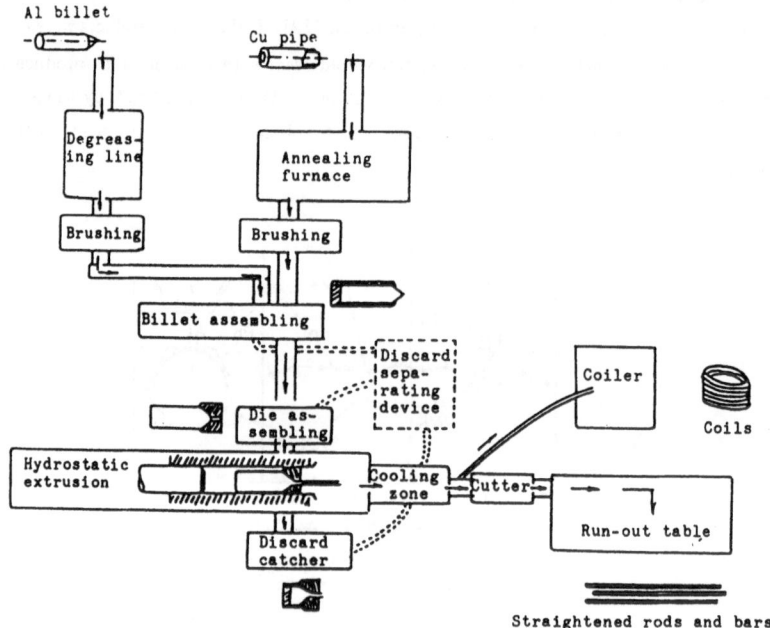

Fig. 31 Production line of copper-clad aluminium wires and bars.

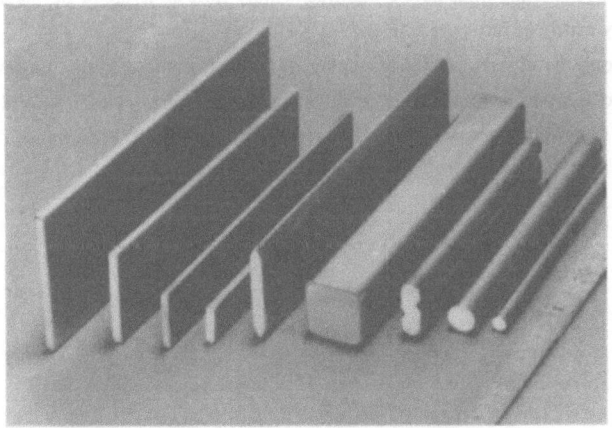

Fig. 32 Appearance of copper-clad aluminium products manufactured by the
hydrostatic extrusion method.

the lead-clad layer will be melted by the generated heat. Therefore, the extrusion ratio is limited to 60 even if hydrostatic extrusion is applied. To extrude down to sizes near those of final products, it is effective to use a small hydrostatic extrusion press.

As a unique example of the commercial use of these wires, the authors' method of producing an ultra-fine honeycomb structure is as follows [1].

To produce single clad wires, round-section lead-clad aluminium billets are hydrostatically extruded to hexagonal-section wires at R.T., and these are then drawn down to their final sizes, retaining their hexagonal-section. In this process, the benefit of hydrostatic extrusion is uniform thickness of the lead-clad layers of the extruded wires both in cross section and longitudinally as shown in Fig. 30.

The hexagonal-section of lead-clad aluminium wires is then cut to a constant length, arranged in order, and bonded all together with epoxy resin. The cores of aluminium are then removed chemically, and a lead honeycomb structure is produced. These honeycomb structures are mass-produced as X-ray collimators for medical scintillation cameras. A collimator and an enlarged view of the structure are shown in Fig. 33.

6.3. Superconducting Wires

Recent increases in the use of high-performance superconducting (SC) magnets for nuclear fusion research, magnetic levitation train systems, etc. demand not only high quality SC wires, but also a reliable method for mass-producing them.

Superconducting wires have a multifilamentary fibre structure of SC material in a high conductivity matrix of very pure copper.

Among the different methods employed, the hydrostatic extrusion process is effective and reliable for the manufacture of such SC composite wires. Nb-Ti alloyed and Nb$_3$Sn compound superconductors are now manufactured by hydrostatic extrusion on an industrial sclae [13], [14].

(a) Appearance of Collimator
$\left(\begin{array}{l}\text{Diameter: 4oo mm, Thickness: 20-30 mm,}\\ \text{Number of holes : 50000 - 100000}\end{array}\right)$

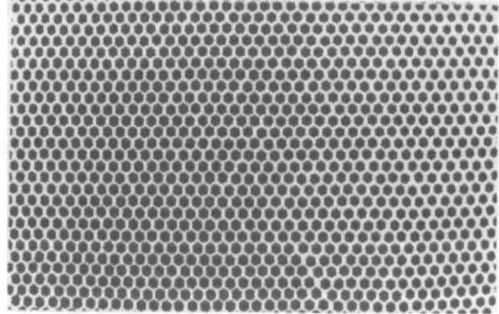

.(b) Enlarged View of Honeycomb Structure
$\left(\begin{array}{l}\text{Hole size : 1.0 - 1.5 mm}\\ \text{Wall thickness : 0.15 - 0.25 mm}\end{array}\right)$

Fig. 33 Lead honeycomb structure using lead-clad aluminium wires manufactured by the hydrostatic extrusion method.

The structure of a multi-filament billet of an alloyed superconductor is illustrated in Fig. 1 (c). Figure 34 shows the relation between extrusion pressure at a steady state and extrusion ratio, and cross-sections of alloyed Nb-Ti SC extrusions are shown in Fig. 35 (a).

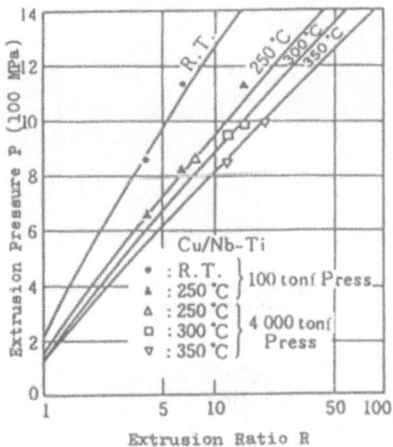

Fig. 34 Relation between extrusion pressure and extrusion ratio of Nb-Ti SC billet.

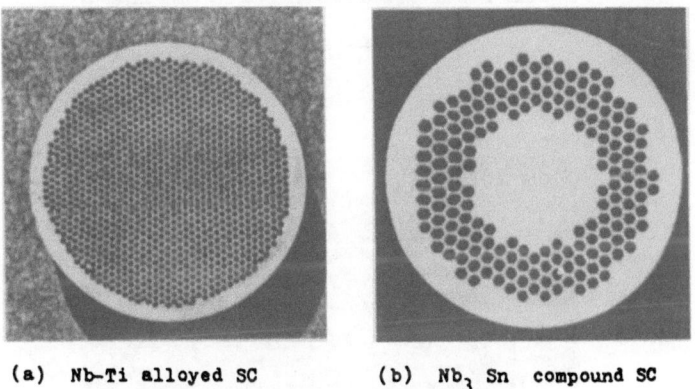

(a) Nb-Ti alloyed SC   (b) Nb$_3$ Sn compound SC

Fig. 35 Cross-sectional photo of extruded SC rods (50$^\phi$).

(a) Appearance of the 4000tf hydrostatic
extrusion press

(b) Appearance of SC billets

(c) Appearance of SC extruded rods

Fig. 36 Production scenery of SC wires by
the 4000tf hydrostatic extrusion
press.

Figure 36 shows the production scenery of SC wires by the 4000tf hydrostatic extrusion press.

In the production of alloyed SC wires, billets with a diameter of about 150mm are extruded down to the sizes 40 – 50mm, thereafter, the extruded rods are drawn to the size of the final products and 1000 – 2000 filaments are contained in a wire.

A flow chart of the production procedure for Nb_3Sn superconducting wire is shown in Fig. 37. When Nb_3Sn superconducting wire is fabricated by the bronze process, triple extrusion of single-, submulti-, and multi-billets is carried out.

The single billet consists of a 13%Sn bronze pipe and a Nb ingot. The submulti-billet is composed of a large number of drawn hexagonal wires surrounded by a Nb diffusion barrier pipe within a copper pipe. The multi-billet is composed of a copper pipe containing a large number of submulti-wire rods tightly packed in a hexagonal arrangement.

Superconducting wire is manufactured by wire drawing, using the draw bench and the bull block, twisting, and final heat treatment for the Nb_3Sn formation. The extruded and

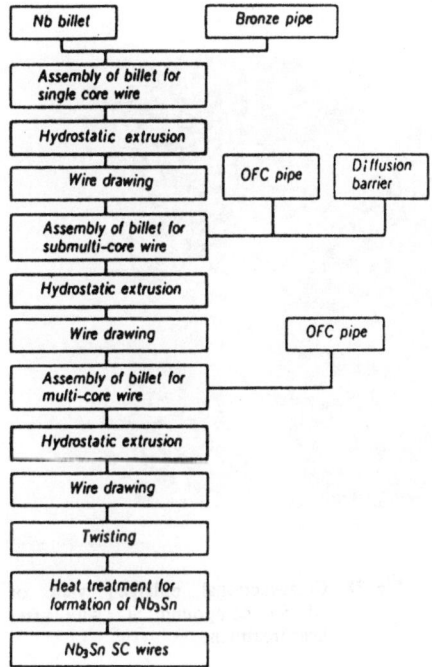

Fig. 37 Manufacturing sequence for Nb_3Sn
SC wire.

drawn wire has a highly uniform arrangement as shown in the cross-sectional view in Fig 35 (b). A cross-section of Nb_3Sn wire after heat treatment is shown in Fig. 38. In this figure an approximately 1 μm thick Nb_3Sn compound is homogeneously formed.

Critical current density (Jc) of Nb-Ti and Nb_3Sn wire obtained by the hydrostatic extrusion process is shown in Fig. 39 as a function of the magnetic field. The Jc of the Nb-Ti SC wire is $1.1 \times 10^5 A/cm^2$ at 8T using the optimum manufacturing process. These are better than those of wires formed by the conventional extrusion process. It is said that working of SC material at high temperature harms its SC properties. The deformation of warm hydrostatic extrusion at $200 - 400°C$ is nearly a cold working for SC materials, which does not harm their properties. The condition of warm hydrostatic extrusion of high reduction produces good bonding between each filament, which provides the excellent workability in the following process on the industrial scale. The uniformity of material flow during extrusion in the longitudinal direction, shown in Fig. 19, maintains the yields during extrusion at a high level, which means that warm hydrostatic extrusion is the most suitable working method for the expensive superconducting materials.

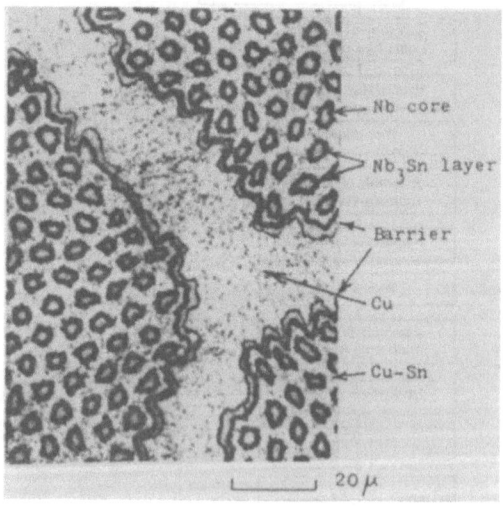

Fig. 38 Cross-sectional enlarged view of Nb_3Sn compound SC wire after heat-treatment.

283

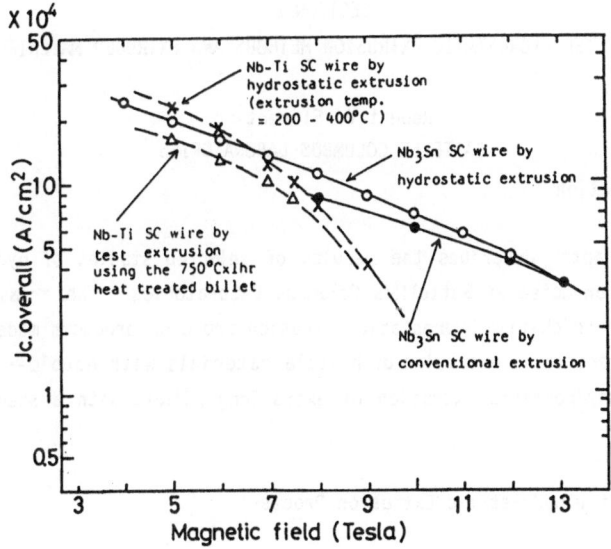

Fig. 39 Relation between magnetic field and critical current density at 4.2 K of superconducting wires.

REFERENCES

1. M. Seido and S. Mitsugi, J. Japan Soc. Tech. Plasticity, 21-238 (1980) 942-948.

2. M. Nishihara, J. High Pressure Inst. Japan, 11-2 (1973) 44-47.

3. B. Avitzur, Metal Forming: Processes and Analysis, McGraw-Hill New York, 1968, p.153.

4. M. Seido, et al., Trans. Japan Soc. Mech. Eng. 48-427 (1982) 378-385.

5. M. Seido, Investigation on Hydrostatic Extrusion of Clad Composite Metals, PhD thesis, Tokyo Institute of Technology, 1984.

6. Y. Yamaguchi, et al., J. Japan Soc. Tech. Plasticity, 15-164 (1974) 724-729.

7. B. Avitzur, Journal of Engineering for Industry, 104 (1982) 293-304.

8. R. Hill and S.J. Tupper, Journal of the Iron and Steel Institute, (1948) 353-359.

9. K. Kubo, Journal of the Society of Materials Science, 20-215 (1971) 35.

10. Y. Matsuura and K. Takase, J. Japan Soc. Tech. Plasticity, 15-157 (1974) 156-165.

11. M. Seido, J. Japan Soc. Tech. Plasticity, 18-198 (1977) 545-552.

12. R. Hogland, et al., Metall, 31 (1977), 515-519.

13. S. Mitsugi, et al., J. Japan Soc. Tech. Plasticity, 23-260 (1982) 909.

14. S. Sakai, et al., ICEC 9/ICMC, (1982) HB-2.

SECTION 2

SELECTED HYDROSTATIC EXTRUSION METHODS AND EXTRUDED MATERIALS

Robert J. Fiorentino

BATTELLE COLUMBUS LABORATORIES

1. Introduction

This chapter describes the results of selected studies of hydrostatic
extrusion conducted at Battelle's Columbus Laboratories. The areas addres-
sed are the thick-film hydrostatic extrusion process, products made by this
process, hydrostatic extrusion of brittle materials with double-reduction
dies, and hydrostatic extrusion of extra long billets with a stepped-bore
container.

2. Thick-Film Hydrostatic Extrusion Process

The thick-film hydrostatic extrusion process* (1,2) is aimed at over-
coming many of the disadvantages of pure hydrostatic extrusion under conditions where a relatively large volume of hydrostatic fluid is typically
used. Among the disadvantages attributed to a large fluid volume are
billet-temperature limitations, occasional difficulty in stick-slip and
billet-stopping control (which can result in unintentional complete extrusion), reduced productivity due to a lower ratio of billet-to-container
volume, and potentially lower total extrusion cycle rates due to greater
fluid-handling time.

A schematic representation of the thick-film hydrostatic extrusion
process is shown in Figure 1. Two key features of the process are:

(1) Hydrostatic fluid is minimized. This is minimized in two loca-
     tions.

   (a) Between billet and container. The radial clearance can be
        less than 1 mm.

---

*   This is also known as the Hydrafilm process, a designation often used
     by Battelle for convenience.

285

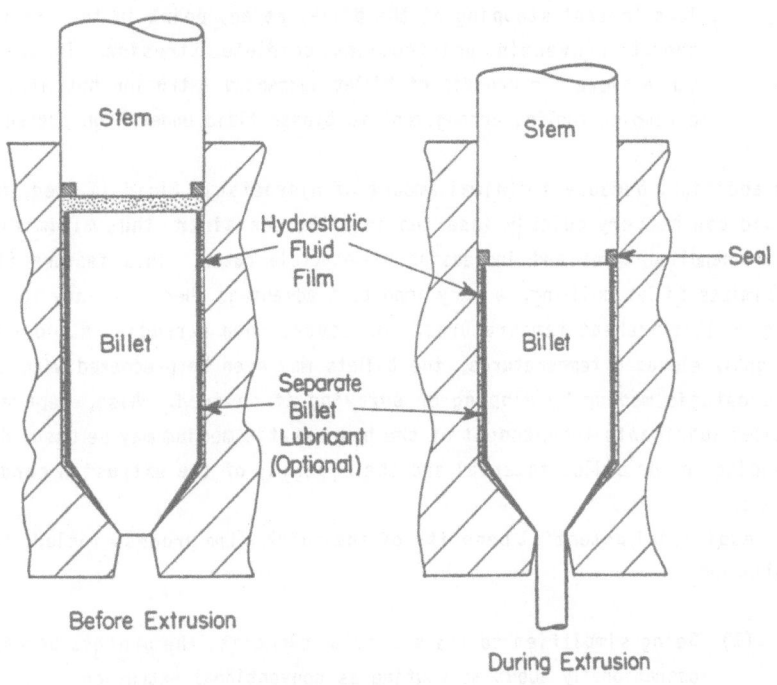

FIGURE 1. A SCHEMATIC OF THE THICK-FILM HYDROSTATIC
EXTRUSION PROCESS

(b) <u>Between billet and stem</u>. Only enough fluid is used to allow enough fluid pressurization to prevent or minimize billet upsetting.

(2) <u>Stem may contact billet</u>. Allowing the stem to contact the billet at least at the start of extrusion permits further minimization of the hydrostatic fluid, prevents or minimizes stick-slip, and allows instant stopping of the billet at any point in the stroke, thereby preventing unintentional complete extrusion. Thus, one can achieve the benefit of billet-augmented extrusion but without a complex tooling arrangement to bypass fluid under high pressure.

In addition, because a minimal amount of hydrostatic fluid is used, the fluid can be very quickly inserted into the container, thus minimizing fluid-handling times and increasing billet-cycle rates. This feature also minimizes billet chilling, a very important advantage when necessary to use high billet preheat temperatures. Moreover, when extruding at room or slightly elevated temperatures, the billets may even be precoated with the hydrostatic medium by dipping or spraying if desired. Also, separate billet lubricants independent of the hydrostatic medium may be used, depending on the billet material and the severity of the extrusion conditions.

Additional potential benefits of the thick-film process include the following:

(1) Being simplified to its essential elements, the process becomes operationally about as routine as conventional extrusion.

(2) The process simplification makes it comparatively easy to convert conventional presses for hydrostatic extrusion.

(3) The thick-film concept can be applied in high-speed mechanical presses and potentially at production rates in the range of 30 to 40 billets/minute.

(4) Hot extrusion at billet temperatures up to 1200 C and beyond is feasible.

2.1 Process Mechanics for Solid Billets

In the thick-film hydrostatic extrusion process, the amount of billet-augmentation stress applied by the ram is a function of the fluid volume. In Case 1 shown in Figure 2, the fluid is only present in the billet-container annulus and none is above (or behind) the billet at the very beginning of the operation (Figure 2a). When the axial billet stress, P_b, applied directly by the ram reaches the billet's yield stress, σ_y, the billet will begin to upset. At that point, the fluid pressure, P_f, would still be essentially zero, neglecting any minor pressurization due to elastic compression of the billet. Further increase in the axial billet stress by the ram would increase the fluid pressure above the deformation zone roughly according to the following:

$$\text{If } P_b = \sigma_y (x), \tag{1}$$

$$\text{then, } P_f = \sigma_y (x-1) = P_b - \sigma_y, \tag{2}$$

where x is some multiple factor of the billet yield strength. Thus, when P_b finally reaches the extrusion pressure, P_e, the fluid pressure would be approximately:

$$P_f = P_e - \sigma_y. \tag{3}$$

Thus, the fluid pressure in this case depends essentially on the difference between the axial billet stress (up to the start of extrusion) and the billet yield strength. The fluid pressure continuously reacts against the tendency of the billet to upset. Furthermore, because the billet-container annular clearance is kept relatively small, the billet cannot upset much even if the fluid pressure were a bit too low to prevent it. At the most, the billet may bulge slightly at mid-length and perhaps make slight contact with the container wall. Such small billet contact in the presence of a highly pressurized fluid lubricant, however, does not result in any significant contribution to container friction.

As the billet extrudes in Case 1, the advancing ram pressurizes the trapped fluid further. At some stage in the ram stroke, the fluid pressure, P_f, would attempt to exceed the billet-end pressure, P_e. At this

288

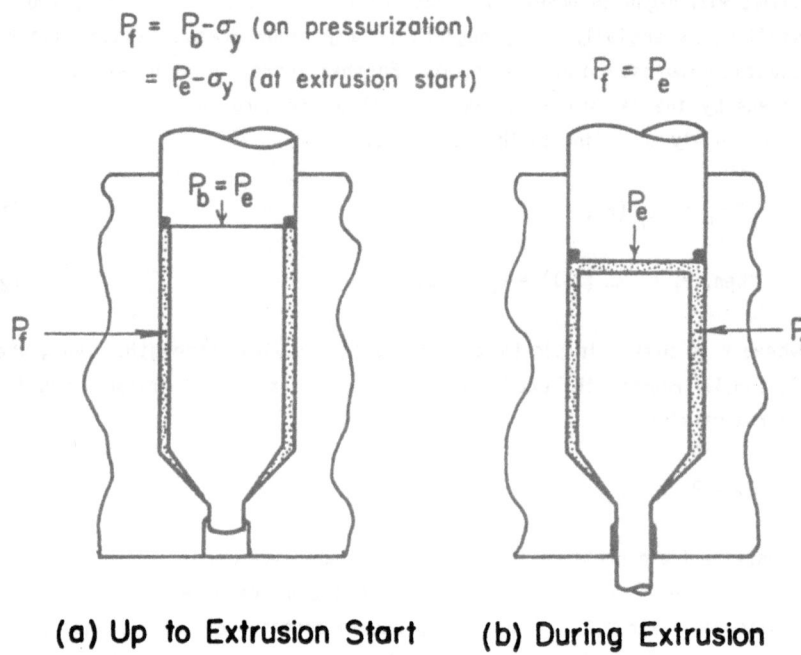

(a) Up to Extrusion Start   (b) During Extrusion

FIGURE 2.   THICK-FILM HYDROSTATIC EXTRUSION OF SOLID BILLETS
WITH NO FLUID HEAD AT START (CASE 1)

point, the ram would tend to separate from the billet (Figure 2b), provided that little or no fluid is lost by extrusion with the product. Thus, the fluid pressure and billet-end pressure would equalize, i.e.,

$$P_e = P_f \qquad\qquad\qquad (4)$$

which is the condition for pure hydrostatic extrusion.

Such a shift to pure hydrostatic extrusion, however, does not present a serious problem of instability or stick-slip. This is mainly because the billet already would be extruding. In pure hydrostatic extrusion with a large fluid volume, stick-slip tends to occur at extrusion breakthrough and, if it does occur, any resulting lubricant breakdown at the billet-die interface may tend to promote stick-slip during extrusion runout as well. However, there is ample evidence (3) that if stick-slip can be prevented at breakthrough, it generally will not develop during runout partly because of hydrodynamic lubrication effects. Thus, it would not be expected to occur during runout in Case 1. Moreover, the fact that the fluid volume is minimal and already fully compressed would further tend to prevent the initiation of stick-slip.

Case 2 differs from Case 1 in that a fluid head is present above the billet at the very start. The fluid head volume, of course, controls the amount of billet-augmentation stress over the range from:

(a) Zero augmentation (pure hydrostatic extrusion) $P_e = P_f$, to
(b) Maximum augmentation (Case 1) $P_e = P_f + \sigma_y$ (from Equation 3).

By controlling the fluid volume carefully, it would be possible to augment the billet stress by an amount approaching the yield stress and still prevent the billet from upsetting. This condition may be desirable to achieve for some applications.

2.2 Process Mechanics for Tubular Billets

Two mandrel arrangements commonly used in hydrostatic extrusion are the fixed and floating designs. The fixed-mandrel design, shown in Figure 3, requires an internal tube to support the mandrel at a fixed distance between the mandrel collar and die. (A fixed mandrel may also be supported

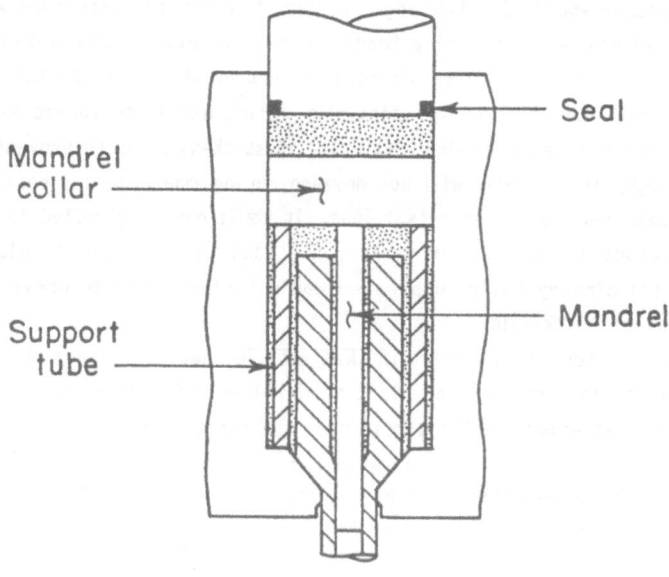

FIGURE 3. INTERNALLY SUPPORTED FIXED-MANDREL ARRANGEMENT
IN PURE HYDROSTATIC EXTRUSION

externally by a separate hydraulic system.) In the floating-mandrel arrangement, the mandrel is supported by bearing directly on the billet itself and moves as the billet extrudes. The floating design is shown schematically in Figure 4 as normally used in "pure" hydrostatic extrusion. (In this case, the word "pure" is used in the traditional sense in that ample fluid exists between (a) the ram and mandrel collar and (b) billet and container wall; however, the billet is necessarily pressure-augmented.) Each method has certain advantages and disadvantages from the standpoint of extrusion pressure requirements, billet-upsetting problems, and tube-wall uniformity. It is obvious from the fixed design that the conditions for pure hydrostatic extrusion prevail. The following discusses the process mechanics for the floating-mandrel design.

In pure hydrostatic extrusion, the main advantages of the floating, over the internally supported, fixed-mandrel arrangement are:

(1) No need for an internal support tube which takes up potential billet volume and can be extremely highly stressed under certain conditions

(2) Generally easier to seal at the mandrel-billet interface

(3) Billet-augmentation pressure is available by an amount proportional to the ratio of cross sectional areas of the unsupported mandrel (A_m) to the billet (A_b), or:

$$P_b = P_e = P_f + P_f \left(\frac{A_m}{A_b}\right) \tag{5}$$

where $P_f \left(\dfrac{A_m}{A_b}\right)$ is the augmenting pressure.

In pure hydrostatic extrusion (shown in Figure 4a), however, the availability of billet-augmentation is not always desirable. This is because of the problem of potential billet upsetting which will occur when

$$P_e - P_f = P_f (A_m/A_b) > \sigma_y \tag{6}$$

292

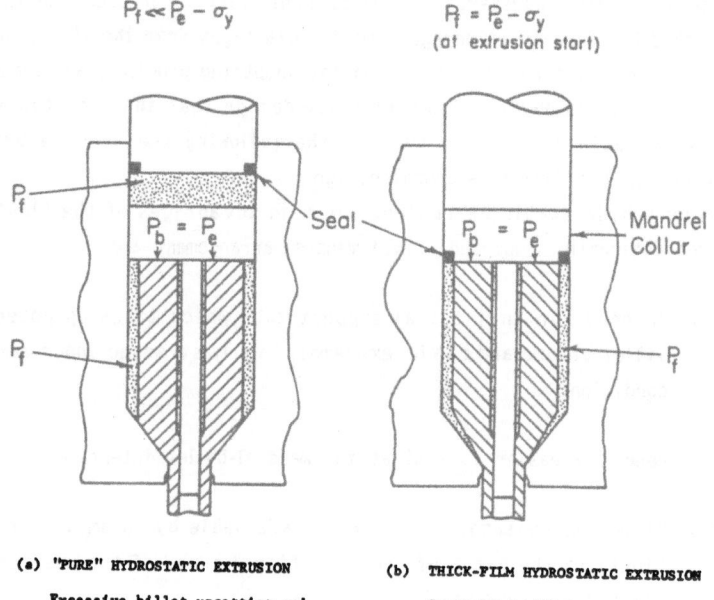

$P_f \ll P_e - \sigma_y$

$P_f = P_e - \sigma_y$
(at extrusion start)

Seal

Mandrel
Collar

P_f

$P_b = P_e$

P_f

$P_b = P_e$

P_f

(a) "PURE" HYDROSTATIC EXTRUSION

Excessive billet upsetting and
container friction can occur

(b) THICK-FILM HYDROSTATIC EXTRUSION

Controlled billet upsetting and
no container friction

FIGURE 4.   COMPARISON OF FLOATING MANDREL IN BOTH "PURE"
AND THICK-FILM HYDROSTATIC EXTRUSION

Thus, upsetting can be prevented in this approach only by controlling the relative cross-sectional areas of the mandrel and billet. This often imposes an unwanted restriction on starting billet dimensions in respect to the mandrel. Furthermore, although billet upsetting may not occur for a given billet/mandrel geometry at a given extrusion ratio, it could readily occur merely on increasing the ratio.

In addition, the ability to use a floating mandrel for extruding relatively thin-walled billets (e.g., OD/wall ratios of about 8 or more) becomes essentially impossible. This is because the unsupported area of the mandrel simply is much too large compared to that of the billet. Attempts have been made to solve this problem by projecting a portion of the mandrel into a recess in the ram, thereby reducing the unsupported area of the mandrel exposed to fluid pressure. However, this approach adds to the complexity of tooling and process control.

One solution appears to be the thick-film hydrostatic extrusion process. As in the case with solid billets, the billet-to-container clearance is minimized so that, if billet upsetting does occur, the extent of upsetting is limited to an acceptable level.

However, this change alone may not be enough. This is because container and die friction may develop to a significant level, especially under conditions where the billet-augmentation stress is high compared to the fluid pressure (e.g., thin-wall tubular billets). The fluid in the annular and wedge regions at the start would merely be displaced into the fluid volume above the mandrel collar. Thus, the benefits of fluid pressure in those regions to minimize container and die friction could be seriously reduced. In fact, this condition could be likened to that of conventional lubricated extrusion, in spite of the presence of hydrostatic fluid behind the billet.

A subtle but significant change that must be made is to alter the position of ram seal to that shown in Figure 4b. Here, the seal is located on the lower surface of the mandrel collar. In so doing, the fluid in the annulus is confined to that region as in the case with solid billets. Thus, the amount of billet-augmentation stress applied by the ram depends only on fluid volume below the mandrel collar and no longer on the relative cross-sectional areas of the mandrel and billet. In other words, for the case where no fluid head is used between the billet and mandrel collar, Equations (1) through (3) apply up to extrusion start. Further extrusion could then result in a change to pure hydrostatic extrusion.

For the case where a fluid head is used at the start between the billet and mandrel collar, the billet-augmentation stress may be varied from zero ($P_e = P_f$) to the maximum ($P_e = P_f + \sigma_y$) by the fluid head volume.

Thus, both a change in seal location and a reduction in billet-to-container clearance are extremely important factors which make it possible in the thick-film process to extend the capabilities of the floating-mandrel arrangement. It now becomes possible to use the floating mandrel regardless of the relative cross-sectional areas of the billet and mandrel. This means, of course, that thin-wall tubular billets can now be extruded just as easily as the thicker-walled sizes.

As a further point, it is evident that the floating-mandrel design, as shown in Figure 4b, is also equivalent to a traveling-mandrel design in which the mandrel is attached directly to the base of the ram. When the traveling-mandrel design is used in the thick-film process, billet-augmented extrusion can be readily used to minimize any problems of stick-slip. This could not be done in pure hydrostatic extrusion where the billet-to-container clearance is large and under conditions that would result in billet upsetting.

2.3 Process Capabilities

The general capabilities of the thick-film hydrostatic extrusion process have been demonstrated in several areas.

2.3.1 Ti-6Al-4V Tubing

Ti-6Al-4V is difficult to extrude because of its high strength (about 915 MPa yield in the annealed condition) and difficulty in lubricating effectively. The thick-film process was applied at warm temperatures to the extrusion of thin-wall tubing from this alloy. (4) The extrusion conditions used were:

    Billet size - 15.7 mm OD x 2.0 mm wall
    Billet 1/d ratio - 6.5:1
    Mandrel design - traveling (floating)
    Mandrel size - 11.3 mm diameter

Die angle - 45 degrees included

Extrusion ratio and pressure - 3:1 (1370 MPa); 5:1 (1800 MPa)

Extrusion size -

    3:1 - 12.8 mm OD x 0.74 mm wall

    5:1 - 12.4 mm OD x 0.5 mm wall

Billet temperature - 815 C

Tooling temperature - 315 C

Ram speed - 34 mm/sec

Hydrostatic medium - high-temperature grease with additives.

A typical extrusion pressure curve obtained in this work is shown in Figure 5. This was for an extrusion ratio of 3:1. The fact that the curve is essentially flat over the entire ram stroke is indicative of very effective lubrication. High breakthrough pressure peaks and stick-slip problems were avoided. Of equal importance is that container friction is essentially nil, in spite of the fact that the billet l/d was 6.5:1, a rather high ratio compared to the 2 or 3:1 levels normally used in conventional extrusion.

Incidentally, if "pure" hydrostatic extrusion had been used instead of the thick-film process, billet upsetting would have occurred for the 5:1 extrusion since the billet-pressure augmentation ratio, P_b/P_f, was calculated to be 2.2:1.

The as-extruded Ti-6Al-4V alloy tubes, shown in Figure 6, were evaluated for dimensional uniformity and surface finish. As an example, the tube extruded at a ratio of 5:1 measured:

OD - 12.4 mm $\pm$0.025 mm

Wall - 0.048 mm $\pm$0.0125.

The concentricity was about $\pm$3 percent on wall and the surface finish was in the order of 0.25 to 0.38 μm CLA. This is in considerable contrast to a wall concentricity of $\pm$10 percent and a surface finish of about 3.8-5.0 μm generally obtained in conventional hot extrusion of this alloy and other high-strength materials.

2.3.2 AISI 4140 Steel Tubing

Pressure-ratio curves for round AISI 4140 steel tubes (5) extruded at 850 C are presented in Figure 7. The pressure requirements for tubes were

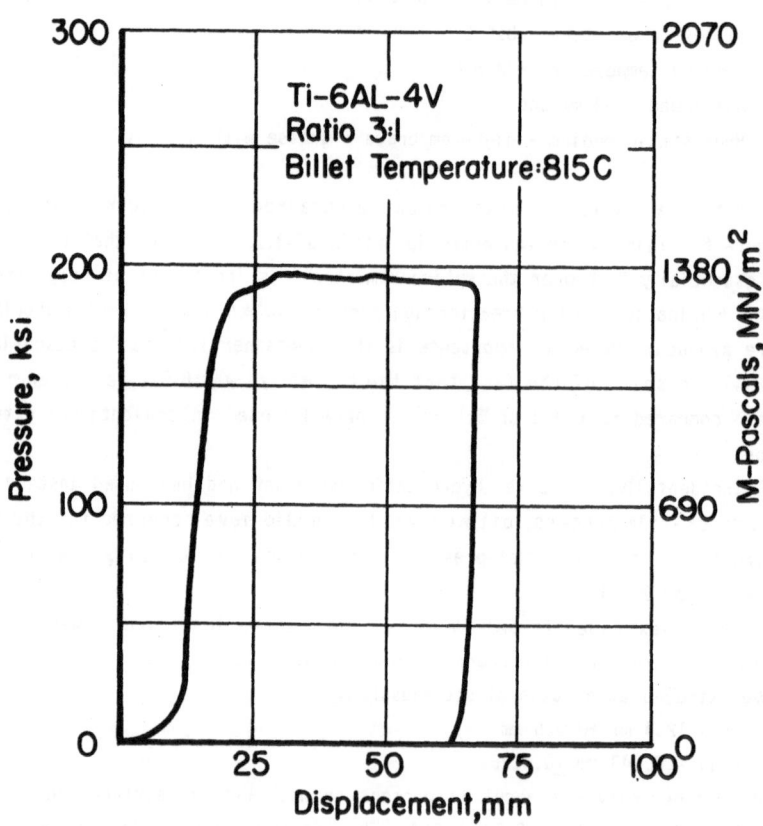

FIGURE 5. EVIDENCE OF EFFECTIVE LUBRICATION (FLAT
PRESSURE CURVE) IN THICK-FILM
HYDROSTATIC EXTRUSION OF Ti-6Al-4V
TUBING AT 815 C

(b) Closeup of tube

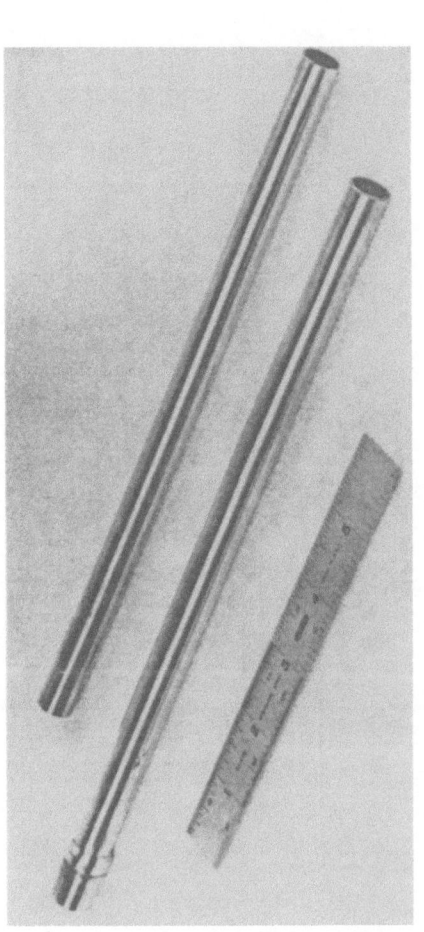

(a) Tubes produced at extrusion ratios of 3 and 5:1

FIGURE 6. Ti-6Al-4V ALLOY TUBING PRODUCED BY THICK-FILM
HYDROSTATIC EXTRUSION PROCESS AT 815 C

298

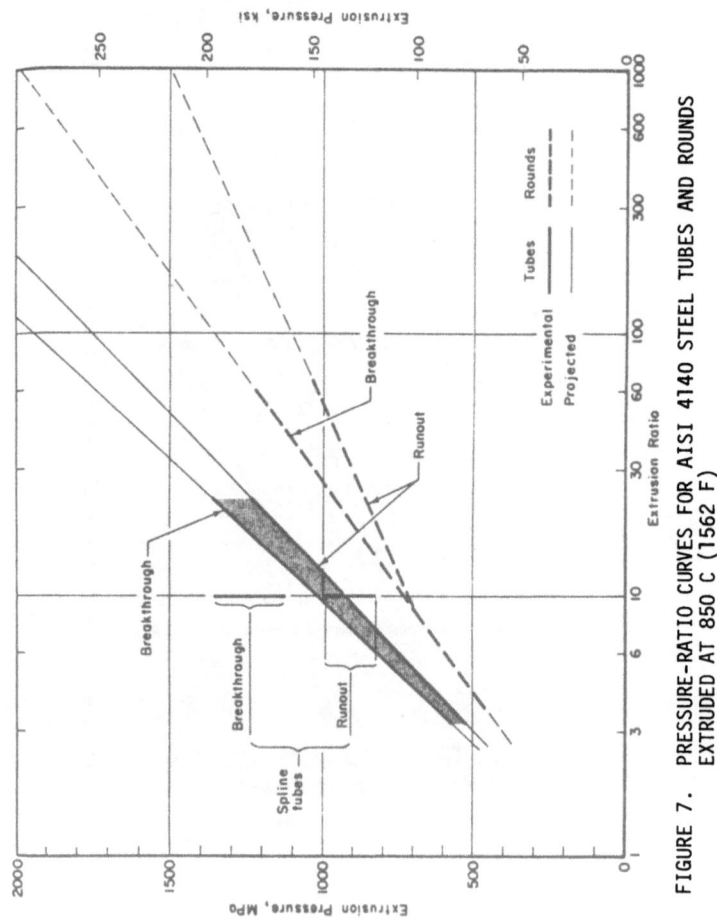

FIGURE 7. PRESSURE-RATIO CURVES FOR AISI 4140 STEEL TUBES AND ROUNDS EXTRUDED AT 850 C (1562 F)

significantly greater than those for solid rounds at the same ratio. Modest breakthrough pressure peaks were encountered and these increased only slightly with ratio. The maximum ratio achieved was about 22:1, producing a tube with a 17.6 mm OD x 2.5 mm wall. A projection of the plot indicates that extrusion ratios of 33:1 and 55:1 may be achieved at pressures of 1500 and 1700 MPa.

Extrusion pressure data for OD-spline tubes are also included in Figure 7. The large pressure peak in this case was largely caused by billet chilling and marginal lubrication.

2.3.3 AISI 316 Stainless Steel Tubing

The pressure-ratio curves for extruding AISI 316 stainless steel tubes (5) at 650 and 850 C are shown in Figure 8. The highest extrusion ratio attempted was 11:1, which produced tubes with a 29.5 mm OD x 2.2 mm wall. Projection of the pressure-ratio curve for 850 C indicates that ratios of 17:1 and 25:1 can be achieved at pressures of 1500 and 1700 MPa. These ratios would be comparable to those achieved in conventional extrusion with glass lubrication at temperatures of 1150 to 1250 C.

2.3.4 Zircaloy-2 Tubing

The pressure-ratio curve for extruding Zircaloy-2 tubes (5) at 700 C is also shown in Figure 8. An extrusion ratio of 57:1 was achieved with this material. However, problems of lubrication breakdown were encountered at ratios much greater than 10:1.

2.3.5 Tubing Concentricity

Typical dimensional data are presented in Table 1 for selected round tubes of different materials extruded at various ratios. In general, the tubes exhibited very uniform dimensions, and there was no significant correlation between dimensional uniformity and either material or extrusion variables. Except for the thin-wall copper tubes extruded at ratios greater than 100:1, the calculated relative eccentricity was ±1.3 percent or less for all the tubes. The eccentricity value for the very thin-wall copper tubes was about 4.5 percent.

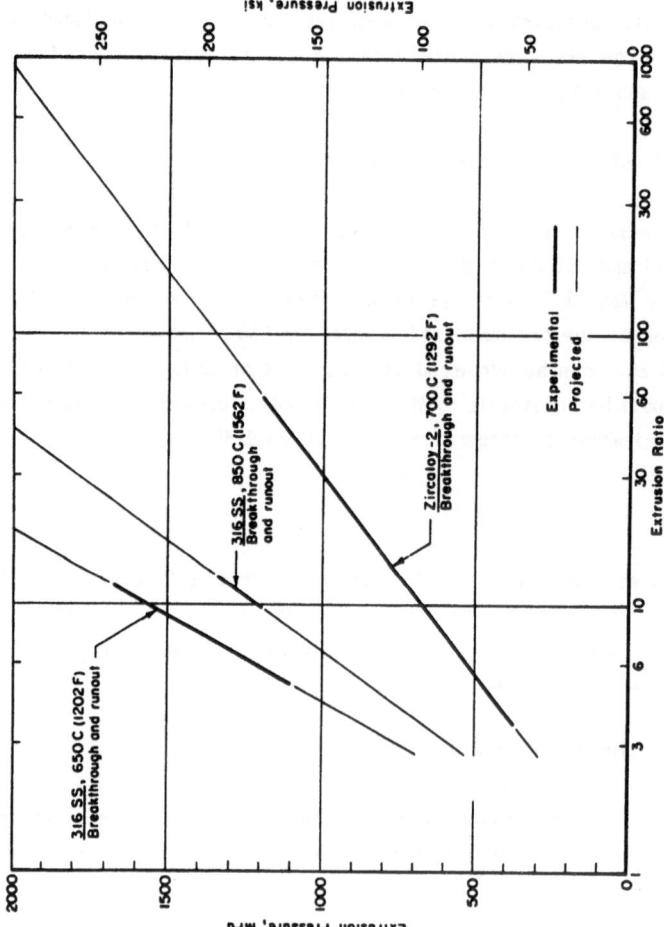

FIGURE 8. PRESSURE-RATIO CURVES FOR EXTRUSION OF AISI 316 STAINLESS STEEL AND ZIRCALOY-2 TUBES AT VARIOUS TEMPERATURES

TABLE 1. DIMENSIONAL DATA FOR SELECTED ROUND TUBES MADE BY THE THICK-FILM PROCESS

Tube Material	Extrusion Temperature, C	Extrusion Ratio	Outside Diameter mm	Extruded Tube Data			Relative Eccentricity Percent (a)
				Wall Thickness, t, mm			
				$t_{max}-t_{min}$	Average		
7075 Aluminum	20	60:1	14.71	0.025	0.98		1.3
7075 Aluminum	200	55:1	14.83	0.025	1.08		1.2
110 Copper	20	22:1	17.63	0.025	2.43		0.5
110 Copper	400	130:1	13.71	0.025	0.50		2.5
110 Copper	750	207:1	13.30	0.025	0.28		4.5
4140 Steel	850	22:1	17.59	0.050	2.44		1.0
316 Stainless Steel	850	11:1	21.39	0.025	4.18		0.3
Titanium	700	11:1	21.53	0.025	4.18		0.3
Zircaloy-2	700	11:1	21.52	0.075	4.16		0.9

(a) Relative eccentricity calculated from: $\dfrac{t_{max}-t_{min}}{t_{max}+t_{min}} \times 100$

2.3.6 Aluminum, Copper, and Steel Round Rod

The thick-film hydrostatic extrusion process has been used to make solid rods from 7075 aluminum, CDA110 copper, and AISI 4140 steel at several temperatures. (6) A summary of the equations representing the breakthrough pressure curves for these materials is presented in the following tabulation.

Billet Material	Billet Preheat Temperature, C	Breakthrough Pressure, MPa
7075 Aluminum	20	110 + 300 ln R
	200	35 + 240 ln R
110 Copper	20	50 + 400 ln R
	400	250 ln R
	750	135 ln R
4140 Steel	750	210 + 395 ln R
		110 + 270 ln R

2.3.7 Copper and Steel Solid Shapes

A variety of shapes shown in Figure 9 were made from several copper and steel alloys by the thick-film hydrostatic extrusion process. (6) Extrusion pressures for selected temperatures and extrusion ratios are presented in Table 2.

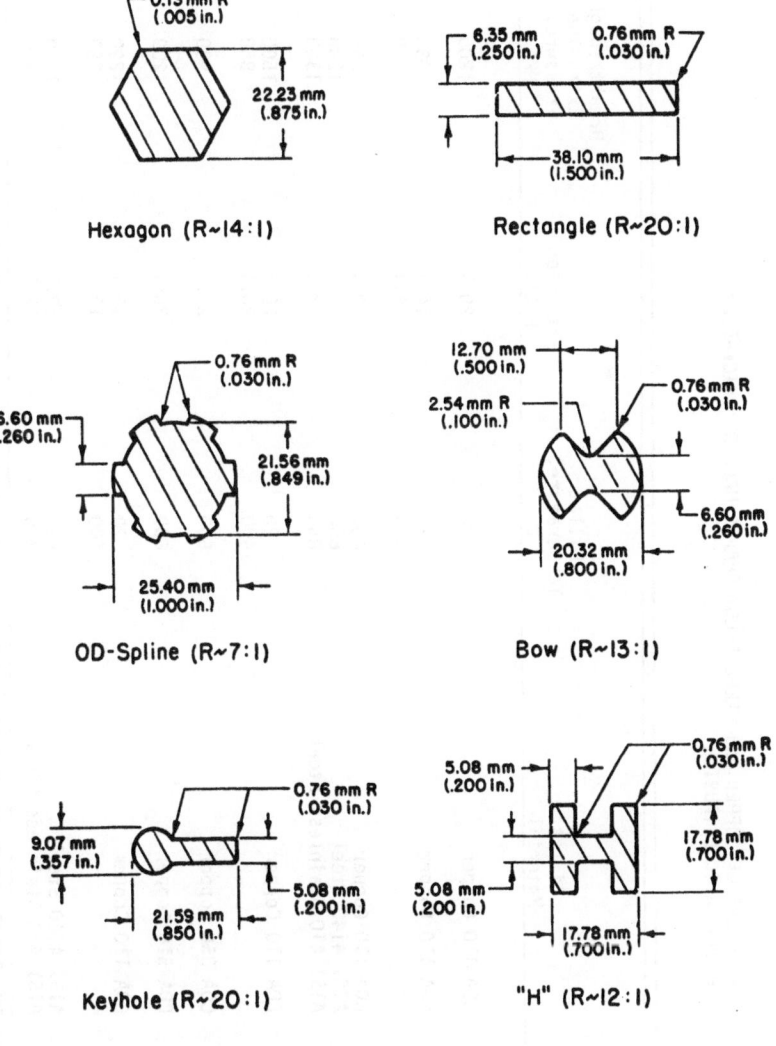

FIGURE 9. SHAPE SECTIONS MADE BY THE THICK-FILM PROCESS

The hexagonal and rectangular sections were extruded
from 88 mm (3.48 in.) billets and the remaining
shapes from 58 mm (2.30 in.) billets.

TABLE 2. EXAMPLES OF COPPER AND STEEL ALLOY SHAPES MADE BY THICK-FILM HYDROSTATIC EXTRUSION

Product Shape(a)	Billet Material	Billet Temperature, C	Extrusion Ratio	Breakthrough Extrusion Pressure, MPa
Rectangle	CDA 110 Copper	20	20:1	1200
Hexagon	CDA 110 Copper	20	14:1	1030
		400	14:1	740
Bow	CDA 110 Copper	400	13:1	830
	AISI 4140 Steel	850	13:1	1280
	AISI 410 Stainless Steel	850	13:1	1320
Keyhole	CDA 110 Copper	20	20:1	1500
		400	20:1	850
	CDA 360 Copper	600	20:1	980
	CDA 510 Copper	600	20:1	209
"H"	CDA 110 Copper	20	12:1	1290
		400	12:1	790
	AISI 4140 Steel	850	12:1	1100
	AISI 410 Stainless Steel	850	12:1	1360

(a) See Figure 9.

3. Hydrostatic Extrusion of Brittle Materials

Many brittle materials are susceptible to circumferential (transverse) and longitudinal surface cracking during hydrostatic extrusion. It is known that such cracking in hydrostatic extrusion can be minimized or prevented by using differential-pressure, fluid-to-fluid extrusion as reported by Pugh (7) and Bobrowsky, et al. (8) In this technique, the billet is hydrostatically extruded from one pressurized fluid into another at a lower pressure, as shown in Figure 10. This approach, however, has several disadvantages which limit its application as a production technique:

(1) Higher tooling and operational costs of the secondary pressure chamber.

(2) Extrusion lengths limited to the length of the secondary pressure chamber

(3) Fluid pressure required to achieve a given reduction in fluid-to-air extrusion must be increased by the amount of counterpressure used in fluid-to-fluid extrusion. This lowers the maximum ratios that could be achieved with a container of a given fluid pressure capability.

In work conducted by Fiorentino, Richardson, and Sabroff, (9) it was found that such brittle materials as beryllium and TZM molybdenum alloy can be cold extruded hydrostatically into air without cracking by using a special die design.

3.1 Double-Reduction Die

The preferred design is a double-reduction die as shown in Figure 11 along with a standard die for comparison. In this die, the major billet reduction is taken at the first or top land and a very small reduction, in the order of 2 percent, at the second land. This second reduction functions in several ways which are depicted schematically in Figure 12. First, it imposes an annular counterpressure, i.e., a longitudinal compressive elastic stress to the extruding product on exit from the first land,

306

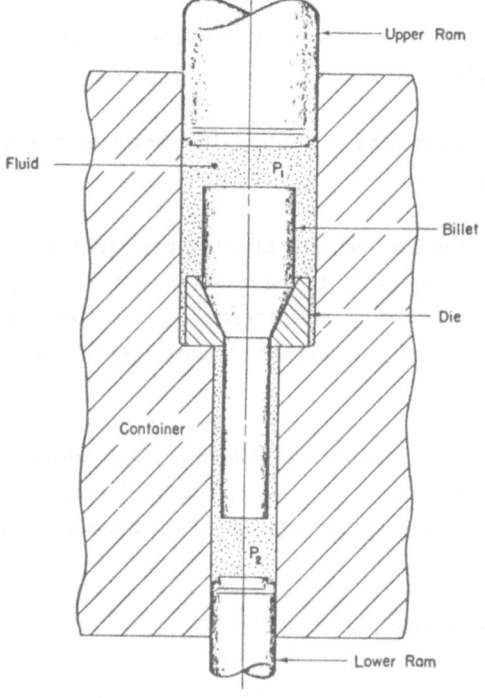

FIGURE 10.   SCHEMATIC OF A COMMON TOOLING ARRANGEMENT FOR
FLUID-TO-FLUID EXTRUSION

307

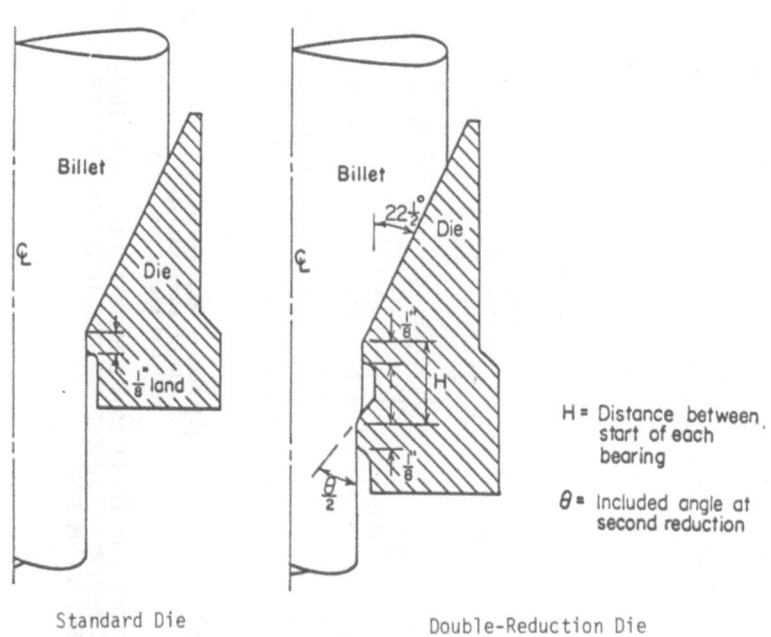

FIGURE 11.   DOUBLE-REDUCTION DIE FOR EXTRUSION OF BRITTLE MATE-
RIALS ALONG WITH STANDARD DIE

308

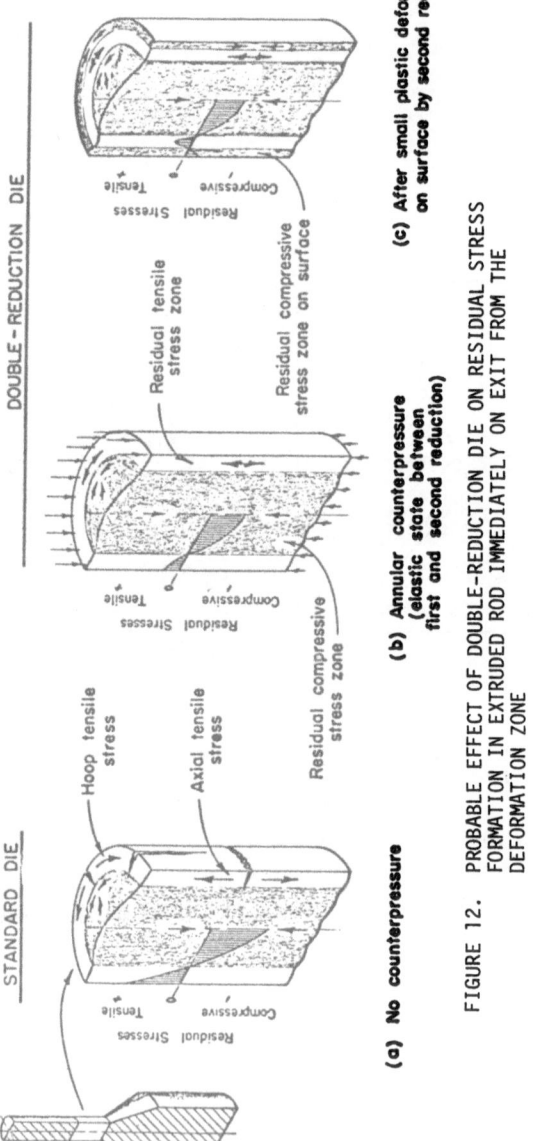

(a) No counterpressure

(b) Annular counterpressure
(elastic state between
first and second reduction)

(c) After small plastic deformation
on surface by second reduction

FIGURE 12.   PROBABLE EFFECT OF DOUBLE-REDUCTION DIE ON RESIDUAL STRESS
FORMATION IN EXTRUDED ROD IMMEDIATELY ON EXIT FROM THE
DEFORMATION ZONE

as shown in Figure 12b. It is believed that this prevents circumferential cracks within and on exit from the first land by reducing the axial tensile stresses, especially those in the outer fibers where the counterpressure is applied. These axial tensile stresses may arise from residual stresses, elastic bending, and land friction. Crack prevention up to this point is controlled mainly, if not completely, by superimposed elastic stresses. Prevention of circumferential cracks on exit from the second land, however, is believed to be associated with a favorable permanent change in the residual stress pattern brought about by the 2 percent plastic deformation. This is illustrated in Figure 12c. Because a light, skin reduction of 2 percent results in mainly a superficial plastic deformation, axial elongation occurs mostly in the outer fibers. On exit from the second land, however, the outer fibers are elastically compressed by the adjacent fibers which are simultaneously put in tension to maintain static equilibrium. This can result in a lower residual axial tensile stress or even a compressive stress in the outer fibers, thus preventing the circumferential cracks.

It is believed that the second reduction functions generally in a similar manner in preventing longitudinal cracks. Such cracking is prevented on exit from the first land because the hoop tensile surface stresses are reduced sufficiently by the annular elastic counterpressure imposed by the second reduction. This is depicted in Figure 12b. On exit from the second land, the longitudinal cracks (as with the circumferential cracks) are prevented by a favorable permanent change in the residual stress pattern brought about by the small plastic reduction on the surface see Figure 12c). This small reduction tends to reduce the residual hoop tensile stress, and may even result in a hoop compressive stress.

With a standard die, however, the extrusion expands elastically and abruptly, immediately on the exit from the die land. This severe stress discontinuity results in appreciable bending of the surface fibers. Such bending can result in longitudinal tensile stresses perhaps close to the elastic limit of the product. These tensile stresses, coupled with the residual axial tensile stresses developed in the extrusion by the deformation process itself, and those introduced by friction on exit from the die land, are believed to be primarily responsible for the initiation of circumferential cracks in the brittle materials.

3.2 Billet Materials and Procedure

The chemical composition and as-received tensile properties of the materials evaluated are given in Table 3.

TABLE 3.   CHEMICAL COMPOSITION AND TENSILE PROPERTIES
OF BILLET MATERIALS

Material	MPa	Y.S.(0.2% Offset), MPa	RA, %	EL, %	Hardness DPN
Be[(a)]	353	255	--	2.5	155
TZM[(b)]	740	620	--	17.0	276

(a)   Be - 1.54 BeO: hot-pressed block

(b)   Mo - 0.42 Ti - 0.1 Zr; wrought and stress relieved

The billets, 44 mm diameter by 150 mm long, were nosed on one end to 45 degrees to mate with the conical entry of the dies. The tapered nose was flush with the start of the die land. The billet surface finish was in the range of 1.5-2.5 µm, CLA. The extrusion ratio was varied by changing the die opening size.

The billets were lubricated with a fused polytetrafluoroethylene (PTFE) layer. The hydrostatic fluid medium used was castor oil. The process used was pure hydrostatic extrusion at room temperature.

3.3 Extrusion Results

3.3.1 Beryllium

The influence of die design on cracking of beryllium during cold hydrostatic extrusion at a ratio of 4:1 can be seen in Figure 13. With the standard die, the beryllium extrusion contains numerous circumferential cracks along its entire length. Longitudinal cracks are also evident.

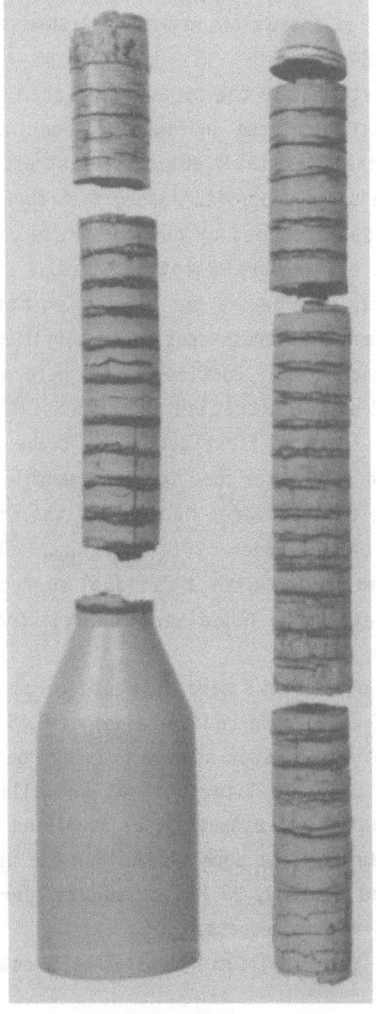 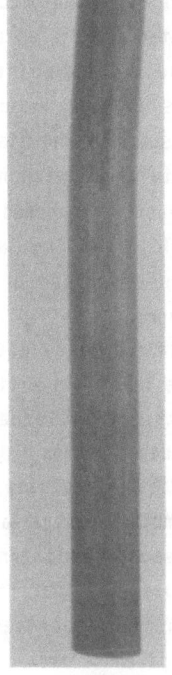

Standard

Double
Reduction

FIGURE 13.   EFFECT OF DIE DESIGN ON SURFACE CRACKING OF
BERYLLIUM DURING FLUID-TO-AIR EXTRUSION

Note the apparent absence of surface cracks
on the deformation zone.

With the double-reduction die, however, both types of cracks are completely eliminated. Metallographic examination up to X800 of transverse and longitudinal cross sections of the beryllium extrusion revealed no longitudinal or circumferential surface cracks. Photomacrographs of a cross section of the billet, deformation zone, and extrusion in both polished and etched conditions are given in Figure 14.

It is of particular interest to note the apparent absence of surface macrocracks in the tapered deformation zone in Figure 13 for the standard die, suggesting sufficiency of the hydrostatic compressive stress component in that region. In spite of this, however, it is seen that surface cracks still occurred in the extrusions produced by both of these dies. This confirms the extreme importance of the development of residual stresses in cracking of the extruded product, once beyond the deformation zone.

Figure 15 is a photomacrograph of a cross-section of beryllium extruded through a standard die at a ratio of 4:1. Portions of the deformation zone near the die orifice point and of the extrusion are shown. Microscopic measurements at X500 indicated that the first detectable crack was 0.8 mm beyond the start of the die land. It would seem to be reasonable to expect cracks to initiate within the die land area because of the high radial compressive stress imposed on the extrusion by the die land. This radial stress would be acting in combination with a residual surface tensile stress being developed in the extrusion immediately on exit from the deformation zone.

With the double-reduction die, cracks were prevented from occurring between the lands, in spite of a sizeable radial relief of 0.25 mm. This indicates the effectiveness of the elastic counterpressure on the outer fibers as depicted in Figure 12b. In later work at Battelle, double-reduction dies without any radial relief between lands were used with equal success. Another desirable feature of the double-reduction die is the fact that the second reduction required only 28 MPa or about 2 percent more fluid pressure over that for the first reduction.

The tensile properties of the beryllium hydrostatically cold extruded at 4:1 are given below:

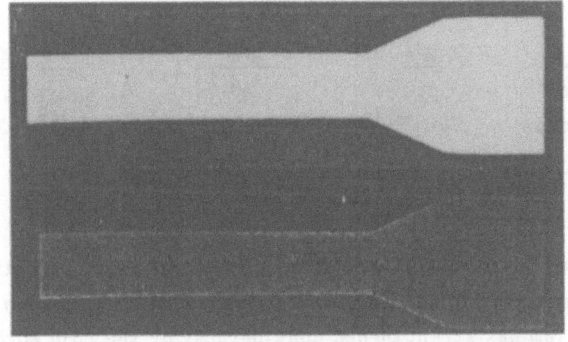

FIGURE 14.   CROSS SECTION OF BERYLLIUM BILLET, DEFORMATION ZONE,
AND EXTRUSION AS PRODUCED WITH A DOUBLE-REDUCTION DIE
IN FLUID-TO-AIR EXTRUSION

TOP polished, BOTTOM etched

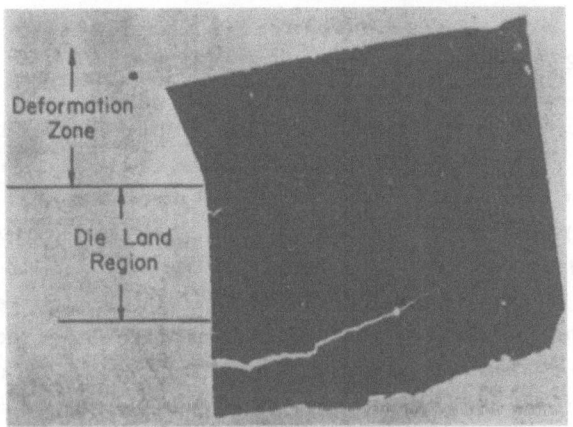

FIGURE 15.   PORTION OF DEFORMATION ZONE AND EXTRUSION SURFACE
OF BERYLLIUM ROD MADE BY FLUID-TO-AIR EXTRUSION AT
A RATIO OF 4:1 WITH STANDARD DIE

Note circumferential cracks in the land region.

Condition	UTS, MPa	0.2% YS, MPa	Elongation (in 2 inches), %
Hot-pressed	352	260	2.5
Cold extruded	790-880[a]	690-745[a]	1.0

(a) Range obtained in four tests.

The tensile specimens were chemically etched after machining to remove surface damage resulting from the machining operation. It can be seen that the beryllium was strengthened considerably by the cold work. This significant strengthening in such a brittle material would seem to indicate that little, if any, structural damage of a weakening nature occurred to the beryllium during cold hydrostatic extrusion under the conditions used. As expected, the ductility decreased, but this would probably increase appreciably on annealing.

3.3.2 TZM Molybdenum Alloy

The effect of die design on cracking of wrought TZM molybdenum alloy is shown in Figure 16. With the standard die, the extrusions produced at ratios of 2.5:1 and 5:1 contained several circumferential cracks at the nose only. These were followed by three hairline longitudinal cracks extending along the remaining length of extrusion.

With the double-reduction die, an apparently crack-free extrusion of TZM was produced at a ratio of 4:1. Again, the pressure required was not appreciably greater (about 5 percent) than that required by a standard die, indicating the pressure to achieve the second reduction was small.

These findings suggest that the principle of counterpressure can be achieved more simply and economically by die design rather than by the approach of fluid-to-fluid extrusion.

Empirical equations relating pressure requirements, P, and extrusion ratio, R, obtained for cold hydrostatic extrusion of beryllium and TZM molybdenum are given below:

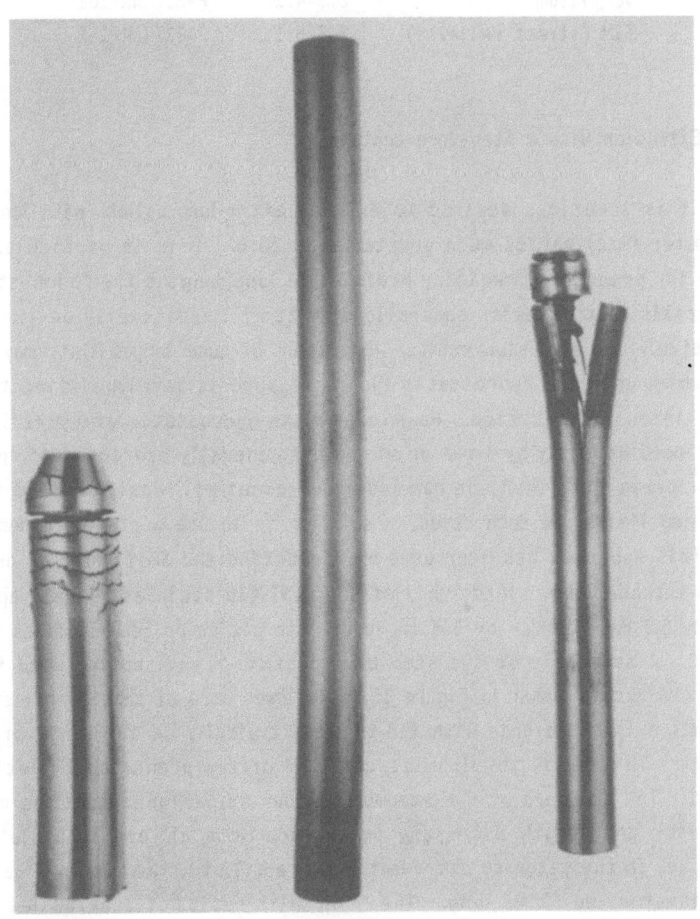

2.5:1	Extrusion Ratio	5:1
Standard	4:1	Standard
die	Double-reduction die	die

FIGURE 16.   EFFECT OF DIE DESIGN ON SURFACE CRACKING OF
WROUGHT TZM MOLYBDENUM DURING FLUID-TO-AIR
EXTRUSION

Material	Extrusion Ratio Range	Fluid Extrusion Pressure, P, MPa
Beryllium	2.5-4:1	P=817 lnR+162
TZM (stress relieved)	2.5-5:1	P=817 lnR+162

4. Extrusion With a Step-Bore Container

It is sometimes desired to extrude extra-long billets with length-to-diameter (l/d) ratios much greater than 20:1. This is particularly the case for producing precision profiles in long lengths from high-strength materials (e.g., steels, superalloys) that will necessarily be limited to relatively low extrusion ratios. Extrusion of such long billets may not be a problem when the hydrostatic fluid pressure is developed directly by a pump-intensifier system. However, if the hydrostatic pressure is to be developed directly by a ram or stem as is generally preferred for production operations, then the ram l/d requirement will easily exceed the ram buckling limit. In such cases, a solution is to use a step-bore container.

This approach has been used by Fiorentino and Smith (10) in order to warm extrude long, slender billets of AISI 416 stainless steel, approximately 6.3 mm diameter by 375 mm long, into pinion or gear profiles at 700-800 C. A schematic of the step-bore container and tooling used to make this product is shown in Figure 17. The lower bore of this container has a minimal radial clearance with the billet - typically in the range of 0.5 to 1.5 mm. This helps to minimize chilling of the preheated billet by the fluid. The top bore of the step-bore liner is designed with the minimum diameter and length necessary to extrude most all of the billet. For example, in the setup to extrude the above billets, the top bore was 19 mm in diameter and 51 mm long. The "hydraulic ratio" of stem/billet cross-sectional areas in this case is only 9:1. Although a high hydraulic ratio obviously is undesirable, a ratio of this level still permitted, in this case, reasonable billet control without excessive stick-slip. Extrusion of billets twice as long can be achieved, of course, by doubling the length of the top bore.

Examples of pinion stock as-warm extruded with this tooling as well as after a single cold-drawing pass are shown in Figure 18. Figures 18 a and

317

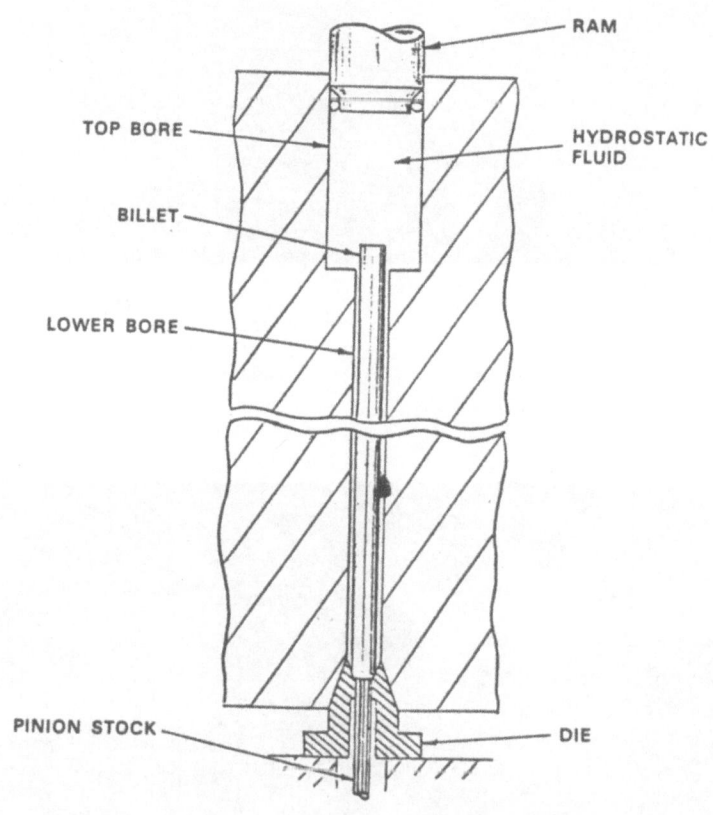

FIGURE 17. SCHEMATIC OF STEP-BORE CONTAINER FOR HYDROSTATIC
EXTRUSION OF HIGH L/D BILLETS

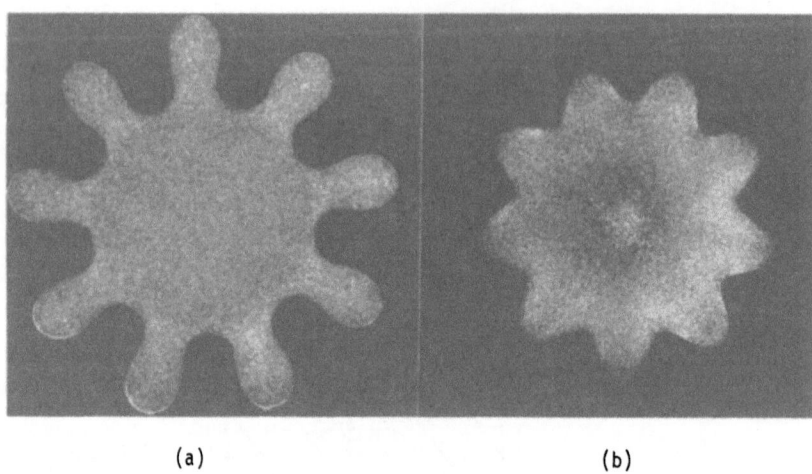

(a)   (b)

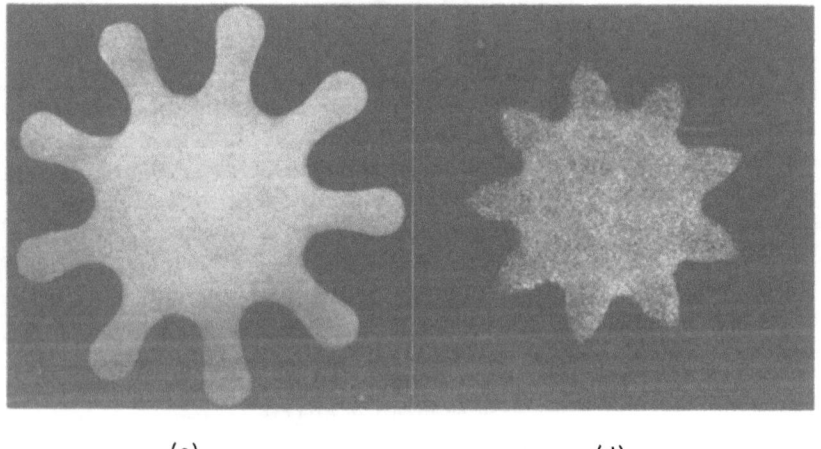

(c)   (d)

FIGURE 18.   CROSS SECTIONS OF AISI 416 STAINLESS STEEL PINION
PROFILES MADE BY WARM HYDROSTATIC EXTRUSION AT
700-800 C [(a) and (b)] AND FOLLOWED BY COLD SIZING
[(c) and (d)]

b show the as-warm extruded cross sections of two pinion profiles made from AISI 416 stainless steel at an extrusion ratio of 3.5:1. The major diameters of the pinions are approximately 5.5 and 4 mm, respectively. Figures 18 c and d show cross sections of the same profiles but after a single cold-drawing reduction of about 20 percent. The purpose of the cold sizing pass is to obtain the precision dimensions and tolerances required (within ± 0.0125 mm) for the application. Typical lengths of the extruded-and-drawn profile depicted in Figure 18 c are shown in Figure 19.

The extrusion breakthrough pressures required for producing the pinion profiles shown in Figures 18 a and b were in the ranges of 1035-1200 MPa and 1310-1450 MPa, respectively.

The step-bore container approach can be used to make a wide range of profiles. Its application will depend for some shapes on the relative economics of potentially competitive processes such as drawing, rolling, and machining.

320

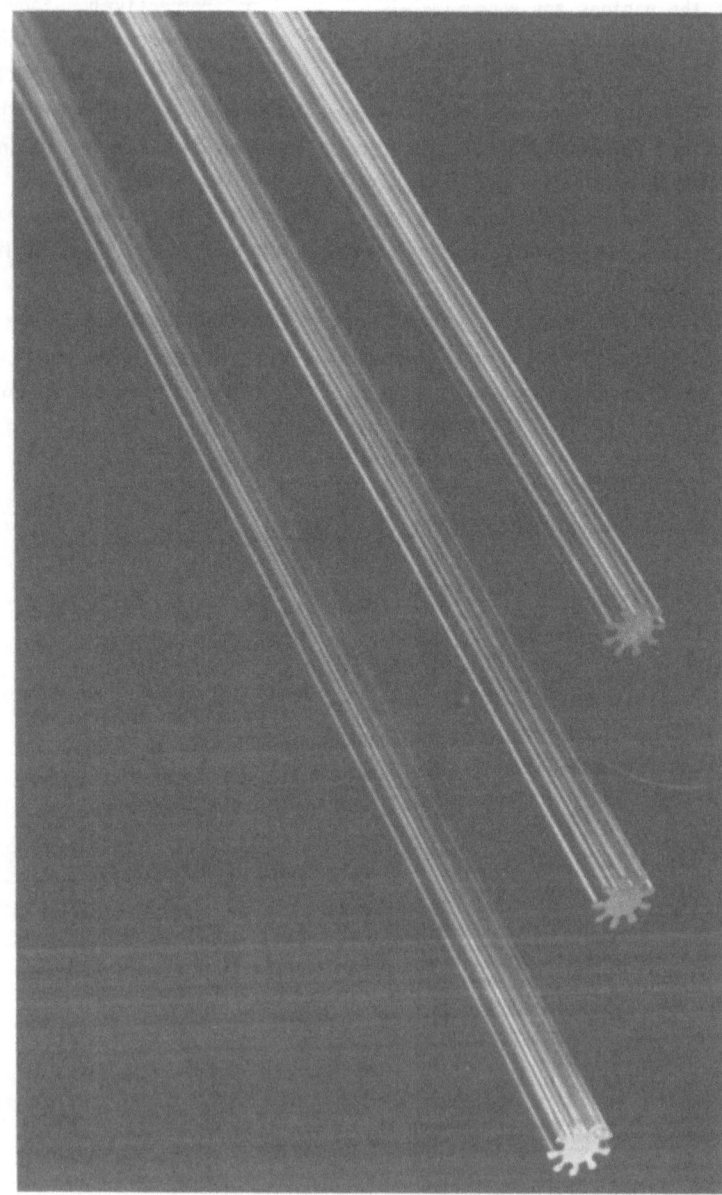

FIGURE 19. LENGTHS OF AISI 416 PINION STOCK MADE BY WARM HYDROSTATIC EXTRUSION FOLLOWED BY A SINGLE COLD-SIZING STEP

321

Acknowledgement

I wish to recognize the special contributions made by my colleague, Mr. E. Garland Smith, Jr. in the research that we conducted jointly on hydrostatic extrusion of precision pinions with a step-bore container and in the research involving Hydrafilm extrusion of tubular and shaped products. I also would like to thank my other colleagues; G. E. Meyer, B. D. Richardson, T. G. Byrer, and A. M. Sabroff; for the many contributions they made to the hydorstatic extrusion technology reported in this chapter. In addition, I wish to thank Mrs. Judith Haas for her assistance in typing the manuscript.

322

References

(1) R. J. Fiorentino, G. E. Meyer, and T. G. Byrer, Some Practical Considerations for Hydrostatic Extrusion, NEL/AIRAPT Conf. on Hydrostatic Extrusion, H. Ll. D. Pugh (Ed.), Mechanical Engineering Publications, London, 1973, pp. 85-93

(2) R. J. Fiorentino, Method of Extrusion, U.S. Patent No. 3,983,730 October 5, 1976

(3) R. J. Fiorentino, B. D. Richardson, G. E. Meyer and A. M. Sabroff, Development of Manufacturing Capabilities of the Hydrostatic Extrusion Process, Technical Report AFML-TR-67-327, Volume I, October 1967, Contract No. AF33(615)-1390

(4) G. E. Meyer, J. A. Houck, T. G. Byrer, and R. J. Fiorentino, Hydrostatic Extrusion of Titanium Alloy Hydraulic Tubing, Interim Report IR243-1(v), Contract F33(615)-71-C-1672, 1972

(5) E. G. Smith, Jr., G. E. Meyer, R. J. Fiorentino, and T. G. Byrer, Hydrafilm Extrusion of Tubular Products, Battelle Columbus Laboratories, December 31, 1975

(6) E. G. Smith, Jr., G. E. Meyer, R. J. Fiorentino, and T. G. Byrer, Hydrafilm Extrusion of Shape Products, Battelle Columbus Laboratories. December 31. 1976

(7) H. Ll. D. Pugh, "Application of High Pressure to the Forming of Brittle Metals", Defense Metals Information Center, Battelle Memorial Institute, Metal Deformation Processing, Volume III, DMIC Report 243, June 10, 1967.

(8) A. Bobrowsky, E. A. Stack, and A. Austen, "Extrusion and Drawing Using High Pressure Hydraulics", Paper No. SP65-33, ASTME Conference, November 1964.

(9) R. J. Fiorentino, B. D. Richardson, and A. M. Sabroff, "Hydrostatic Extrusion of Brittle Materials", Metal Forming, September 1969, pp. 243-252

(10) R. J. Fiorentino and E. G. Smith, Jr., Hydrostatic Extrusion of Precision Pinions, U.S. Army ManTech Journal, 5, No. 3., 1980, pp. 3-10.

Section 3

FINE WIRES

K. OSAKADA

HIROSHIMA UNIVERSITY

1. Introduction

In hydrostatic extrusion, the billet is not limited to straight rod, for even a coiled wire can be processed. Since a drawing stress is usually applied from the die exit to the product when a wire is extruded, the process is called hydrostatic extrusion-drawing [1] or product-augmented extrusion [2].

The reduction of area in each pass in the conventional wire-drawing process must be small to avoid tensile fracture. On the other hand, the drawing stress in the hydrostatic extrusion-drawing process can be small as it only controls the speed of wire, so that a very large reduction is attainable in a single pass. However, the volume of wire processed in each operation is limited by the volume of the high-pressure container. Thus, manufacturing of fine wires of diameters less than 0.1 mm by hydrostatic extrusion-drawing is considered to be advantageous because a very long wire can be accommodated in a container of limited volume.

2. Equipment

Billet wires are supplied in the form of coils with or without bobbins. A coil without a bobbin, self-standing coil, is more suitable for industrial use than a bobbin-wound coil because a longer wire can be kept in a container with a limited inner diameter. However, a great care must be taken to prepare a self-standing coil so as to be smoothly uncoiled from its bore without tangle during extrusion.

For threading wire into the throat of the die and for setting the coiled wire into the container, it is convenient to use a die unit which is prepared in advance with a die, a coiled wire with threaded end, lubricant and a heating device if necessary. Fig.1 illustrates two examples of die unit for coils with and without a bobbin [3]. The bottom end of each unit is covered by a thin vinyl film to isolate the lubricant from the pressure transmitting fluid.

An experimental apparatus for wire extrusion is shown in Fig.2 [3,4].

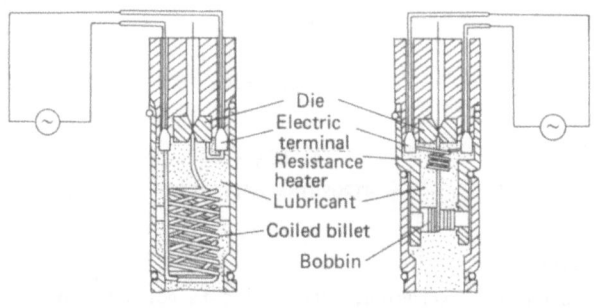

(a) Free-standing coil with direct heating method

(b) Bobbin-wound billet with indirect heating method

Fig.1 Details of die unit.

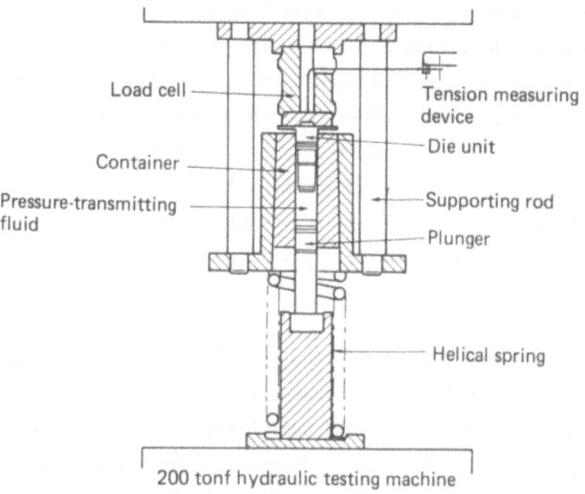

Fig.2 Experimental apparatus for hydrostatic extrusion-drawing

The plunger is compressed upwards to retain the lubricant within the container and the billet wire is positioned in the die unit. The speed or tensile stress of the extruded wire is controlled by using a tension measuring device, tachometer and a winding drum driven by a synchronous motor. The experimental data presented below were collected on this apparatus.

Fig.3 illuotrates a commercial extrusion machine which was designed on the line of drawing machine [5]. A coiled billet is inserted directly into the high pressure container. The extruded wire is drawn up by the capstan with a speed up to 1000 m/min and is wound up on the bobbin. The high pressure is generated by two plunger type pumps (700 kg/cm^2) and two intensifiers, and the pressure is automatically maintained at a constant value up to 15,000 kg/cm^2.

Fig.3 Commercial hydrostatic extrusion–drawing machine

3. Extrusion Pressure in Cold Hydrostatic Extrusion
(1) Effect of Diameter of Billet

The relation between the total extrusion pressure of fine copper wire and extrusion ratio is shown on a semi-log scale in Fig.4. The total extrusion pressure denotes the sum of the internal pressure and the drawing stress. For comparison, the results for copper rod extruded through a tool steel die of throat diameter 10 mm are also plotted. It is clear that fine wires require higher extrusion pressures (1.3 - 1.7 times) than does the conventional billet. This size effect is attributed to the difference in frictional stress, i.e., the frictional stress is considered to increase as the diameter of billet decreases.

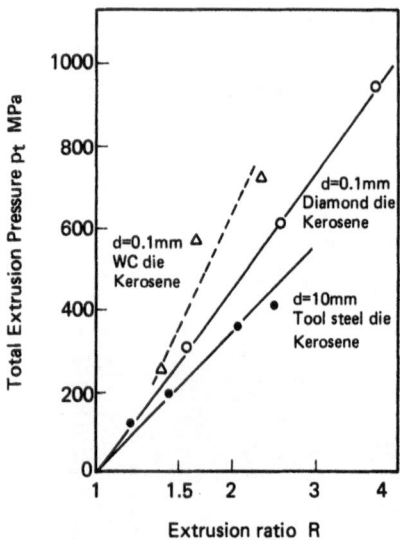

Fig.4 Effect of extrusion ratio on total extrusion pressure

It is also seen from Fig.4 that a higher extrusion pressure is needed to extrude fine wire through the carbide die than through the diamond die. This suggests that diamond causes lower friction with copper than does tungsten carbide.

(2) Effect of Speed

Fig.5 shows the relationship between extrusion speed and total extrusion pressure. The self-standing coiled billet and bobbin-wound billet are extruded through a carbide die of included die angle $2\alpha = 30°$ at an extrusion ratio $R = 1.4$ using kerosene and machine oil as lubricant. The bobbin-wound billet is found to produce a marked speed effect due to the viscous resistance acting on the rotating bobbin when machine oil is used. The effect of a low viscosity lubricant such as kerosene ($2cP = 0.02g/cm^2sec$) should be used when a rotating bobbin is employed.

The effect of extrusion speed is also observed for the self-standing coiled billet, in the case of the viscous lubricant. Hayashi et al.[6] have reported that the increase in extrusion pressure due to viscous drag acting on the surface of the billet is inversely proportional to its diameter. For

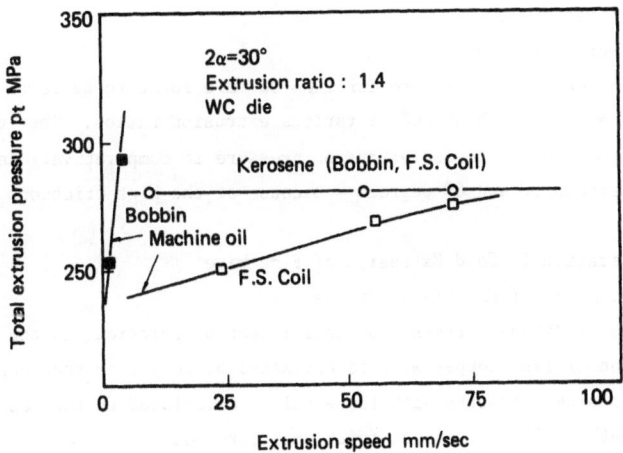

Fig.5 Effect of extrusion speed on total extrusion pressure for fine wire

industrial conditions, i.e., at a higher extrusion speed, lubricants of very low viscosity would be required, even when a self-standing coiled billet is used.

(3) Effect of Drawing Stress

The correlation between the internal pressure and the drawing stress is shown in Fig.6. As the internal pressure increases, the drawing stress decreases. The total extrusion pressure, the sum of the internal pressure and the drawing stress, is almost constant or increases slightly with

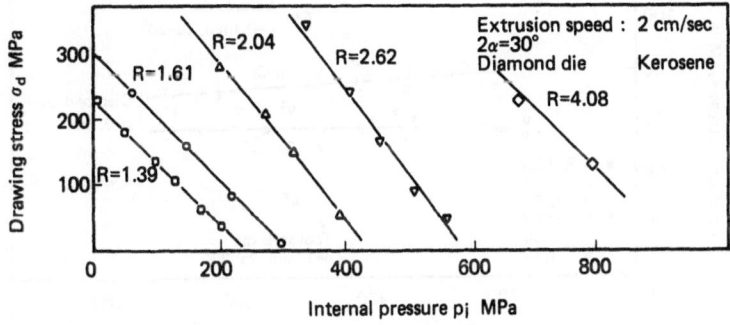

Fig.6 Correlation between drawing stress and internal pressure

internal pressure.

(4) Effect of Die Angle

The extrusion pressure for $2\alpha = 30°$ are found to be lower than those for $2\alpha = 15°$ by about 20% at various extrusion ratios. The optimum die angle to give the lowest extrusion pressure is comparatively large - possibly between 30 and 60 degrees - because of the high friction.

4. Lubrication in Cold Extrusion of Fine Wire

(1) Estimation of Frictional Stress

The frictional stress, or coefficient of friction, in the hydrostatic extrusion of fine copper wire is estimated by comparing the measured value of extrusion pressure with those values calculated on the basis of upper bound method [7] for various frictional stresses.

In Fig.7, the estimated frictional stresses are plotted against contact pressure. These results are obtained at room temperature for the combination of diamond die and kerosene lubricant. Some results for fine wire extrusion with a carbide die and for rod extrusion with a tool steel die are also illustrated for comparison. It will be seen that the frictional stress varies only a little with contact pressure, i.e. with drawing stress and extrusion ratio, for a constant die angle, and increases with die angle. The frictional stress in fine-wire extrusion is almost three times as great as that in rod extrusion.

(2) Mechanism of Lubrication

There are two mechanisms for trapping liquid lubricant between the billet and the tool surfaces; the hydrodynamic and geometrical mechanisms.

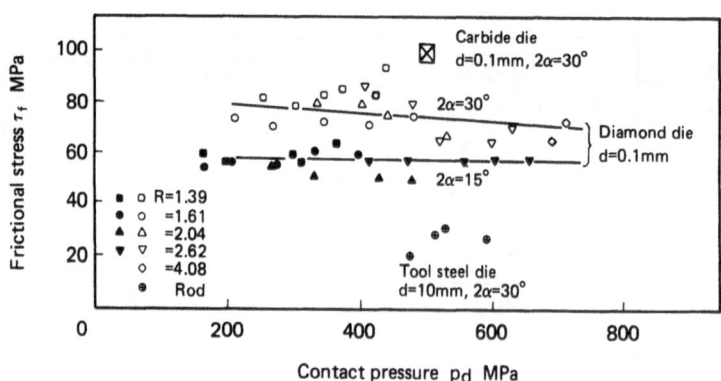

Fig.7 Effect of contact pressure on estimated frictional stress

In hydrodynamic lubrication, the film thickness of trapped lubricant is essentially dependent on its viscosity. As explained above, a very light lubricant such as kerosene is likely to be used in fine wire extrusion, so that lubrication by the hydrodynamic mechanism is not expected to work well.

It has been found that rods with surface irregularities due to machining cause low extrusion pressures in hydrostatic extrusion because the lubricant is trapped between the die surface and the concave part of the surface roughness. The film thickness of trapped lubricant by this mechanism can be estimated by using a simple surface model.

At the corner of the die entry, the billet is deformed before contacting the die surface and a rounded corner is observed as shown in Fig.8 (a). The shape of the ripple on the surface is changed as Fig.8 (b) while the surface is passing along the round corner. The lubricant is trapped into the recessed part when the peaks of the ripple contact the die surface.

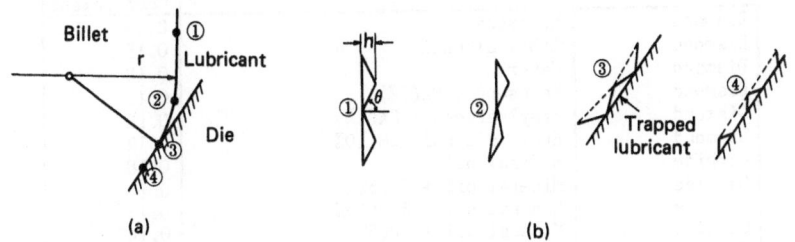

Fig.8 Trapping of lubricant in surface irregularities

From a simple calculation of geometry, the mean thickness of trapped lubricant film is given by;

$$t = 0.5\{(h + r\sqrt{1 - H})\sqrt{1 - H} - r\} (h/r + \sqrt{1 - H}) , \qquad (1)$$

where $H=(h^2\tan^2 \theta)/r^2$. The film thickness decreases as the radius of curvature decreases. This mechanism has been confirmed qualitatively in hydrostatic extrusion of rod [8].

The radius of curvature changes almost in proportion to the diameter of the wire, whereas the roughness is not related to the size of billet. Thus the thickness of lubricant film due to geometrical trapping becomes thinner with the decreasing diameter of the wire.

(3) Effect of Additives in Lubricant

Since the lubricant film caused by the hydrodynamic or geometrical mechanisms is not thick in fine wire extrusion, a possible way of decreasing the frictional stress over the contact area is to form a chemical layer of boundary lubrication on the surface.

To increase the effect of the chemical properties of a lubricant, extrusion of fine copper wire is carried out using lubricants with various additives. As shown in Table 1, the organic base lubricants (kerosene and ethyl alcohol) give lower coefficients than the inorganic base lubricant (water). The additives for boundary lubrication (stearic acid and an epoxy monomer) are effective when they are mixed with ethyl alcohol. The extreme pressure additives (zinc dialkyl dithiophosphate and chlorinated paraffin) show marked effects especially when they are mixed with a boundary lubricant.

Table 1 Effect of additives on friction in hydrostatic extrusion of fine wire

Die Material	Lubricant	Friction Coefficient
Diamond	Kerosene	0.15
Diamond	Ethyl alcohol	0.15
Diamond	Water	0.18
Diamond	Kerosene + SA(1%)	0.15
Diamond	Ethyl alcohol+ SA(1%)	0.12
Diamond	Ethyl alcohol+ EM(10%)	0.12
Carbide	Mineral oil	0.18
Carbide	Mineral oil + SA(5%)	0.16
Carbide	Mineral oil + ZDD(10%)	0.13
Carbide	Mineral oil + CP(5%)	0.14
Carbide	Mineral oil + SA(5%) + ZDD(10%)	0.08

2 =30, R=1.69 for diamond die and R=1.77 for carbide die.
SA: Stearic acid, EM: Epoxy monomer, ZDD: Zinc stearic acid,
ZDD: Zinc dialkyl dithiophosphate, CP: Chlorinated paraffin.

4. Warm Hydrostatic Extrusion

Two heating methods have been proposed (see Fig.1): (a) the direct method by passing an electric current between the carbide die and the self-standing coil, and (b) the indirect method by an electric resistance heater located in front of the die. The indirect method needs more heat energy than the direct method.

Fig.9 illustrates the temperature dependence of extrusion pressure, where the pressure decreasing gradually with temperature. When the carbide die is used, stick-slip motion of the billet occurs at temperatures above 170 ° C and smooth extrusion does not continue (hatched part in the figure).

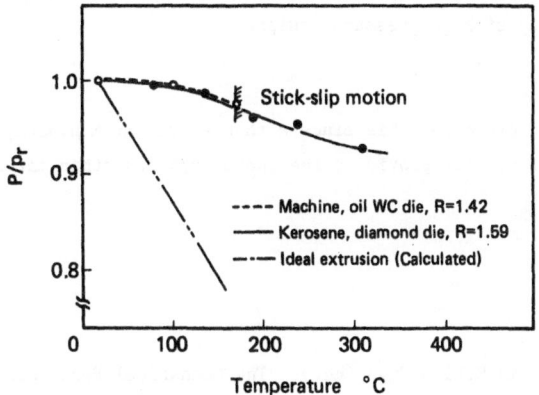

Fig.9 Effect of temperature on relative extrusion pressure

The diamond die with a rotating bobbin does not show this marked stick-slip motion, even at up to 350°C.

The experimental extrusion pressure at 300°C is lower than that at room temperature by only 8%, whereas the flow stress is lower by about 40% [9]. This result suggests that the coefficient of friction increases with temperature, owing to the loss of boundary lubrication in the oil at elevated temperatures.

The additives are found to be effective in reducing the friction coefficient in warm extrusion. It has been established that the lubricant with the extreme pressure additive and the boundary lubrication additive presents the best result up to 350°C.

5. Concluding Remarks

Since the wire drawing technique well established in industry, mass production of fine wire will continue to be carried out by drawing. The hydrostatic extrusion process can give a large reduction in a single pass with compact equipment, therefore this method is likely to be used for small batch production of uncommon or precious metals. Another advantage of this process over the wire-drawing process is that heating of wire is effective to extend the working limit, differently from drawing in which the weakening of heated wire after die exit reduces the working limit. Consequently, fine wires of less ductile metals as beryllium [10] and of hard metals may be manufactured by hydrostatic extrusion-drawing. In any case, successful operation depends on the lubrication technique in addition to the reliability of high pressure equipments.

Acknowledgement

The author expresses his sincere thanks to Mr.M.Hinata, Sumitomo Electric Industries, for providing the photograph and other materials used in this section.

REFERENCES

1 H.Ll.D.Pugh, in H.Ll.D.Pugh (Ed.), The Mechanical Behaviour of Materials under Pressure, 1970, Elsevier, p.492.

2 J.M.Alexander and B.Lengyel, Hydrostatic Extrusion, Mills and Boon, London, 1971.

3 K.Osakada and R.Asada, J. Mech. Work. Tech., 1, 1977/78, pp.277-290.

4 K.Osakada, R.Narutaki and K.Minami, J. Jpn. Soc. Tech. Plasticity (in Japanese), 15, 1974, 737-743.

5 Sumitomo Electric Industries, Pamphlet.

6 M.Hayashi, M.Yokota, T.Kondo and M.Hinata, Proc. 4th Int. Conf. High Pressure, AIRAPT, 1974.

7 K.Osakada and Y.Niimi, Int. J. Mech. Sci., 17, 1975, pp.241-254.

8 K.Osakada, J. Jpn. Soc. Tech. Plasticity (in Japanese), 12, 1971, pp.313-321

9 J.F.Alder and V.A.Phillips, J. Inst Metals, 83, 1954/55, pp.80-86.

10 J.C.Uy, B.D.R.Richardson, T.S.Felker and R.J.Fiorentino, Wire Industry, 40, 1972, pp.40-43 and pp.110-112.

SECTION 4

POLYMERS

N. INOUE

SCIENCE UNIVERSITY OF TOKYO

1. Introduction

The primary interest in applying the techniques of hydrostatic extrusion to the deformation processing of polymeric material is in improving the mechanical properties of the material, especially the tensile modulus and strength. The theoretical modulus of a perfect crystal of polyethylene is reported to be in the range of 240 - 340 GPa [1], and the theoretical strength of molecular chains is calculated to amount to 19 GPa [2]. Various means were tried to achieve this. They are tensile drawing [3], shear crystallization from solution [4], capillary extrusion [5], high pressure injection moulding [6], solid state extrusion [7], die drawing [8], radial compression [9], and so forth. The strengthening mechanism proposed for the crystalline polymer, extruded or drawn in one direction is the following [10]. Deformation processing causes the orientation of the molecular chains. The oriented polymer consists of oriented crystallites called lamellae, with the c-axis aligned in the direction of extrusion or drawing, which are connected by oriented tie-molecules. The stiffness and strength of the material are related to the state and number of the tie-molecules. If they are tight in tension, the modulus is high. If they are large in number, the strength is high.

The tensile modulus of the oriented polymer is primarily dependent on the deformation ratio, i.e. the ratio of the cross sectional area of a billet to that of a product, of the orientation process, irrespective of the particular method employed for processing. Quite a number of researchers have worked on the linear polyethylene, aiming at reaching the theoretical modulus by raising the deformation ratio. The modulus vs. deformation ratio relation obtained so far for this material is approximated by a bilinear line on the logarithmic scale in Fig. 1. The highest modulus obtained used to be around 70 GPa, which is close to that of aluminum and approximately 1/3 of the theoretical modulus. In the last few years, however, the gap between the theoretical and experimental values was almost

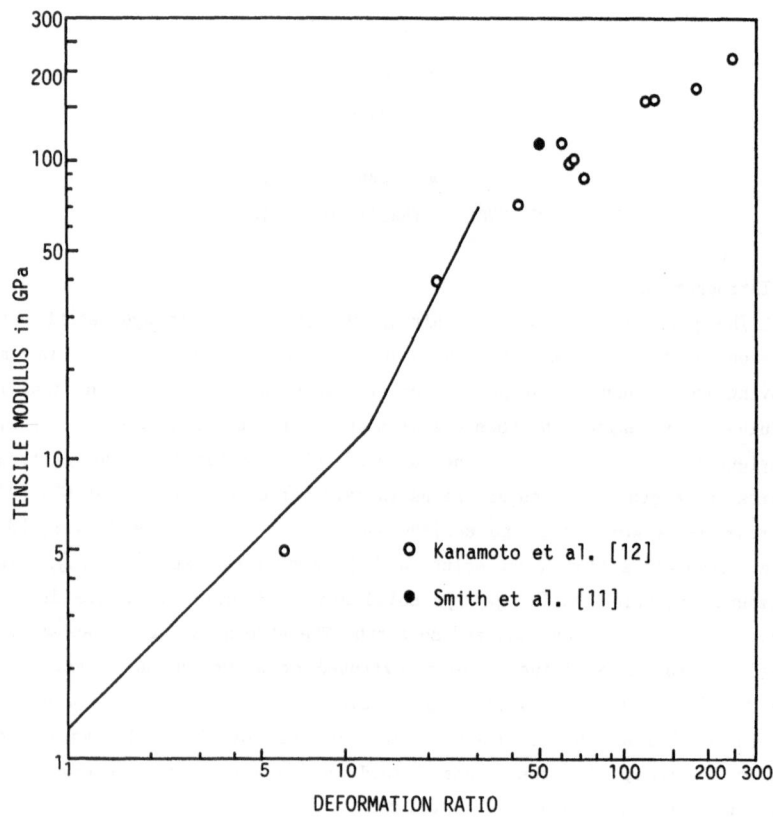

Fig. 1. Tensile modulus vs. deformation ratio relation of high-density polyethylene.

closed up by the use of rather unusual means. Thus Smith et al. [11] worked with the dried gel of ultrahigh molecular weight polyethylene, prepared by quenching its semi-dilute solution, and drew it up to the deformation ratio of 50. The highly drawn material exhibited the tensile modulus of 115 GPa and strength of 4.7 GPa. Quite recently, Kanamoto et al. [12] employed a novel technique of two-stage drawing and obtained a highly drawn ultrahigh molecular weight polyethylene with the tensile modulus of 227 GPa. The polyethylene used was Hizex Million 240 M ($M_v = 2 \times 10^6$) of Mitsui Petrochemical Co., Ltd. Solution-grown crystals of this material were isothermally precipitated from its dilute solution in xylene, which was kept at 85°C for 20 h. Sedimented mats of solution-grown crystals were obtained by slowly filtering the crystal suspension followed by drying it in vacuo at 50°C. One or more of the mats were placed between two split billet halves of normal molecular weight polyethylene. The JX-20 (M_w = 6.7 x 10^4) of Mitsubishi Petrochemical Co., Ltd. was employed for this pourpose. The whole assembly was coextruded, imparting the extrusion ratio of 6 to the sandwiched crystals. A thin film, 2 - 4 mm in width and 3 - 10 mm in length, was subsequently drawn in tension at near 115°C with a very low rate. The film showed an excellent drawability and was drawn at almost a constant load. The total deformation ratio imparted to the starting material amounted to 247. This unusual drawability is due, the authors claim, to the preparation of the favorable morphology of the crystallites to pull out the tie molecules.

Polymers are normally processed in a molten state. The solid state processing, however, has some merits. First, the energy for processing is saved. There is no need to heat up the work to a high temperature. Secondly processing time is saved. Polymers have an extremely low conductivity and because of this it takes a long time to heat up and subsequently cool down the work. Cold working eliminates the process of the heating cycle. The cold working of plastics, however, has not been extensively employed in industry because of their dimensional instability. A large amount of spring-back due to the low modulus, combined with a seemingly never-ending recovery of dimensions due to viscoelasticity, makes cold forming unattractive for practical use. With the advent of ultrahigh modulus polymers, however, a new era of the solid-state forming of plastics is being opened.

Since solid-state forming has been widely employed in the field of metalworking, quite a variety of techniques have been developed for this purpose. Most of them are readily available for the deformation processing of plastics. Hydrostatic extrusion is one of them, which is featured by absence of friction between the billet and container wall, and also by extremely low friction between the billet and die. This is particularly favorable for polymer processing because keeping the extrusion pressure low is the key to the success in polymer extrusion. Moreover, from the industrial point of view a continuous process is preferred and this may be realized by using the techniques of hydrostatic extrusion. It is not without reason that quite a lot of research has been recently done in the field of hydrostatic extrusion of polymers.

When compared with the mechanical properties of metallic materials those of polymeric materials are highly dependent on strain, strain rate, pressure, and temperature. Due consideration must be paid to these features in designing the hydrostatic extrusion of plastics. In the following, recent findings in this field are reviewed, problems are identified, and trends delineated.

2. Plastic Deformation of Polymers

When pulled in uniaxial tension polymers, either crystalline or a-morphous, yield at a certain load. Normally there is a peak in a load – elongation curve and it is called a yield point. If there is no distinct peak, a proof stress at 0.2 – 2 % offset is employed as a yield strength in a similar manner as in metals. When not cold worked, polymers can be considered isotropic. The yield strength in compression of isotropic poly-mers is higher than that in tension. Under the triaxial state of stresses the Mises' yield criterion can be used for the isotropic polymer, but the pressure dependent yield criterion of Raghava et al. [13] seems to better represent the yield behavior. In its most general form, the proposed cri-terion for macroscopic yielding is expressed as:

$$(\sigma_1 - \sigma_2)^2 + (\sigma_2 - \sigma_3)^2 + (\sigma_3 - \sigma_1)^2 + 2(C - T)(\sigma_1 + \sigma_2 + \sigma_3) = 2CT \quad (1)$$

where σ_1, σ_2, and σ_3 are the three principal stresses while C and T repre-sent the absolute values of the compressive and tensile yield strengths respectively, measured at atmospheric pressure. If C = T then Eq. 1 reduces to the usual form of the Mises' criterion. The influence of the hydrostatic portion of the applied stress state is introduced by the quantity (σ_1 + σ_2 + σ_3). The C/T ratio is always greater than 1 in the isotropic polymer and is normally around 1.3.

Once plastic deformation sets in, the lamellae in the crystalline polymer are transformed into a fibrous structure and the amorphous com-ponent becomes also oriented. Orientation in polymers is a phenomenon of great technical and theoretical importance. Recent researches on the structure and properties of oriented polymers are reviewed in an excellent monograph edited by Ward [14]. The oriented polymers are anisotropic. The anisotropic yield criterion first proposed by Hill [15] has been modified to account for differences in tensile and compressive yield strengths in a given direction [16], as shown in the following:

$$H(\sigma_1 - \sigma_2)^2 + F(\sigma_2 - \sigma_3)^2 + G(\sigma_3 - \sigma_1)^2 + K_1\sigma_1 + K_2\sigma_2 + K_3\sigma_3 = 1 \qquad (2)$$

where all six coefficients of the stress terms are defined in terms of the tensile and compressive yield strengths T_1, T_2, T_3; C_1, C_2, C_3 respectively in the directions of the principal axes of anisotropy:

$$H + G = (C_1T_1)^{-1}, \quad F + H = (C_2T_2)^{-1}, \quad G + F = (C_3T_3)^{-1},$$

$$K_1 = (C_1 - T_1)(C_1T_1)^{-1}, \quad K_2 = (C_2 - T_2)(C_2T_2)^{-1}, \quad K_3 = (C_3 - T_3)(C_3T_3)^{-1} \qquad (3)$$

When a tensile specimen of the polymeric material is pulled in tension, normally a peak in load, followed by necking [17], is observed. Subsequently the specimen elongates under an almost constant load. This means the polymer is approximately linearly workhardening. There is no general formula applicable to every polymer to represent its stress – strain behavior. For crystalline polymers, however, the true stress vs. true strain relationship proposed by Maruyama et al. [18] seems to well represent the experimental data, especially of high-density polyethylene, its blend with n-paraffin, and nylon 6. It is given by

$$\ln (\sigma / \sigma^*) \ln (\varepsilon / \varepsilon^*) = -c \qquad (4)$$

where σ is the true tensile stress, ε true tensile strain, and σ^*, ε^*, c are material constants. The positive constant c is characteristic of polymer species, being independent of melt index, temperature, deformation rate and the blending ratio of paraffin to polyethylene.

Since the plastic deformation of polymers is dependent on pressure p, temperature t, and strain rate $\dot{\varepsilon}$, efforts have been expended to include these parameters in the stress – strain equation. From a series of tensile testings at pressures up to 300 MPa, temperatures up to 100°C, and strain rates up to 1 sec^{-1} with high-density polyethylene, the following true stress σ vs. true strain ε relation is proposed [19]:

$$\sigma = [5.66 \times 10^{-4}\, p\, (43.1 - \ln \dot{\varepsilon})(430 - t) - 3.29 \times 10^{-2}\, t\, (6.70 \times \ln \dot{\varepsilon})$$
$$+ 0.29 \times \ln \dot{\varepsilon} + 39.3][\, 1 + 0.38(e^\varepsilon - 1)^{1.15}] \qquad (5)$$

3. Effects of Hydrostatic Pressure

The stiffness, yield, and fracture of polymers are greatly affected by pressure, as well as temperature. This is in contrast to metals whose stiffness and yield are practically independent of pressure. It has been suggested that a pressure dependent modulus is a logical consequence of the theory of finite strains [20]. The principal reason why the modulus changes more drastically with pressure for polymers than for metals is simply that the ratio of the imposed pressure p to the initial tensile or shear modulus, E_0 or G_0 respectively, is much greater. Birch [21] has shown that if a small axial or shear strain is superimposed on a material subject to finite volume strains, the Young's and shear moduli, E and G respectively, will both be functions of pressure. His equations can be written in the simplified form

$$E(p) = E_0 + 2(5 - 4\nu)(1 - \nu)p, \quad G(p) = G_0 + 1.5(3 - 4\nu)(1 + \nu)^{-1}p \qquad (6)$$

If it is assumed, as a first approximation, that the Poisson's ratio ν is independent of pressure, then the Birch Eqs. 6 predict that the modulus will vary linearly with pressure. A comparison of computed values with experimentally observed ones has been made for some crystalline and amorphous polymers [22]. The agreement is quite good.

The yield criterion of isotropic polymers is already given in Eq. 1, but there is another form of the criterion due to Sternstein and Ongchin [23], which also represents the pressure dependence of the yield behavior very well and is widely used. It assumes that at yield the octahedral shear stress is a linear function of the surrounding pressure p. In uniaxial tension or compression the tensile or compressive yield strength, T_p or C_p respectively, is given by

$$T_p = T + kp, \quad C_p = C + k'p \qquad (7)$$

where T and C are the tensile and compressive yield strengths at atmospheric pressure respectively. The pressure coefficients k and k' are given in Table 1 together with references.

The ductility of polymers is affected by the surrounding pressure also. From the various experimental data reported to date concerning the influence of pressure on the fracture behavior of polymers it appears possible to draw some general conclusions [28]. If the polymer is brittle and fracture occurs at low applied tensile stress as a result of high

Table 1 Yield strengths and their pressure coefficients.

Material	T in MPa	C in MPa	k	k'	Ref.
Polyethylene	22.1		0.095		[24]
Polypropylene	37.3		0.206		[24]
Nylon 6	53.0		0.037		[25]
Acryronitril-butadiene-styrene copolymer	42.2		0.115		[26]
Polytetrafluoroethylene	8.97	14.1	0.080	0.094	[20]
Polyvinyl chloride	62.1	75.9	0.135	0.20	[24]
	49.1		0.095		[25]
Cellulose acetate	41.4	48.3	0.196	0.21	[26]
Polyimide	103	106	0.062	0.074	[27]
Polysulfone	90.3	101	0.065	0.097	[24]

localized stresses associated with the presence of surface flaws or crazes, the imposition of hydrostatic pressure will tend to close the microcracks and inhibit crack propagation. Hence when the pressure reaches some transition value, shear yielding and plastic deformation will occur rather than brittle fracture. The value of the transition pressure will depend on the polymer, on the medium if the polymer is not adequately protected, on the temperature, and on the strain rate. This is the case with polymethyl methacrylate [29] or polystyrene [30].

If, on the other hand, the polymer is ductile and develops considerable plastic deformation when tested in tension at a given strain rate and at atmospheric pressure, then the nominal tensile strain to fracture will probably reduce with increasing pressure, since cold drawing will become more difficult as a result of the decreasing free volume and the increasing restrictions on molecular mobility. Quite different from the situation with metals the application of pressure does not necessarily result in the enhancement of ductility.

The pressure dependence of the mechanical properties has been predicted theoretically on the basis of the free volume approach by Ferry and Stratton [31]. They derived the pressure analog of the well-known WLF equation [32] in the form

$$\ln a_p = (B/2.303\ f_0)(p - p_0)[f_0/K_f - (p - p_0)]^{-1} \tag{8}$$

where a_p is the ratio of the relaxation times at pressure p to those at a reference pressure p_0; B is a constant assumed to be of the order of unity; f_0 is the fractional free volume at the reference pressure; and K_f is the isothermal compressibility of the free volume. On the other hand, tensile measurements on elastomers by Patterson [33] under superposed hydrostatic pressure clearly showed that the effect of increasing pressure on Young's modulus was similar to that of decreasing temperature, i.e., glass-like behavior is encountered above a transition temperature. The existence of a pressure-induced transition in polymers has been discussed by various authors and now appears to be universally accepted. As an example, Sauer et al. [22] made stress - strain measurements on polypropylene at various hydrostatic pressures, and from the observed break in the modulus - pressure curve they estimated the pressure induced shift of T_g to be about 20°C/kbar. Combination of the usual isobaric measurements at atmospheric pressure as a function of temperature with isothermal measurements as a function of pressure allows, in principle, all the molecular parameters required by the free volume theory to be determined unambiguously. By considering the bulk modulus to be linearly related to pressure, an extension of Eq. 8 to include pressure effects has been attempted by Fillers and Tschoegl [34].

4. Mechanics of Hydrostatic Extrusion

Polymers creep at room temperature under a constant true tensile stress. When pulled in tension, majority of the thermoplastic polymers stretch under a constant load. These characteristics are well reflected in the extrusion pressure p vs. nominal extrusion ratio R_n relation. The nominal extrusion ratio is defined as the ratio of the cross-sectional area A_1 of the billet to that of the die opening area A_2. The slab method, one of the elementary ways of analyzing the extrusion process, readily gives the p - R_n relation when a uniform deformation of the billet with negligible interfacial friction is assumed:

$$p \propto \ln R_n \qquad \text{for constant true stress} \qquad (9)$$

$$p \propto (R_n - 1) \quad \text{for constant load} \qquad (10)$$

When the polymer is extruded at a low rate of strain of the order of 10^{-1} s^{-1} the first type of the p - R_n relation is obtained [35]. When the strain rate is high and is of the order of $1 \ s^{-1}$, on the other hand, the second type appears [36].

If the pressure dependence of the yield strength is taken into consideration by way of Eq. 7, Eq. 10 can be modified as

$$p = T_n \, (R_n - 1)/(1 - k) \tag{11}$$

where T_n is the normal yield strength in tension at atmospheric pressure. The pressure p' in the billet-die interface is given by

$$p' = T_n \, R_n \, /(1 - k) \tag{12}$$

If a Coulomb friction in the billet-die interface is assumed, with the coefficient μ of friction, the extrusion pressure of Eq. 11 is further modified as

$$p = (1 + \mu \cot \alpha) \, T_n \, (R_n - 1)/(1 - k)(1 - \mu \tan \alpha) \tag{13}$$

where α is the semi-cone angle of the die. If the pressure p_r to overcome the redundant work [37]:

$$p_r = 4 \, T_n \, \tan(\alpha/3)/\sqrt{3} \, (1 - k) \tag{14}$$

is introduced, the extrusion pressure, with the pressure dependence of the yield strength, friction, and redundant work taken into consideration, is given as

$$p = [T_n/(1 - k)][(4/\sqrt{3})\tan(\alpha/3) + (1 + \mu \cot \alpha)(R_n - 1)/(1 - \mu \tan\alpha)] \tag{15}$$

In the hydrostatic extrusion of a sheathed billet of polymethyl methacrylate at room temperature and at strain rate in the range of $2 - 2.8 \, s^{-1}$ with a conical die of $\alpha = 20°$ good agreement was obtained between the experimental extrusion pressure and Eq. 15 in the range of $R_n = 1 - 2.5$, by assuming $\mu = 0.01$ [38]. The nominal yield strength in tension at atmospheric pressure is a function of temperature and strain rate. In hydrostatic extrusion with a conical die the strain rate increases towards the die exit and so does the temperature because of the adiabatic heating of the billet. The yield strength decreases with the rise in temperature and increases with the strain rate. It is one of the basic assumptions of the analysis above described that this strength does not vary to an appreciable degree during the whole process of extrusion. The yield strength T_n in Eq. 15, therefore, can be referred to that of the temperature and strain rate of the billet either at the die entry or at the die exit.

To be more accurate, the flow stress of the billet should be specified as a function of the coordinate along the axis because the mean stress, strain, strain rate, and temperature vary along it and the flow stress is dependent on them. This is partially done with the hydrostatic extrusion of linear polyethylene and polyoxymethylene [39]. Assuming a constant temperature throughout the process and neglecting the pressure dependence of the flow stress but considering its strain rate dependence, good a-greement is obtained between the theoretical and experimental p - R relation. The range of nominal extrusion ratio studied was 1 - 20 for linear polyethylene and 1 - 10 for polyoxymethylene. The assumed value of μ to obtain the best fit was 0.03 for linear polyethylene and 0.08 for polyoxymethylene.

The analyses above presented are based on the equilibrium of stresses and can be called a lower bound approach of the limit analysis in the theory of plasticity. An upper bound approach has been also attempted to analyze the hydrostatic extrusion process of high-density polyethylene [40]. A radial flow [41] is employed as the basis for estimating the power required for internal deformation, redundant work, and friction. The flow stress σ_f was assumed to increase exponentially with increasing strain, linearly with hydrostatic pressure, and logarithmically with strain rate:

$$\sigma_f = [kp + m \ln \dot{\varepsilon} + n] \exp (\varepsilon) \tag{16}$$

where m and n are constants. The variation of pressure p through the die was approximated from the zero-friction equilibrium stress distribution in the die as

$$p = \sigma_f [\frac{2}{3} - \ln (A_2/A)] \tag{17}$$

where A is the cross-sectional area of the billet in the deformation zone. The two constants m and n in Eq. 16 were determined from tensile tests. The pressure coefficient k of the flow stress was assumed to be 0.094 (Table 1). By taking the coefficient of friction to be 0.02 an excellent agreement was obtained between the computed and experimental extrusion pressures up to $R_n = 7$ for $\alpha = 10°$. Both of the computed and experimental extrusion pressures assumed their minimum at $\alpha = 10°$.

In the analysis stated above a radial flow was assumed. Whether or not this is true can be checked by visioplasticity analysis. Round billets of high-density polyethylene with grids on a meridian section was hydro-

statically extruded to study the deformation pattern [42]. The range of
nominal reductions R'_n in area studied was 20 - 85 %, that of temperatures
25 - 120°C, that of strain rates at die exit 0.9 - 3 s^{-1}, and that of semicone angles of the extrusion die 20 - 45°. The strain and strain-rate distributions were calculated. Assuming Levy-Mises constitutive equations a
finite-difference analysis was performed to obtain the stress and temperature distributions. The experimentally determined dependence (Eq. 5) of
flow stress on pressure, temperature, and strain rate was employed in the
analysis. The friction along the billet-die interface was obtained along
with the mean-stress distribution. The results show that the flow pattern
is most strongly affected by R'_n and is far from uniform when $R'_n > 50$ % even
at room temperature and with a small semicone angle of 20°C. The main
feature of the flow pattern is the following: (1) In the entry zone the
plastic flow begins at the periphery of the billet. The core lags behind,
making the transverse grid lines concave toward the die exit. (2) As the
deformation of the billet proceeds in the deformation zone the peripheral
part is retarded with respect to the core. When the billet comes out of
the die, the transverse grid lines are convex toward the die exit. (3) When
the die angle is small and the reduction high, a flow pattern with double
peaks, which was observed in the high-rate high-temperature extrusion of
aluminum and titanium alloys and designated as a double-maximum flow
pattern [43], appears (Fig. 2).

Another process variable, which influences the flow pattern most, is
the die angle. For the flow pattern to be considered uniform the semicone
angle must be less than 20°. It is rather surprising to see that the
temperature and strain rate do not significantly affect the flow pattern.

The result of the stress analysis shows that, unlike the hydrostatic
extrusion of metals [44], the billet temperature is higher on the surface
than in the core. It reaches its maximum at the die exit. This explains
the fact that a trace of melting is sometimes observed on the extrudate
surface. A common defect in the extrusion of metal wires is central burst,
which is explained by the occurrence of triaxial tension along the billet
axis [45]. In the extrusion of polymers, however, the zone of triaxial
tension is located on and near the surface. This explains the fact that
cracks in polymeric extrudates are always observed to begin on the surface.
The friction in the billet-die interface is not constant throughout the
deformation zone. It increases from a very low value at the entry zone,
reaches its maximum, and decreases toward the die exit.

statically extruded to study the deformation pattern [42]. The range of
nominal reduction in area R'_n in area studied was 20 - 85 %, that of temperatures 25 - 120°C, that of strain rates at die exit 0.9 - 3 s^{-1}, and that

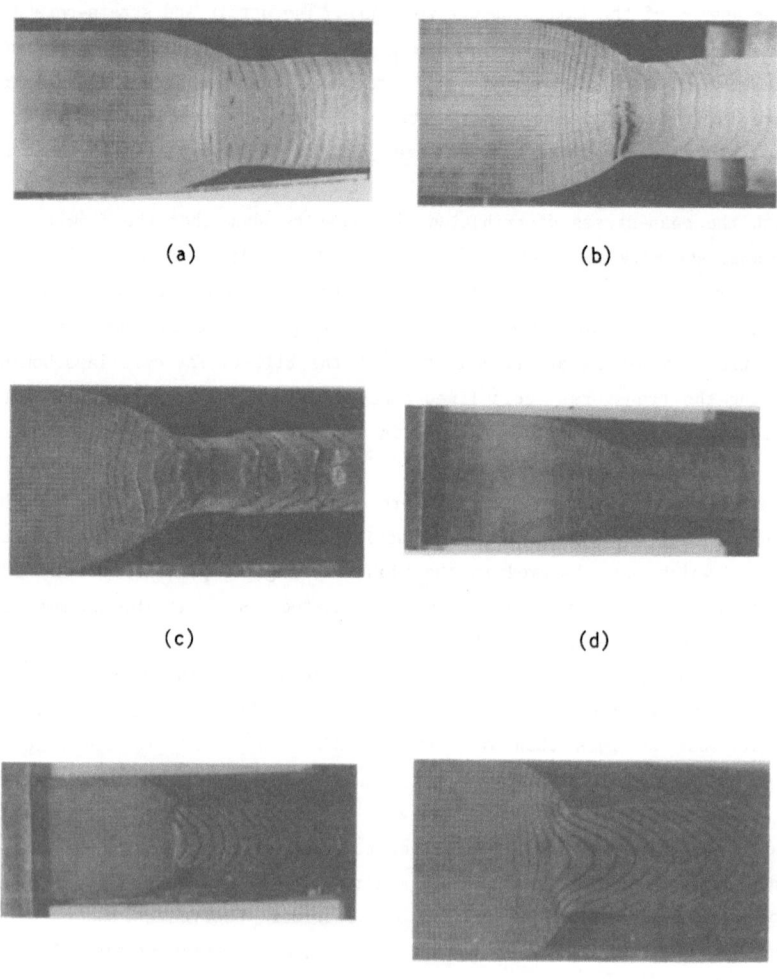

(a)   (b)

(c)   (d)

(e)   (f)

Fig. 2. Deformation patterns of high-density polyethylene billets hydrostatically extruded at various temperatures (t), reductions (R'_n) in area,
and die semicone angles (α).
(a) t = 25°C, R'_n = 66%, α = 20°, (b) t = 25°C, R'_n = 76%, α = 20°,
(c) t = 25°C, R'_n = 85%, α = 20°, (d) t = 50°C, R'_n = 76%, α = 20°,
(e) t = 50°C, R'_n = 76%, α = 30°, (f) t = 50°C, R'_n = 76%, α = 45°.

345

5. Lubrication

The isothermal hydrodynamic theory of hydrostatic extrusion, developed by Wilson [46] for metallic billets, was applied to the hydrostatic extrusion of high-density polyethylene [47] under the following
assumptions: (1) The die is rigid and of a conical form. (2) The billet
material is rigid up to the yield point. The deformation after yield is
a uniaxial tension under pressure, and the elastic components of strains
are neglected. (3) The deformation is uniform and the surface velocity
of the billet is inversely proportional to the square of its diameter,
because no volume change occurs during the plastic deformation. (4) When
the billet material flows in the uniaxial tension, the flow stress is
constant till the strain reaches a certain amount, and then the material
begins to workharden. During the workhardening the nominal flow stress
assumes a constant value. (5) The lubricant film thickness is thin compared with the billet diameter. (6) In the billet-die interface the
Reynolds equation for steady incompressible flow applies. (7) An isothermal condition exists in the lubricant film, and the billet temperature
is kept at the inlet temperature until workhardening sets in. During work-
hardening the billet temperature is kept at the outlet temperature. (8)
The lubricant viscosity is Newtonian and exponentially increases with
pressure. (9) The whole process of hydrostatic extrusion may be divided
into four zones, that is, the inlet, the perfectly plastic, the work-
hardening and the outlet zones (Fig. 3). The pressure gradient in the work

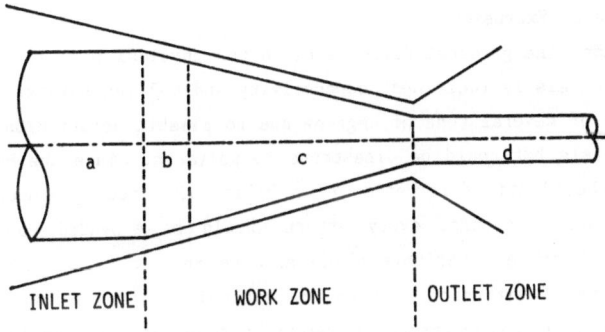

Fig. 3. Divided zones of hydrostatic extrusion process. (a) Inlet,
(b) perfectly plastic, (c) workhardening, and (d) outlet zones.

zones (b) and (c) is so small that its effect on the lubricant flow may be neglected and the pressure gradients in the inlet and outlet zones are assumed to become zero at the edge of the work zones.

Good agreement between the theory and experiment was obtained for the room-temperature hydrostatic-extrusion pressure of high-density polyethylene with the conical die with a 20° semicone angle for R_n = 1.2 - 7. The film thickness calculated, however, was extremely thin when compared with that for metallic billets mainly because of a low level of extrusion pressure. A similar observation was made in the hydrostatic extrusion of unsheathed polymethyl methacrylate [48]. The unsheathed billet was extruded at a temperature of 90°C up to R_n = 3.5. A friction coefficient in the range of 0.1 - 0.2 was obtained, suggesting that the condition of boundary lubrication prevailed. Sand-blasting the surface of the billet effectively reduced the initial billet-die friction.

In hydrostatic extrusion of polymers castor oil is predominantly employed as a pressure medium because of its high viscosity. The theory of hydrodynamic lubrication predicts a thicker lubricant film for a higher fluid viscosity, a higher rate of extrusion, and a smaller die angle. Thus, Nakayama and Kanetsuna [49] found that the extrusion pressure using castor oil and glycerine as the pressure medium was lower than that using water and ethanol. Silicone oil gave a higher extrusion pressure [40] and the coating of the billet with a graphite lubricant had little effect on the extrusion pressure.

6. Process Variables

6.1 Rate of Extrusion

Perhaps the greatest differencre in the physical properties of polymers and metals is their heat conductivity and melting points. A temperature rise of several tens of degrees due to plastic deformation would not affect metals but could be disastrous to polymers. Since the heat generated by the plastic deformation of a billet is directly proportional to the extrusion pressure, every effort should be expended to lower the pressure. A process variable which affects the extrusion pressure most is the rate of extrusion v. When the extrusion ratio is low, e.g. R_n = 2 - 3 for a room-temperature extrusion of high-density polyethylene, and 5 for the similar extrusion at 100°C, the rate of extrusion has practically no effect on the extrusion pressure. The rate can be varied over a wide range without changing the pressure very much. When the extrusion ratio is higher, e.g. 4 - 10 for a room-temperature extrusion of high-density polyethylene, the extrusion pressure increases linearly with ln v up to a certain point, rather abruptly drops a certain amount, and then increases again linearly with ln v (Fig. 4). In an effort to find out the cause of the pressure drop, three thermocouples were embedded in the high-density polyethylene billet at various distances from its axis and the temperature rise was directly monitored during the deformation process of hydrostatic extrusion [50]. No significant rise in temperature was observed during the first stage in Fig. 4. When the extrusion rate v was further increased, the temperature started to rise due to the adiabatic heating of the billet. The pressure drop in stage II in Fig. 4 was found to be due to the rise in temperature. In stage III the pressure drop due to the temperature rise is overcome by the pressure rise due to the rate increase. When the extrusion ratio is further increased, e.g. 20 for the hydrostatic extrusion of linear polyethylene [51], only a slight increase in the rate of extrusion causes a large increase in the extrusion pressure and the extrusion is possible only at an extremely low rate of strain, say 1 $\%s^{-1}$.

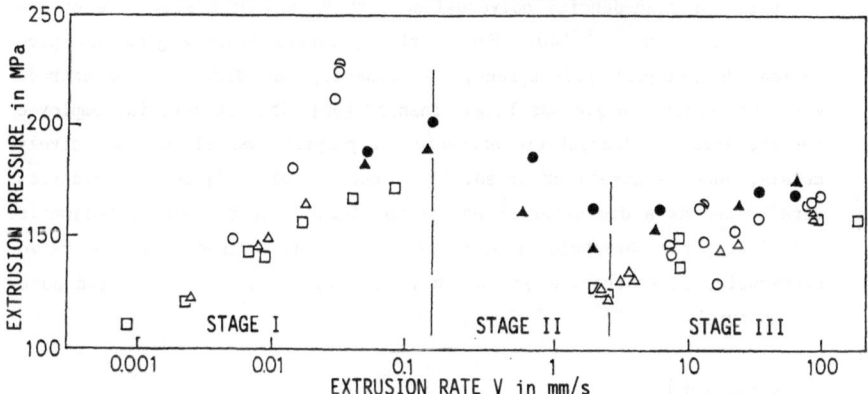

Fig. 4. Relation between extrusion pressure and extrusion rate in the hydrostatic and ram extrusions of high-density polyethylene with nominal extrusion ratio 7 at room temperature, using several dies with large and small openings and made of steels and plastics.

6.2 Temperature

The main effects of the extrusion temperature in the extrusion of polymers are on (1) the maximum extrusion ratio attainable, (2) the stiffness of the extrudates, and (3) the dimensional stability of the product. For crystalline polymers with high degrees of crystallinity increasing the temperature allows lower pressures to be used. Consequently, the maximum extrusion ratio attainable is raised. For amorphous polymers, on the other hand, raising the temperature reduces the maximum extrusion ratio [52]. This is also true for the crystalline polymers with low crystallinity [53]. For the purpose of improving the stiffness of the product a lower extrusion temperature is always preferred [54]. When a polymeric billet is exposed to a certain temperature during the hydrostatic extrusion process, the extrudate shows an improved dimensional stability up to that temperature. This means the higher the extrusion temperature the better the dimensional stability of the product at elevated temperatures [55].

6.3 Die Angle

The conical die has been exclusively employed by researchers of the hydrostatic extrusion of polymers. The semicone angles used range from 5° to 90°. There is an optimum angle for reducing the extrusion pressure. For example, it was found to be 10° in the room-temperature hydrostatic extrusion of high-density polyethylene with R_n = 7 at a slow rate of extrusion of 1 mm mn^{-1} [40]. For brittle polymers lower angles are preferred. High-impact polystyrene, for example, was difficult to extrude when the semicone angle was larger than 5° [56]. The die material employed for the study of hydrostatic extrusion of polymers was almost exclusively metals, such as steels or brass. In a recent study [57] certain plastics were tested as a die material and it was found that the use of polymeric dies lowers the threshold pressure and raises the maximum extrusion ratio attainable. Polycarbonate and nylon 6 were found to be best suited as a die material.

6.4 Forward Pull

Unlike the hydrostatic extrusion of metals the application of excess pressure is a cause of failure. One of the ways of reducing the extrusion pressure is to apply a haul-off load, that is, to pull the billet forward.

In the push-pull extrusion of high-density polyethylene at 120°C with the deformation ratio of 30 the extrusion rate is reported to increase substantially with the application of tension in the range of 0.02 - 0.04 GPa exerted by the attached loads, especially in the lower range of the extrusion pressure of 0.13 - 0.17 GPa, where the rate enhancement was up to tenfold [58]. On the contrary, the application of the back pressures in the hydrostatic extrusion of polymers always turned out to be deleterious, and the maximum extrusion pressure attained was reduced [59].

6.5 Coextrusion

The visioplasticity analysis [60, 61] reveals there is a shearing zone in the peripheral part of a billet, where the shearing stresses are high especially in the vicinity of the die exit. Stresses and strains are uniform and extensional in the core. This suggests that if only the core is utilized for the purpose of cold drawing a more uniform deformation may be expected and the maximum attainable extrusion ratio raised. This practice is called coextrusion or two-phase extrusion. Thus a round bar of high-density polyethylene housed in a coaxial tube of lead was successfully extruded at temperatures of 100 - 125°C to the deformation ratio of 55 [62]. Moreover, if the billet is split into halves along a meridian plane the effect of alleviating the shearing stresses may be enhanced. An attempt was made to sandwich a thin sheet of high-density polyethylene with two halves of a round bar of the same material and extrude them together. An ultra-thin film obtained at the deformation ratio of 30 was transparent and exhibited dead bend, its tensile modulus having been 30 GPa [63]. A further study revealed that when the yield strength of the tube material is similar to that of the core material, the diameter of the core must be less than that of the die opening in order to accomplish a successful coextrusion. This was experimentally proved with a polyethylene billet sheathed with lead.

6.6 Crystal Morphology

In order to produce an ultra-drawn morphology of high-density polyethylene it is postulated [64] that it is important to prepare certain crystal morphology prior to proceeding with hydrostatic extrusion. Thus Capiati et al. [64] compacted the material at a temperature just below the melting point and subsequently melted it at atmospheric pressure. After the temperature was lowered below the melting point a series of careful compression pressurizations were performed to accomplish the desired level of crystallization. The procedure was continued until the maximum density was reached and the material was ready for hydrostatic extrusion. By the use of this technique for billet preparation the deformation ratio over 50 was made possible. Capaccio et al. [65] maintain that crystallinity per se does not have a primary effect in determining the rate of local deformation except in the case of polymers of very low average molecular weight and narrow molecular weight distribution. A dominant role is played by the broad feature of the sample morphology as detected by optical microscopy. They explained these results on the basis of the concept of a network structure whose nature is affected by the annealing treatment to an extent which depends on the degree of coupling between adjacent crystalline regions in the isotropic undrawn polymer.

6.7 Melt Flow Index and Adiabatic Heating

Gibson and Ward [66] conducted hydrostatic extrusion of linear polyethylene having a range of different molecular weights at 100°C and concluded that the behavior correlated well with the melt flow index. The maximum extrusion ratio attainable increases with the melt flow index, reaches its maximum when the index is in the region of 5 - 10, and then begins to decrease. They also tested dies with various diameters at the opening. If the diameter is small and the rate of extrusion is moderate, it is possible that the process is isothermal, most of the heat of deformation being conducted out of the deformation zone by the tooling. From a dimensional analysis they proposed that the level of adiabatic heating is determined by the parameter vd/K, where v is the rate of extrusion, d the diameter of die opening, and K the thermal diffusivity of the billet material in the transverse direction. This parameter represents the ratio of the amount of heat transported out of the deformation zone by bulk movement of the material to that removed by transverse conduction. A low value of this number would correspond to the isothermal situation while a high value would approximate to the adiabatic case.

6.8 Composites

In metallurgy, multi-phase materials are common and techniques such as alloying or dispersion hardening are employed to control and modify the structure and properties of the materials. Kubat et al. [67] hydrostatically extruded a blend of linear polyethylene with a water-soluble

material consisting of sodium lignosulphate (Wanin S), and Kubat and
Manson [68] worked with blends of linear polyethylene with carbon black,
camphor and dimethylterephthalate in the hope of improving the mechanical
properties of the extrudates. Compared to the unfilled material, the addition of carbon black did not have any significant influence on the extrusion pressure, while a substantial reduction in pressure was noted with
other materials as the second phase. They concluded that the composites
containing the compatible substances could be processed at a significantly
lower pressure level compared to the unfilled polyethylene.

7. Properties of Product

7.1 Elastic Modulus

The tensile modulus of the extrudate is determined primarily by the
deformation ratio, but it is also dependent on the grade, molecular
weight, morphology, extrusion temperature, and extrusion rate. Thus the
tensile modulus increases almost linearly with the increase in the deformation ratio, but with the same amount of deformation ratio the product
with a lower extrusion temperature has a higher modulus. Hope and Parsons
[69] hydrostatically extruded linear polyethylene billets of various diameters at a nominal extrusion temperature of 100°C and $R_n = 10$, and found
that the modulus of the product with the 25.1 mm diameter was approximately half that with the 2.50 mm diameter. At high product diameters extrusion occurs in a predominantly adiabatic thermal regime, causing a
temperature rise in the billet. It is mandatory to heat up the billet to
a temperature just below the melting point in order to obtain a high deformation ratio, but every effort must be made to prevent the temperature
rising during the deformation processing if a higher value of product
modulus is desired. From this point of view a slower rate of extrusion
is preferred. In the preparation of superdrawn filaments of isotactic
polypropylene the product modulus is almost halved by increasing the rate
by a factor of 10 [70]. The product modulus is also dependent on the initial morphology of the billet. Thus in the solid state extrusion of
ultrahigh molecular weight polyethylene an extrudate from the chain-folded
morphology gave approximately 1/10 of the modulus of the chain-extended
one at the deformation ratio of 5, and the compacted powder almost 1/2
of the melt-extruded material [71]. When the tensile modulus obtained
under the optimum condition is plotted against the deformation ratio, Fig.
5 is obtained as an average of the modulus data from various sources.

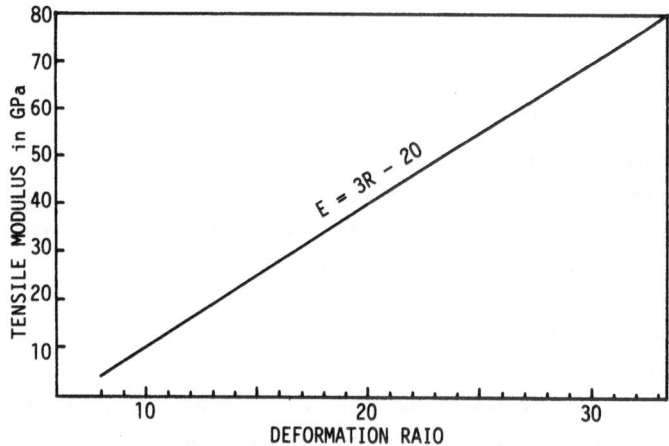

Fig. 5. Tensile modulus vs. deformation ratio for high–density polyethylene.

Modulus measurements are mostly done for the tensile modulus in the direction of extrusion by way of three-point bending. Moduli in the transverse direction and in torsional shear are expected to change little with the deformation ratio as is experimentally proved with polyoxymethylene [72].

Normally the stiffness of the extrudates is little affected by the amount of hydrostatic pressure applied during the deformation processing. For fiber-reinforced plastics, however, the situation is different. The solid-phase forming of glass-fiber reinforced polyoxymethylene was studied using the techniques of hydrostatic extrusion and die-drawing [73]. Although considerable orientation of the fibers and matrix occurred in both processes, there were very appreciable differences between the stiffness of the products in each case. For hydrostatic extrusion, the moduli of the glass-filled products were significantly higher than those of the unfilled products. In die-drawing, however, the absence of hydrostatic pressure allowed debonding of the fibers from the matrix to occur, accompanied by extensive void formation. The fibers were, consequently, no longer effective as a reinforcing phase: the moduli of the glass-filled products were no higher than those of the unfilled products.

7.2 Strength

Kojima et al. [74] studied the influence of deformation ratio and extrusion temperature on tensile strength. They found that the tensile strength approached a limit at high deformation ratios, independent of extrusion temperature over the range 110 - 134°C. The maximum obtainable tensile strength is around 0.4 GPa [75] with the deformation ratio of around 30. A similar value of the axial tensile strength is reported [76] for the high-density polyethylene tube, hydrostatically extruded with R_n = 15. The tensile strength in the circumferential direction, i.e. the hoop strength, is almost halved when R_n = 15 [77]. In an effort to improve tensile strength in the hoop direction Nakayama et. al. [78] tried a forming method of combined expansion and extrusion and obtained high-density polyethylene tubes with improved strength in both axial and hoop directions. This is the strengthening of tubes by way of biaxial drawing, but Austen in the United States applied a similar process to obtain a biaxially drawn sheet of polypropylene [79]. Thus he first produced polypropylene tubes with improved strength in both axial and hoop directions by way of hydrostatic extrusion, and then made a slit in the tube axially, flattened the slit cylinder, and finally produced helmets out of the biaxially drawn sheets of polypropylene.

A mathematical study was made of the collapse problem of thick-walled cylinders made of anisotropic polymers subjected to either internal or external pressure [80]. A remarkable increase in limit pressure was found, owing to the pressure dependence of yield strengths. When the direction of the highest tensile yield strength of the transversely isotropic material is aligned with the maximum principal stress axis, the limit pressure increases with the degree of anisotropy. On the other hand, if the direction of the highest tensile yield strength is aligned with other principal stress axes, the limit pressure decreases with the increase in the degree of anisotropy. Fig. 6 shows the relation between the limit internal pressure and the diameter ratio of thick-walled cylinders made of orthotropic plastics with tensile yield strength in the circumferential direction n times as much as that in other principal directions of orthotropy. The ratio of the yield strength in compression to that in tension of the isotropic starting material is assumed to be 1.3.

7.5 Elongation at Fracture

Normally in the hydrostatic extrusion of polymers the elongation

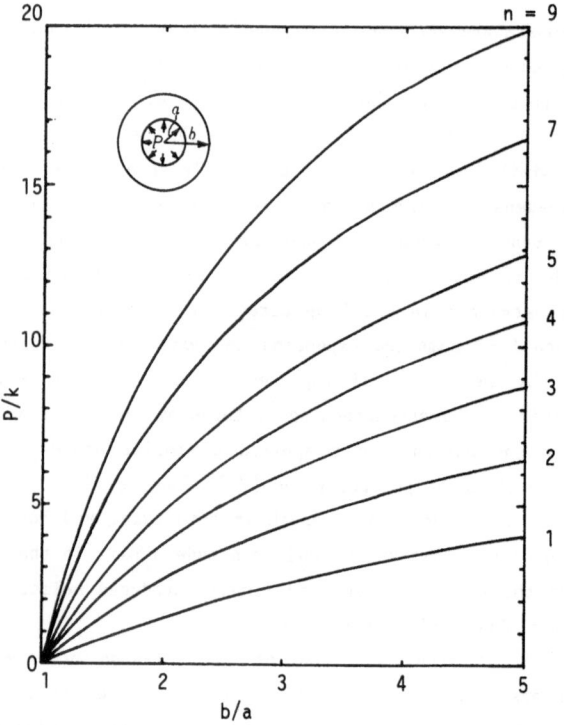

Fig. 6. Limit internal pressure P/k vs. diameter ratio b/a of thick-walled cylinders made of orthotropic plastics with the tensile yield strength in the circumferential direction increased by the factor of n from the isotropic material with the yield strength in compression 1.3 times as much as that in tension. k: yield strength in shear of the isotropic material. a, b: inner and outer radii respectively.

at fracture of the extrudate is less than that of the starting material. There is a class of polymers which are brittle at room temperature and under atmospheric pressure because the state of cold drawing is intercepted by fracture. Polymethyl methacrylate and polystyrene are typical examples of those polymers. The extrudates of those normally brittle polymers behave quite differently. Typical nominal stress – strain curves obtained with polymethyl methacrylate specimens are shown in Fig. 7 [81]. The elongation at fracture remarkably increased with the reduction and reached 45 % in terms of engineering strain with the extrudate of 60 % reduction in area, whereas the virgin samples broke at a few percent strain without showing any sign of yield or flow. The stress – strain curves showed no yield drop, and the post-yield flow stress was almost constant when expressed in terms of nominal stress. In Fig. 8 a smooth curve is shown through the stress – strain curves obtained at atmospheric

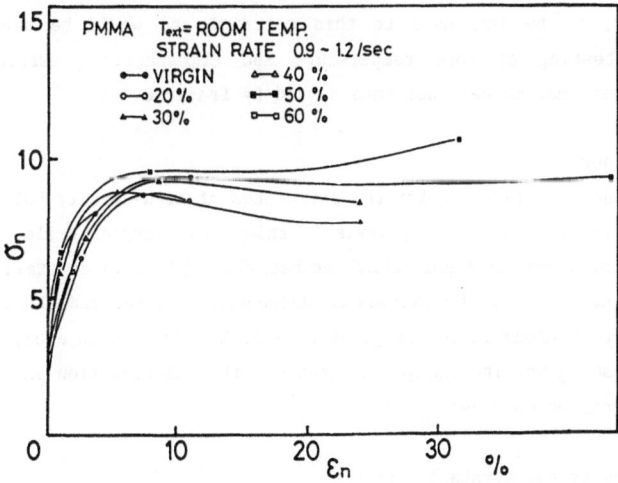

Fig. 7. Nominal stress σ_n – strain ε_n curves of polymethyl methacrylate at room temperature and atmospheric pressure.

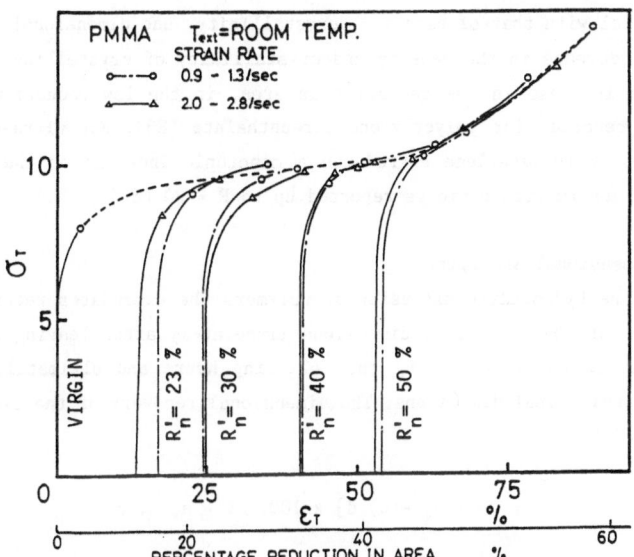

Fig. 8. True stress σ_t – strain ε_t curves of polymethyl methacrylate at room temperature and atmospheric pressure.

pressure subsequent to the various high-pressure prestrains by hydrostatic extrusion, yielding an average stress - strain curve which falls on and extends the stress - strain curve for tests conducted entirely at atmospheric pressure. The stress - strain curve in Fig. 8 may be interpreted, therefore, to be intrinsic to this material and would be seen in the tensile testing at room temperature and atmospheric pressure if its tensile deformation were not intercepted by fracture.

7.4 Hardness

Hardness is reported for the extrudates of high-density polyethylene, polypropylene, nylon 6, polyvinyl chloride, acrylonitrile-butadiene-styrene copolymer, and polymethyl methacrylate [82]. In crystalline polymers the hardness of the extrudates decreases with the reduction in area, reaches its minimum in the range of 20 - 60 % reduction in area, and tends to increase again. The change in hardness with the reduction in area, however, is rather marginal.

7.5 Density and Crystallinity

Density measurements have been made on the products of hydrostatic extrusion [82]. The variation of density with the reduction in area is in parallel with that of hardness, crystallinity, and dimensional stability. The decrease in the density and crystallinity of crystalline polymers with the increase in the reduction in area, in the low reduction range, is also reported for polyethylene terephthalate [83]. For ultra-oriented high-density polyethylene extrudates a monotonic increase of the density with the deformation ratio is reported up to R = 30 [84].

7.6 Dimensional Stability

In the hydrostatic extrusion of polymers the extrudates recover some fractions of their original dimensions immediately after leaving the die, and continue to do so during the following hours and ultimately settle down to their final dimensions. The dimensional recovery of the extrudates is defined by

$$\gamma_i = [(d_i - d)/d] \times 100, \quad i = a, b, c \tag{18}$$

where d is the diameter of the billet and the suffixes a, b, and c refer to the stages right after the extrusion, when the dimension has stabilized

at room temperature and at 100°C, respectively. They vary with the reduction in area as shown in Figs. 9 - 11 for high-density polyethylene
(PE), polypropylene (PP), nylon 6 (Ny6), polyacetal (PA), polyvinyl
chloride (PVC), acrylonitrile-butadiene-styrene (ABS), polymethyl methacrylate (PMMA), and polycarbonate (PC). The dimensional recovery of
crystalline polymers increases with the reduction in area, presenting a
common problem in the cold working of polymers, but it decreases rather
rapidly after the 60 % reduction in area has been reached. This fact is
worthy of particular attention when cold working of polymers is undertaken. When the deformation ratio is extremely high, e.g. 30 for the
ultra-oriented high-density polyethylene, no dimensional recovery is observed even at temperatures close to the melting point [84].

7.7 Thermal Expansion

The thermal expansion behavior of a series of oriented linear polyethylene, the product of hydrostatic extrusion, has been measured over
the temperature range - 130 to + 70°C [85]. The hydrostatic extrusion was
performed at a temperature of 100°C with the deformation ratios ranging
from 5 to 20. The coefficient of thermal expansion in the direction of
extrusion is negative and its absolute value increases with the deformation ratio, approaching that of the crystal unit cell. The coefficient
of thermal expansion in the transverse direction, on the other hand, is
positive and nearly constant, irrespective of the deformation ratio. It
is nearly equal to the average of those in the crystal a- and b-axis directions.

7.8 Thermal Conductivity

The thermal conductivity of hydrostatically extruded linear polyethylene samples has been measured parallel (K_a) and perpendicular (K_t) to the
extrusion direction over a wide temperature range [86,87]. At -173°C the
ratio K_a/K_t is about 20 and K_a has a value comparable to that of stainless
steel. At this temperature K_a is determined by the degree of crystal conti-
nuity and the conductivity is considered to be interpreted on the basis
of the thermal analog of the modified Takayanagi model [88]. It is also
reported that the low-temperature thermal conductivity of one of the large-
scale extruded polyoxymethylene samples was greater in the axial direction
than in the transverse direction [89]. The polyoxymethylene products obtained by hydrostatic extrusion exhibited enhanced properties in the axial

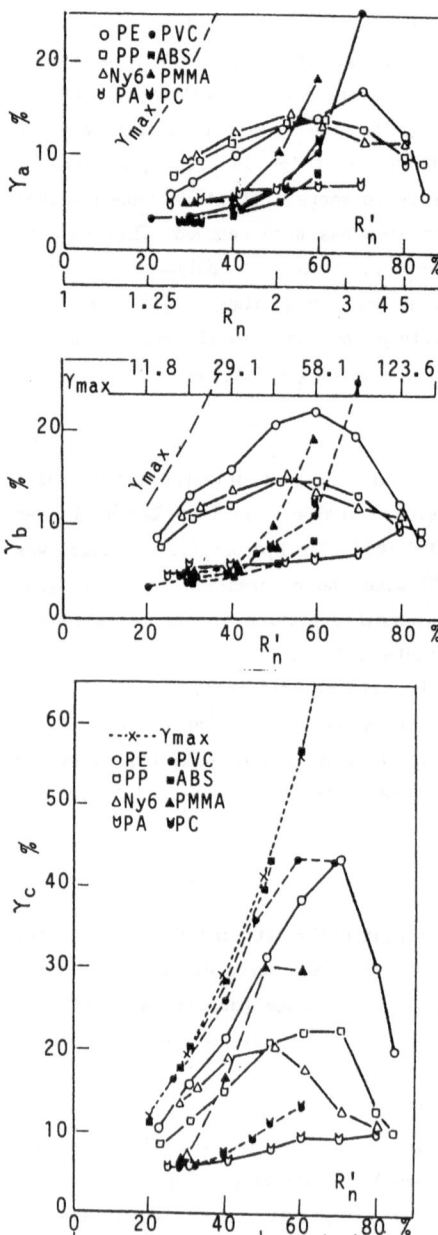

Fig. 9. Relation between recovery percentage and reduction percentage of extrudates, monitored right after extrusion. Curve γ_{max} plots the complete recovery up to virgin-billet size.

Fig. 10. Relation between recovery percentage and reduction percentage of extrudates, monitored when stabilized at room temperature. Curve γ_{max} plots the complete recovery up to virgin-billet size.

Fig. 11. Relation between thermal recovery percentage and reduction percentage of extrudates, monitored when stabilized at 100 ± 3°C in oven. Curve γ_{max} plots the complete recovery up to virgin-billet size.

extrusion direction, with roughly similar properties to the original isotropic material in the transverse direction [72].

8. Conclusions

Cold or warm forming of polymers has the advantage of improving the
mechanical properties by way of molecular chain orientation, which is
comparable with the workhardening of metals and alloys in metalworking.
It is essential to extend the polymers without shear, and the process of
hydrostatic extrusion is best suited for this purpose. Increase in elastic
moduli, yield and fracture strengths is expected as well as improvements
in ductility and thermal properties.

The hydrostatic extrusion process allows a high reduction in area
in a single pass, and eliminates a heating-and-cooling cycle. This is especially effective in reducing the working process and time, particularly
for processing large-diameter rods or thick sections of polymers, which
are notorious for their low thermal conductivity.

Although the technique of hydrostatic extrusion is already established as an industrial process for metalworking, its application to
polymers is still in the research and development stage. Much more work
still requires to be done to elucidate the mechanical behavior of polymers
under pressure, possible reactions of polymers with pressure medium, mathematical analyses of hydrostatic extrusion process, die design, or the
optimization of pressure, temperature, or the rate of extrusion.

Acknowledgment

The present author is indebted to his research associates, Dr. T.
Nakayama and Mr. T. Ariyama, and his students at the Science University
of Tokyo for their unlimited assistance in preparing the manuscript.
Thanks are also due to numerous authors, whose important works are freely
cited in the monograph. He also wishes to give his sincere thanks to Prof.
Avitzur of Lehigh University for supplying information on the current
state of the industrial application of hydrostatic extrusion to polymers
in the United States.

REFERENCES

1 I. Sakurada, T. Ito, and K. Nakamae, J. Polymer Sci., C, 15(1966)75.
2 D.S. Boudreaux, J. Polymer Sci., Phys. Edn., 11(1973)1285.
3 G. Cappacio and I.M. Ward, Polymer Engg. Sci., 15(1975)219.
4 A. Zwijnenburg and A. J. Pennings, Coll. Polymer Sci., 254(1976)868.
5 R.S. Porter, J.H. Southern and N. Weeks, Polymer Engg. Sci., 15(1975)213
6 K. Djurner, J. Kubat and M. Rigdahl, Polymer, 18(1977)1068.
7 K. Imada and M. Takayanagi, Int. J. Polymeric Materials, 2(1973)89.
8 A.G. Gibson and I.M. Ward, J. Materials Science, 15(1980)979.
9 P.D. Griswold, R.S. Porter, C.R. Desper, and R.J. Farris, Polymer Engg. Sci., 18(1978)537.
10 A. Peterlin, Polymer Engg. Sci., 19(1979)118.
11 P. Smith, P.J. Lemstra, and H.C. Booij, J. Polymer Sci., Phys. Ed., 19(1981)877.
12 T. Kanamoto, A. Tsuruta, K. Tanaka, M. Takeda, and R.S. Porter, Polymer Journal, 15(1983)327.
13 R. Raghava, R.M. Caddell, and G.S.Y. Yeh, J. Materials Science, 8(1973) 225.
14 I.M. Ward (Ed.), Structure and Properties of Oriented Polymers, Applied Science, London, 1975.
15 R. Hill, Mathematical Theory of Plasticity, Clarendon Press, Oxford, 1950.
16 R.M. Caddell and J.W. Kim, Int. J. Mech. Sci., 23(1981)99.
17 J.W. Hutchinson and K.W. Neale, J. Mech. Phys. Solids, 31(1983)405.
18 S. Maruyama, K. Imada, and M. Takayanagi, Int. J. Polymeric Materials, 1(1972)211.
19 T. Nakayama and N. Inoue, Proc. 59th National Congress of Japan Society of Mechanical Engineers, No. 810 - 11(1981)68.
20 J.A. Sauer, D.R. Mears, and K.D. Pae, European Polym. J., 6(1970)1015.
21 F. Birch, J. Applied Physics, 9(1938)4.
22 J.A. Sauer and K.D. Pae, Coll. Polymer Science, 252(1974)680.
23 S.S. Sternstein and L. Ongchin, Polymer, Preprints, 10(1969)1117.
24 J.A. Sauer, K.D. Pae, and S.K. Bhateja, J. Macromol. Sci. - Phys., B8 (3 - 4)(1973)631.
25 K. Ohji, K. Ogura, T. Yoshimura, and K. Kuroda, J. Soc. Materials Science, Japan, 23(1974)26.
26 K.D. Pae and J.A. Sauer, Engineering Solids under Pressure, Inst. Mech. Eng., London, 1971.
27 S.K. Bhateja and K.D. Pae, J. Polymer Science, B, 10(1972)531.
28 J.A. Sauer, Polymer Engg. Sci., 17(1977)150.
29 K. Matsushige, S.V. Radcliffe, and E. Baer, J. Polymer Science, Polymer Physics Ed., 14(1976)703.
30 K. Matsushige, S.V. Radcliffe, and E. Baer, J. Materials Science, 10 (1975)833.
31 J.D. Ferry and R.A. Stratton, Kolloid - Z., 171(1960)107.
32 J.D. Ferry, Viscoelastic Properties of Polymers, John Wiley and Sons, New York, 1970.
33 M.S. Patterson, J. Applied Physics, 35(1961)176.
34 R.W. Fillers and N.W. Tschoegl, Trans. Soc. Rheology, 21(1977)51.
35 S. Onogi, K. Matsumoto, H. Kawamura, K. Kubo, N. Utsumi, and T. Matsushita, Kobunshi Ronbunshu, 38(1981)315.
36 T. Nakayama and N. Inoue, Bulletin of Japan Society of Mechanical Engineers, 20(1977)688.

37 H.Ll.D. Pugh, in H.Ll.D. Pugh (Ed.), The Mechanical Behaviour of Materials Under Pressure, Hydrostatic Extrusion, Applied Science, London, 1971, p.401.
38 N. Inoue, T. Nakayama, and T. Ariyama, J. Macromol. Sci. - Phys., B 19(3)(1981)543.
39 P.D. Coates, A.G. Gibson, and I.M. Ward, J. Materials Science, 15(1980) 359.
40 R. Gupta and P.G. McCormick, J. Materials Science, 15(1980)619.
41 B. Avitzur, Handbook of Metal-Forming Processes, Wiley Interscience, New York, 1983.
42 N. Inoue, T. Nakayama, M. Takano, and Y. Wakayama, Proc. 4th International Conference on Production Engineering, Tokyo (1980)746.
43 C.P. Hinesley and H. Conrad, Materials Sci. Engg., 12(1973)47.
44 M. Nishihara, M. Noguchi, T. Matsushita, and Y. Yamaguchi, Preprint, MTDR Conf., (1977).
45 L.F. Coffin and H.C. Rogers, Trans. ASM, 60(1967)672.
46 W.R.D. Wilson and J.A. Walowit, J. Lubrication Technology, Trans. ASME, 93(1971)69-74.
47 N. Inoue and T. Nakayama, Proc. 1982 Joint Conference on Experimental Mechanics, (1982)849.
48 P.S. Hope, I.M. Ward, and A.G. Gibson, J. Materials Science, 15(1980) 2207.
49 K. Nakayama and H. Kanetsuna, Kobunshi Kagaku, 31(1974)321.
50 M. Hakamada, N. Kaneko, T. Nakayama, and N. Inoue, Proc. 61st National Congress of Japan Society of Mechanical Engineers, No. 830 - 10(1983) 98.
51 A.G. Gibson, I.M. Ward, B.N. Cole, and B. Parsons, J. Materials Science, Letters, 9(1974)1193.
52 T. Nakayama and N. Inoue, Proc. 58th National Conference of Japan Society of Mechanical Engineers, No. 800 - 11(1980)86.
53 W. Yagihashi, T. Nakayama, and N. Inoue, Proc. 61st National Conference of Japan Society of Mechanical Engineers, No. 830 - 10(1983)92.
54 P.S. Hope, A.G. Gibson, and I.M. Ward, J. Polymer Science, Polymer Physics Ed., 18(1980)1243.
55 T. Nakayama and N. Inoue, Proc. 56th National Conference of Japan Society of Mechanical Engineers, No. 780 - 13(1978)140.
56 N. Inoue and T. Nakayama, Proc. Spring Meeting of Japan Society for Technology of Plasticity, (1976)233.
57 N. Kaneko, M. Hakamada, A. Ohkubo, T. Nakayama, and N. Inoue, Proc. 33rd Joint Conference for the Technology of Plasticity, (1982)481.
58 T. Shimada, A.E. Zachariades, M.P.C. Watts, and R.S. Porter, J. Applied Polymer Science, 26(1981)1309.
59 K. Nakayama and H. Kanetsuna, J. Applied Polymer Science, 23(1979)2543.
60 T. Nakayama, N. Inoue, and Y. Wakayama, J. Japan Society for Technology of Plasticity, 21(1980)1108.
61 T. Nakayama, N. Inoue, and M. Takano, J. Japan Society for Technology of Plasticity, 22(1981)654.
62 J. Niki, S. Tasaka, and S. Miyata, Kobunshi Ronbunshu, 38(1981)307.
63 A.E. Zachariades, P.D. Griswold, and R.S. Porter, Polymer Engg. Sci., 19(1979)441.
64 N. Capiati, S. Kojima, W. Perkins, and R.S. Porter, J. Materials Science, 12(1977)334.
65 G. Capaccio, T.A. Crompton, and I.M. Ward, Polymer Engg. Sci., 18(1978) 533.
66 A.G. Gibson and I.M. Ward, J. Polymer Science, Polymer Physics Ed., 16(1978)2015.

67 J. Kubat, J.-A. Manson, and M. Rigdahl, J. Materials Science, 17(1982) 901.
68 J. Kubat and J.-A. Manson, J. Materials Science, 17(1982)910.
69 P.S. Hope and B. Parsons, Polymer Engg. Sci., 20(1980)589.
70 W.N. Taylor, Jr. and E.S. Clark, Polymer Engg. Sci., 18(1978)518.
71 A.E. Zachariades, M.P.C. Watts, and R.S. Porter, Polymer Engg. Sci., 20(1980)555.
72 P.D. Coates and I.M. Ward, J. Polymer Science, Polymer Physics Ed., 16(1978)2031.
73 P.S. Hope, A. Richardson, and I.M. Ward, Polymer Engg. Sci., 22(1982) 307.
74 S. Kojima and R.S. Porter, J. Polymer Science, Polymer Physics Ed., 16(1978)1729.
75 W.T. Mead, C.R. Desper, and R.S. Porter, J. Polymer Science, Polymer Physics Ed., 17(1979)859.
76 D.M. Bigg, E.G. Smith, M.M. Epstein, and R.J. Fiorentino, Polymer Engg. Sci., 18(1978)908.
77 P.S. Hope, S. Henderson, B. Parsons, and I.M. Ward, J. Mechanical Working Technology, 5(1981)223.
78 K. Nakayama, H. Kanetsuna, M. Kida, T. Masaoka, S. Nakamura, H. Kiyono, and S. Nakata, Proc. 34th Joint Conference for the Technology of Plasticity, (1983)361.
79 A.R. Austen and D.V. Humphries, U.S. Patent No. 4,282,277, August 4, 1981.
80 N. Inoue, J. Japan Society for the Technology of Plasticity, 25(1984) 227.
81 T. Nakayama and N. Inoue, Proc. 2nd International Conference on Mechanical Behavior of Materials, (1976)1305.
82 T. Nakayama and N. Inoue, Trans. Japan Society of Mechanical Engineers, 42(1976)3126.
83 M. Ito, J.R.C. Pereira and R.S. Porter, J. Polymer Science, Polymer Letters Edition, 20(1982)61.
84 A. Tsuruta, T. Kanamoto, and K. Takeda, Polymer Engg. Sci., 23(1983)521.
85 A.G. Gibson and I.M. Ward, J. Materials Science, 14(1979)1838.
86 S. Burgess and D. Greig, J. Phys. C: Solid State Phys., 8(1975)1637.
87 A.G. Gibson, D. Greig, M. Sahota, I.M. Ward and C.L. Choy, J. Polym. Sci., Polym. Lett. Edn., 15(1977)183.
88 M. Takayanagi, H. Harima and Y. Iwata, Mem. Fac. Eng., Kyushu Univ., 23(1963)1.
89 C.L. Choy and D. Greig, J. Phys. C, 10(1977)169.

AUTHOR INDEX

SUBJECT INDEX

Shaped bar, 14

Shear factor, 145

Shrinkage, compound, 205

Size effect, 326

Slab method, 21

Slip line field method, 21

Sodium lignosulphate, 351

Softening phenomenon, 135

Solid rod, 302

Sound extrusion, condition for, 74,
254

Special alloy, 113

Special metal, 113

Speed, effect of, 326

Spring-back, 335

Static design, 203

Static fluid supporting, 206

Steady state problem, 34

Steel, 4,14,111

  18% nickel maraging, 195,208

  air-melted, 200

  AISI 316 stainless, 299

  AISI 416 stainless, 316

  AISI 4140, 295,302

  alloy, 155

  carbon, 155,159

  nickel-chromium-molybdenum, 194,
    202,208

  tool, 195

  5% chromium, 195,200,208

  vacuum-remelted, 200

Stem, 185,217,219

  hollow, 186

  solid, 186

  speed, 83,150

Stem-part seal, 181,182

  gland type, 181,182

  piston type, 181,182

Stick-slip motion, 41,66,74,85,94

Stick-slip phenomenon, 239

Strain, distribution, 35

  equivalent, 8,25

  rate, 24

    deviatoric, 30

Stream function method, 28

Strength, buckling, 217

  fatigue, 196,200,202,213

    reversed, 198

  tensile, 353

  theoretical, 333

  ultimate tensile, 202

  yield, 202

Stress, 22

  amplitude, 198

    allowable, 198

  axial, 203

  buckling, 218,219

  deviatoric, 23

  distribution, 35

  frictional, 328

  hydrostatic, 23

  intensity factor, range of, 209

  maximum shear, 208,213

  maximum tangential, 208

  mean, 211

    compressive, 198

  radial, 203

  residual, 18,41,205,309,312

    compressive, 204,215

  state, 17

  tangential, 203

Strip drawing, 45

Structure, feature of, 230

Surface cracking, circumferential,
305

  longitudinal, 305